SOIL WATER AND NITROGEN

Developments in Plant and Soil Sciences

Volume 1

Also in this series

2. J. C. Brogan, ed., Nitrogen Losses and Surface Run-off from Landspreading of Manures. Proceedings of a Workshop in the EEG Programme of Coordination of Research on Effluents from Livestock, held at The Agricultural Institute, Johnstown Castle Research Centre, Wexford, Ireland, May 20–22, 1980.
    ISBN 90-247-2471-6
3. J. D. Bewley, ed., Nitrogen and Carbon Metabolism. Symposium on the Physiology and Biochemistry of Plant Productivity, held in Calgary, Canada, July 14–17, 1980.
    ISBN 90-247-2472-4

In preparation:

R. Brouwer, O. Gasparikova, J. Kolek and B. C. Loughman, eds., Structure and Function of Plant Roots. Proceedings of the Second International Symposium, held in Bratislava, Czechoslovakia, September 1–5, 1980.

Series ISBN 90-247-2405-8

Soil Water and Nitrogen
in Mediterranean-type Environments

edited by

JOHN MONTEITH
University of Nottingham, Nottingham, U.K.

and

COLIN WEBB
The International Center for Agricultural Research in the Dry Areas,
Aleppo, Syria

Selected reviews, reprinted from *Plant and Soil* Vol. 58 (1981)

1981

SPRINGER-SCIENCE+BUSINESS MEDIA, B.V.

Soil Water and Nitrogen in Mediterranean-type Environments, based on a workshop, organized by The International Center for Agricultural Research in the Dry Areas, which was sponsored by The International Center for Agricultural Research in the Dry Areas, and The United Nations Development Programme

This volume is listed in the Library of Congress Cataloging in Publication Data

ISBN 978-94-015-0320-4   ISBN 978-94-015-0861-2 (eBook)
DOI 10.1007/978-94-015-0861-2

Preface

ICARDA has the serious and urgent responsibility for increasing the quantity and availability of food in the extensive North Africa-West Asia region, and therefore must give high priority to optimising the use of soil water and nitrogen which are considered to be among the main limiting factors to production.

To obtain the knowledge to further this aim, and with the help of the United Nations Development Program (UNDP), ICARDA is developing a special project on 'Increasing the Fixation of Soil Nitrogen and the Efficiency of Soil Water Use in Rainfed Agricultural Systems in the Countries of North Africa and Western Asia'.

In planning this project, ICARDA has called on the expertise of leading scientists in a number of countries who can give the benefit of their experience, and advise on methods and priorities on which this specific research can be soundly based and conducted.

To provide a forum for the presentation and exchange of such information, ICARDA, with the help of UNDP, invited a number of these scientists to a week-long workshop at Aleppo in January 1980. This workshop gave rise to valuable discussions which culminated in several recommendations by which the project will be guided. The organising committee comprised Drs. J. Begg (Australia), P. Cooper and D. Gibbon (ICARDA), P. Dart (Australia), G. J. Koopman and J. McWilliam (ICARDA Board of Trustees), A. Kassam (England) and P. Vlek (U.S.A.).

ICARDA's operating committee at the workshop included Drs. D. Gibbon, A. Allan, P. Cooper and R. Stewart, Messrs. F. Jabri and C. Webb, and Misses M. Boyagi and L. Brahimsha. Thanks are expressed to these scientists and also to the committees which provided such a smooth and effective organisation for the meeting.

However the influence of a good workshop does not finish with a set of recommendations. This book, compiled from the review papers presented at this meeting, provides a continuing record of the current state of knowledge about soil, water and nitrogen which are so fundamental to life itself. A paper covering the important area of nitrogen fixation was included in the conference proceedings but could not be obtained in manuscript form in time.

While of particular interest to the North Africa-West Asia region and other Mediterranean-type areas which occur in practically every continent, this publication has a much wider application. It contains the very principles by which soil, water and nitrogen can be studied in every environment. Thus it will provide a very useful reference book throughout the world.

It is particularly satisfying that this publication which is due very much to ICARDA's initiative, will make such a wide and valuable contribution to knowledge. I trust that it will continue to benefit people throughout the world until, inevitably, it will be up-dated by information which scientists will continue to provide.

H. S. DARLING,
December, 1980   Director-General,
ICARDA

Editorial note

We should like to bring to the attention of our readers that papers on techniques have also been published in the special edition of *Plant and Soil*, volume 58.

We wish to acknowledge specially the help of several people who contributed to this publication. They are Dr. Shawki Barghouti, leader of ICARDA's Training and Communications Program; Dr. Richard Stewart, especially for his relationships with the publishers; Miss Leila Brahimsha for typing services; and Mr. Hassan Khairallah who designed the figures.

December, 1980   JOHN MONTEITH
                                                             COLIN WEBB

Contents

Authors

Participants in the ICARDA/UNDP ·vorkshop were as follows.

Dr. *Floyd Bolton* is an associate professor of crop science at Oregon State University, Corvallis, Oregon, USA, where he is a research leader in dryland cereal production in Oregon. He obtained his Doctor of Philosophy degree at Colorado State University, Colorado, USA.

In 1963–65 he was an instructor in agronomy, and agronomy research leader in Ethiopia on a contract with USAID and Oklahoma State University. Later, he spent several years (1969–73) as a crop production agronomist in Ankara, Turkey, with the Rockefeller Foundation wheat research and training center.

He has also been a consultant for the Ford Foundation in Western Iran; for USAID in Tunisia: and FAO in Egypt.

Dr. *John Burford* is a soil chemist with ICRISAT, Hyderabad, India. He obtained his Doctor of Philosophy degree at the University of Adelaide, Adelaide, South Australia, Australia.

He did post-doctoral work at Iowa State University, Iowa, USA, in 1969–71 and was a Research Fellow at the University of Reading, England in 1971–74, before spending four years at the Litcombe Laboratory.

He joined the Farming Systems Research Program at ICRISAT in 1978.

Dr. *A. E. Fischer* is a principal research scientist of CSIRO's Division of Plant Industry at Canberra, Australia. He obtained his Doctor of Philosophy degree at the University of California, Davis, California, USA, after a study of the mechanism of stomatal opening and the role of potassium uptake. Before that, he did research on wheat agronomy at Wagga Wagga, NSW, Australia.

During the late 1960's, he examined the effects of water stress on wheat in Professor R. O. Slatyer's group at the Australian National University. This was followed by five years as a wheat physiologist/agronomist at CIMMYT, Mexico, under Dr. N. E. Borlaug. Since 1975, he has concentrated on wheat physiology and zero tillage research with CSIRO's Division of Plant Industry.

Dr. *David Gibbon* was leader of the Farming Systems Research Program at ICARDA, 1977–80, where his work included a study of existing farming systems, studies of the Syrian economy, cropping systems and crop/livestock systems research, and cooperative programs with other programs and outside agencies.

He was on contract with ICARDA from the School of Development Studies, University of East Anglia, Norwich, England, with an interdisciplinary team of scientists. He returned to England at the end of the ICARDA/UNDP workshop to resume his post as senior lecturer in Development Studies.

Dr. Gibbon's experience includes lectureships in agronomy at Leeds University, the University of Dar es Salam, and agronomic and farming systems research in Trinidad, Tanzania, Botswana, Sudan and the United Kingdom.

Dr. *Hazel Harris* is a research fellow in the Department of Agronomy and Soil Science at the University of New England, Armidale, NSW, Australia. She did both her undergraduate and Doctor of Philosophy studies at this university.

Her research interests include plant-environmental interactions and adaptation, particularly to moisture and temperature.

Dr. *Douglas Johnson* is a plant physiologist, specialising in forage and range plant improvement for the United States Department of Agriculture, at Utah State University, Logan, Utah, USA.

He obtained his Doctor of Philosophy degree at Utah State University after a study of range ecology. His current work is concentrated on plant improvement for the semi-arid rangelands of western United States of America.

Dr. *Amir Kassam* has been a consultant in agricultural ecology and land resources evaluation since 1976. After graduating as Master of Science at the University of California, California, USA, he obtained his Doctor of Philosophy degree at the University of Reading, England.

In 1971–74, he was a research fellow in the agronomy department of the Institute for Agricultural Research, Samaru, Northern Nigeria. He was head of the cereal physiology section, ICRISAT, Hyderabad, India for two years before taking up his present consultancy.

Dr. *Rafael Novoa* is leader of the ecology program of the INIA agricultural research institute in Santiago, Chile. He obtained his Doctor of Philosophy degree at the University of California, Davis, California, USA, after studying as an agronomist at the University of Concepion, Chile, and as a plant physiologist at the University of Paris, France.

His experience includes a soil survey, research on the response of tomatoes to fertilizers, and modelling of the nitrogen metabolism of the higher plants.

Dr. *Joe T. Ritchie* has worked since 1966 with the Soil and Water Conservation Research Division of The Agricultural Research Service, USDA. He was previously employed at the Texas Agricultural Experimental Station and at Iowa State University. He has broad interests in agronomy, soil science, and microclimatology and is well known for his contributions to the analysis and prediction of evaporation from crops in the field with special reference to row crops and to the relation between actual evaporation and soil water content.

Dr. *Richard Smith* is a senior lecturer in agronomy at the University of New England, Armidale, NSW, Australia. He obtained his Doctor of Philosophy degree at the University of Western Australia after a study of grazing systems agronomy, and did a post-doctoral year at the University of California, Davis, California, USA, on computer simulation.

His main interests include systems agronomy, applications of computer simulation to agriculture, computer scheduling of irrigation, and studies in crop adaptation.

Dr. *Neil Turner* is a principal research scientist at CSIRO's Division of Plant Industry at Canberra, Australia. He obtained his Doctor of Philosophy degree at the Waite Agricultural Research Institute, University of Adelaide, Adelaide, South Australia.

His experience includes a long period as plant physiologist at the Connecticut Agricultural Experiment Station, New Haven, Connecticut, USA; nine months as visiting research fellow to the Department of Botany, University of Aberdeen, Aberdeen, Scotland; and a consultancy with UNESCO at the UNDP Center for Soil Water Management, Haryana Agricultural University, Hissar, India.

Dr. *Herman van Keulen* is a research officer at the Center for Agrobiological Research, Wageningen, Netherlands. He obtained his Doctor of Philosophy degree at the Agriculture University, Wageningen.

. Highlights of his experience include a study of primary production and water use efficiency of natural pastures in semi-arid regions in the framework of cooperative research between the Netherlands and developing countries, and the development and application of systems analysis in agricultural systems.

From 1975 to 1977, he participated in a bilateral cooperation program between the Netherlands and Indonesia.

Dr. *Paul Vlek* is a soil scientist with the International Fertilizer Development Center at Muscle Shoals, Alabama, USA. He obtained his Doctor of Philosophy degree at Colorado State University, Colorado, USA, after a study of soil physical chemistry.

His experience includes research with acid sulphate soils, molybdenum, soil chemistry, and nitrogen in flooded soils.

Prologue: context and scope

J. R. McWILLIAM

ICARDA Board of Trustees

Agriculture directly controls the economic and social life of nearly two-thirds of the people of North Africa and western Asia; the region in which our efforts at this conference will be focused. Improved levels of economic development, political stability and human well-being in this area, will depend to a large extent upon how rapidly the low-income countries of the region can significantly improve the productivity of their rural sectors.

Progress in this respect in much of the region has been sluggish and erratic. Farm production has grown, but per capita production has remained the same and in most countries the incomes of rural workers have remained well below their urban counterparts, and have been unresponsive to improvements in GNP. As a result, there has been a continuing migration to the cities and the adoption of more appropriate production technology at the farm level has been slow.

It is ironic, and to a degree humbling, that we should be here at the end of the second millennium AD giving evidence to the people of a region that first developed stable agriculture about 10 000 years ago.

The earliest agricultural economies such as this in the Fertile Crescent of West Asia developed in dry environments and their major problems then, as now, centered around their efforts to farm in an environment characterised by a highly variable and chronically deficient rainfall.

To cope with this, they developed quite sophisticated dry farming systems, utilizing fallowing and various water harvesting techniques and irrigated systems draining water from rivers and wells including the famous ganats. With these systems, they grew crops such as wheat, barley, peas, lentils, chickpeas and cucurbits which are still the staple crops of the region. These farmers were also some of the first farmers to experience problems of soil fertility decline, erosion and siltation and widespread salinization.

We have much to learn from their 10 000 years of agricultural experience and I hope something to give in return to help restore this region to a more stable and productive condition.

Any attempt to improve the agriculture of these low-income countries by the transfer of technology from the West, must go hand in hand with efforts by the national governments and others to provide additional inputs, including price incentives, improved agricultural infrastructure, better credit and marketing facilities and support for research and extension to assist in adapting the technology for use under local conditions.

The International Research Centers, and ICARDA in particular, have an important catalytic role to play in this effort to improve the pace and direction of agricultural development. They have concentrated their efforts in the area of research and technology with a focus, in the case of ICARDA, on the development of more productive and stable dry-farming systems for the region.

In this approach, there has been a special concern for the cereal and legume crops which constitute the basis of the crop rotation in the region. Also a special effort is being made to devise ways of integrating legume-based ley pastures into the cropping rotation to improve fertility and provide additional feed for the sheep and goat populations which represent an important source of cash flow in many rural communities.

In the predominantly rainfed agricultural systems which constitute the largest and poorest sector of the farming community, soil and water resources, which are function of the environment, largely determine the level of production and ultimately the overall economic growth of the region.

This workshop has been designed to focus on these two basic resources and, in particular, to analyse the more effective use of soil water and nitrogen in the dry farming systems of the region. The need to develop a clearer scientific perspective on this problem, and to explore alternative approaches has been motivated by ICARDA's desire to mount a significant research program in this area.

In any approach to this problem, one of the first requirements will be to review and analyse the information available on the agroclimatic resources of the region, especially that relating to the land and water resources. As part of this review, there is a need for long term historical weather records which help to identify climatic patterns and establish the agricultural potential of the region.

From such an analysis, it will be apparent that the agricultural potential is largely determined by the incidence of winter rainfall, modified by radiation and temperature and the soil conditions, especially those pertaining to soil depth and fertility. The inherent variability and interactive nature of these factors demands a prescriptive analysis to help devise more efficient strategies for utilizing water and nitrogen in the rainfed farming systems. The principles underlying such strategies should have generality for comparable agro-ecological zones and will need to be developed from models of crop growth and yield, based on sound experimental data and validated to ensure reality.

The development of these dynamic models simulating the most relevant components of the cropping system and ultimately the more complex farming system, will require an appropriately focused research effort to understand the fluxes of water and nitrogen and to develop functional relationships to describe their interactions in the soil-plant-atmosphere continuum.

To appreciate the role of soil water, there is a need to predict the duration of the potential growing season, as measured by available soil water and the distribution of water with depth and time. Also information is needed on the consumptive use of water by the various crops during their life cycles and the consequences of water stress during critical stages of their development.

Similar information is required for nitrogen and, in particular, there is a need to quantify the nature and extent of the nitrogen transfers in the various soil systems and to construct realistic nitrogen balance sheets for alternative cropping systems. The levels of available nitrogen derived from all sources, including endogenous reserves, N_2 from legume symbiosis and fertilizer nitrogen inputs where appropriate, must be equated with their contribution to crop development and ultimately through interactions with water and other nutrients to the yield and quality of major crops of the region.

Although a greater awareness and understanding of the constraints imposed by limitations in water and nitrogen by itself may not contribute to yield improvement, the development of predictive models of these particular subsystems can provide guidelines for the plant breeders, whereby they can sharpen their appreciation of the responses needed to improve the adaptation (yield and stability) of crop and pasture species in the region.

Also information on the fluxes of water and nitrogen throughout the course of the growing season of the crop and during the fallow cycle can suggest ways of optimising the use of these key resources through crop and soil management; thereby reducing the risk associated with farming in these difficult environments.

Both of these endeavours, the genetic improvement of the crops and the development of improved agronomic procedures to make more effective use of water and nitrogen, are important steps in the development of better farming systems for the region. The adoption of this improved technology by farmers, however, may be more difficult if these innovations are not built into the structure of the existing farming systems and seen as one component of a wider process of rural development.

Most of you at this meeting appreciate this point, and are fully aware of the difficulties involved in the introduction of alternate technology in developing countries. The International Agricultural Research Centres, including ICARDA, see this as their primary responsibility along with the development of research capacity among the agricultural scientists of the region.

This workshop, which has been convened by ICARDA, is part of this general objective of adapting the best of current agricultural technology to serve the needs of agriculture in this region.

Within this general objective, our task at this workshop is quite specific; namely to review what is known and what is relevant to know about the dynamics of water and nitrogen in the farming systems of the region and to establish what scientific approaches and research techniques will be the most appropriate to use in achieving this objective.

Finally, it is hoped that the collective wisdom of this group might be harnessed to assist in defining research priorities, and to suggest how these might be developed into a coherent research program. I hope we are successful in this endeavour, and we appreciate your willingness, under the circumstances, to contribute to this special workshop.

1. Climate, soil and land resources in North Africa and West Asia

A. H. KASSAM

Swinscoe, Ashbourne, Derbyshire, England

The international institutes concerned with agricultural research appreciate the need for inventories of physical resources relevant to the use of land for crop production. Such inventories can be used to plan research programmes and in the interpretation of records from the field. This paper reviews the climatic and soil resources in the region served by ICARDA within which the term 'Mediterranean' has been applied to a wide range of environments.*

In North Africa the countries covered are Morocco, Algeria, Tunisia, Libya, Sudan and Egypt; in West Asia, they are Jordan, Saudi Arabia, People's Democratic Republic of Yemen, Yemen Arab Republic, Oman, United Arab Emirates, Qatar, Bahrain, Lebanon, Israel, Syria, Turkey, Iraq, Kuwait, Iran and Afghanistan.

In a recent report from the Food and Agriculture Organisation (FAO, 1978), the methodology for assessing potential land use is based on the following procedures:

i   selection and definition of the type of land utilization, i.e. crop and product, production type, input level;
ii compilation of a climatic inventory including phenological requirements and the response of photosynthesis to temperature and radiation;
iii assembly of information on the soil requirements of the crop;
iv compilation of a quantitative climatic inventory (1: 5 million scale) based on climate (characterizing temperature differences) and length of growing periods (characterizing period when water and temperature permit crop growth);
v   assembly of a soil inventory, by countries, from the FAO/Unesco Soil Map of the World (1: 5 million scale);
vi overlay of the climatic inventory on the soil map and area measurement of resultant climate: soil units, the so-called *agro-ecological zones*;

---

* The review is based on the information generated by the Agroecological Zones Project in the Land and Water Development Division, FAO. The material on North Africa is presented in FAO (1978), and on West Asia in FAO (1979).

vii calculation (from v and vi) of national areas for soil units (by slope class, texture class and phase) by major climates and growing period zone (30-day intervals);

viii matching climatic inventory (iv) with the crop climatic requirements (ii) and, where the climatic requirements of the crop are met, calculations of constraint-free crop yields by growing period zones;

ix matching the soil requirements of the crop (iii) with the soil units, slope classes, texture classes and phases of the soil map, by rating soil limitations (agro-edaphic constraints);

x compilation and rating of the various agro-climatic constraints to crop production occurring in the various major climates and growing period zones;

xi application of the agro-climatic constraints (x) to the constraint-free crop yields (viii) to derive attainable crop yields, by growing period zones;

xii *agro-climatic suitability classification* of each growing period zone according to expected crop yields (xi);

xiii application of the soil limitation ratings on the agro-climatic suitability classification of each growing period zone according to the soil composition of the zone, to arrive at the *land suitability classification*, i.e. areas of land variously suited to the production of the crop (i.e. the type of land utilization under consideration).

The title of this paper covers a good deal of territory and has a scope that is too broad to allow detailed treatment of specific information. Further, the countries within ICARDA's mandate cover tropical, subtropical and temperate thermal climates. Consequently, the analysis of the land resources is presented as a reconnaissance. It uses static crop-environment models while resorting to subjective rules where necessary. However, many of the concepts and principles used are independent of scale, and they highlight the problems of developing rational and quantitative approaches to the evaluation of land resources.

BASIC CLIMATES AND MEDITERRANEAN-TYPE ENVIRONMENTS

All global classifications of climates based on temperature recognize four basic thermal climates (Fig. 1). In the thermal tropics, monthly mean daily temperatures at Sea Level in all months are above 18 °C; in the thermal subtropics, one month or more has Sea Level mean temperatures below 18 °C but all months are above 6 °C; in the thermal temperate zone, one month or more (up to a maximum of nine months) has monthly mean temperatures below 6 °C; and in the thermal

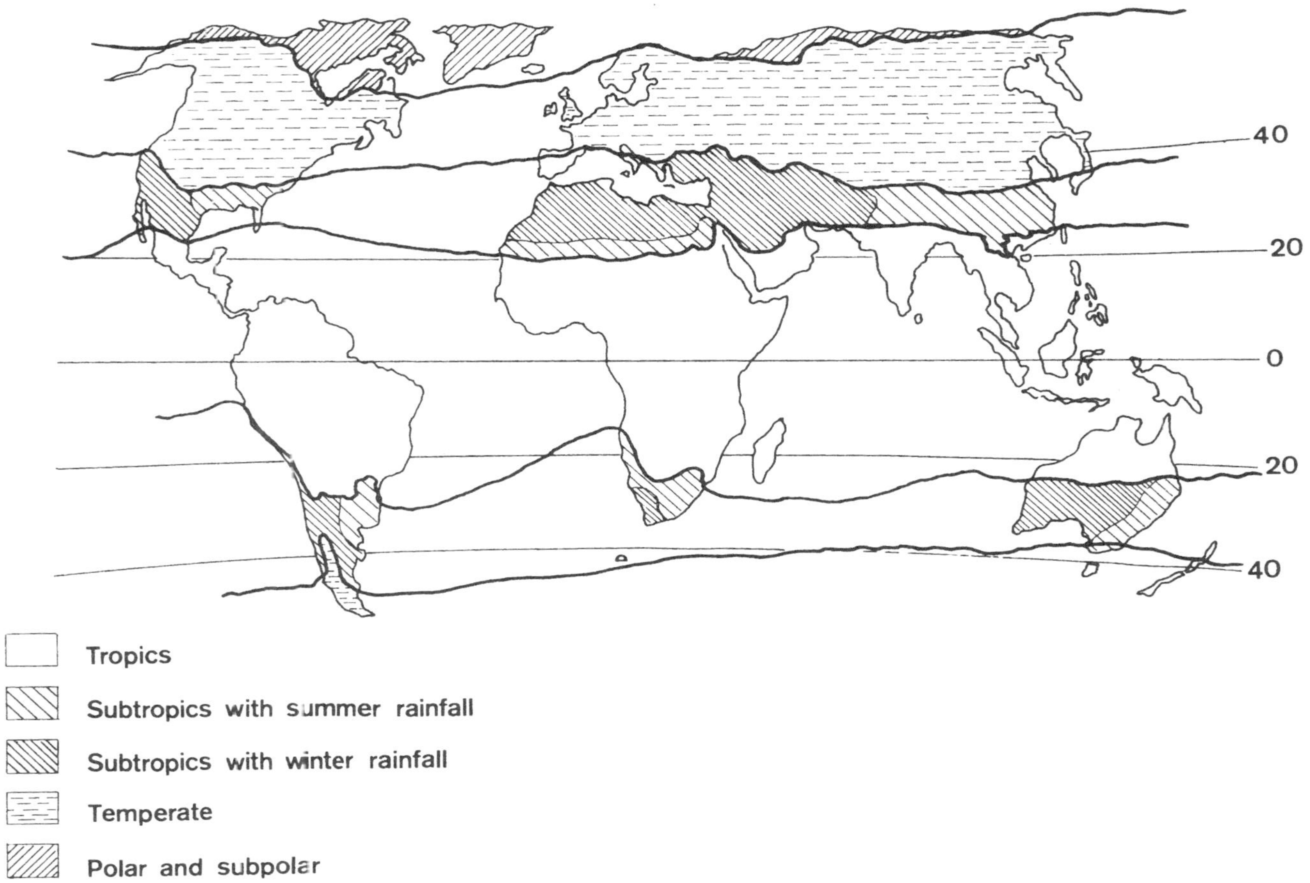

3

Tropics

Subtropics with summer rainfall

Subtropics with winter rainfall

Temperate

Polar and subpolar

Fig. 1. Basic thermal climates and distribution of subtropics with winter rainfall.

polar region more than nine months have monthly mean temperatures below 6 °C.

Within each basic thermal climate there are warm and cool variants due to latitude and altitude; and at a given latitude and altitude, there are oceanic and continental variants due to the land: sea configuration. Further, because of the variation in the amount of precipitation, each basic thermal climate has arid and humid variants. However, the precipitation may be concentrated in either the warmer part or the cooler part, so that each thermal-moisture combination has 'summer' and 'winter' variants.

It is the areas in the subtropics with winter rainfall that define a Mediterranean climate. Such areas are found in all continents and their global distribution is shown in Fig. 1. However, the true Mediterranean environment has a mild winter and hot dry summer with winter precipitation. Therefore the Mediterranean climate proper is always seasonally arid and never found at a great distance from the sea or at high altitude.

Table 1. Major variations of the Mediterranean climates in the thermal subtropics

Altitude	Oceanic	Continental
<1500 m	Warm oceanic Mediterranean	Warm continental Mediterranean
	Thermal regime:	*Thermal regime:*
	Cool or cold winter with warm or hot summer.	Cool or cold winter with warm or hot summer.
	Annual fluctuations <25 °C	Annual fluctuations >25 °C
	Moisture regime:	*Moisture regime:*
	Humid to desert	Semi-arid to desert
>1500 m	Cool oceanic Mediterranean	Cool continental Mediterranean
	Thermal regime:	*Thermal regime:*
	Cold winter with cool summer.	Cold winter with cool summer.
	Annual fluctuations <25 °C	Annual fluctuations >25 °C
	Moisture regime:	*Moisture regime:*
	Subhumid to desert	Semi-arid to desert

Temperature (°C)				*Rainfall (mm)*				
Cold:	<5	Warm:	18 to 30	Humid:	1000 to 2000		Arid:	100 to 200
Cool:	5 to 18	Hot:	>30	Subhumid:	600 to 1000		Desert:	<100
				Semi-arid:	200 to 600			

If the thermal and moisture conditions in the subtropical winter rainfall areas are considered together with the factors of latitude, altitude and continentality, it is obvious that the number of Mediterranean-type environments can be very large. Further, because the subtropics are bounded by the tropics on their warmer margins and by the temperate climate on their cooler margins, it is possible to experience tropical and temperate variants of some of the subtropical Mediterranean-type environments. These are referred-to as temperate Mediterranean and tropical Mediterranean climates.

A summary of the main classes of Mediterranean climates in the subtropics is given in Table 1. The largest single continuous area of such climates lies around and to the east of the Mediterranean Basin, stretching from the Atlantic Coast in the west to Afghanistan in the east. It merges into the temperate climates in the north in Turkey, the subtropical summer rainfall and tropical Sahelian climates to the south and the tropical monsoon climates to the east in Pakistan, northern India and south west Nepal.

ICARDA recognizes two major agro-climatic zones in the winter rainfall areas: a 'lowland' zone of less than 1200 m altitude, and a 'plateau' zone of more than 1200 m (Darling, 1979). This distinction reflects the fact that, somewhere between 1000 m and 1500 m altitude, thermal and moisture regimes become recognizably different from those prevailing in the lowlands, and the difference in climate is presumed to have important ecological consequences in terms of how the land is farmed.

Because agro-climatic zones merge into each other, there is a transition area between zones. As more information on climate becomes available, it would be possible to define quantitatively and more precisely the agro-climatic zones relevant to ICARDA's numerous activities.

CROP AND LAND UTILIZATION TYPE

In North Africa and West Asia more than 75 per cent of the cultivated area with winter precipitation is occupied by rainfed wheat grown as a winter crop. The land resources of the region are therefore assessed in relation to their suitability for the production of rainfed winter wheat.

Having selected the crop, it is necessary to define the conditions under which it is to be grown. Without such a definition, the evaluation is not valid, because suitability for a crop varies considerably according to the circumstances under which it is produced.

For example, lands with slopes of more than 14 per cent, or of a very stony nature, are not normally suited to mechanical cultivation, but can be cultivated

6

with hand tools. On the other hand, very heavy soils, such as Vertisols, cannot be cultivated with hand tools, but are suited to mechanized cultivation. Thus a description of the circumstances of cultivation is vital to any sound assessment of land resources. Combined description of such things as crop, produce, inputs and technical expertise, serve as the land utilization type of the evaluation (Beek, 1978).

The crop is considered at two levels of inputs – low and a high. The attributes of the two input levels are shown in Table 2, and form the basis of the definition of the land utilization type employed in the assessment. In a regional analysis, it is not possible to collate the data needed for a specific definition of land utilization type. Accordingly, two generalized major land utilization types have been considered in the assessment, approximating to conditions of low inputs and high inputs. The two input levels can be visualized as representing two points on a production: input curve, corresponding to no (or few) on-farm capital inputs, and a high (not unlimited) level of capital inputs.

Table 2. Attributes of land utilization types considered

Attribute	Low Inputs	High Inputs
Crop and product	Winter wheat (either bread or durum) for dry grain production	
Production method	Rainfed production, no irrigation or water importation. Sole cropping only, no multiple (sequential or mixed) cropping.	
Market orientation	Subsistence production	Commercial production
Capital intensity	Low	High
Labour intensity	High, including uncosted family labour	Low, family labour costed, if used
Power source	Manual labour with hand tools	Complete mechanization including harvesting operations
Technology employed	Local cultivars. No (or insufficient) fertilizer, no chemical pest, disease or weed control, fallow periods for both water and nutrient accumulation	High yielding cultivars. Adequate fertilizer application, chemical pest, disease and weed control, fallow periods for water accumulation only
Infrastructure requirement	Market accessibility not essential, inadequate advisory services	Communications and market accessibility essential, high level of advisory services
Land holding	Small, sometimes fragmented	Large, consolidated
Income level	Low	High

Only bread or durum wheat is considered, produced from water which falls on the land as precipitation. The water may be used for wheat production, with or without a fallow period depending on its amount. Wheat production involving flood water, and water that is imported either by natural or artificial means, are not considered in the assessment.

Only the areas less than 1500 m above Sea Level are considered i.e. the low altitude warm oceanic and continental Mediterranean environments (Table 1). The areas above 1500 m have a snow cover for more than three months, and generally do not have a thermal growing period during the time when there is precipitation so that where wheat is cultivated, it grows and matures on the water from the snow melt. This system is different from that in the low altitude Mediterranean environments where the wheat crop develops and matures more on the 'current' water balance account than on the 'deposit' account.

RESPONSE TO CLIMATE

To enable crops and types of land use to be matched to land qualities, the climatic inventory must be compiled in a form that permits the interpretation of climate in terms of its suitability for the production of crops in question. Appropriate attributes of the crops dictate what variables are to be taken into account in the compilation of the climatic inventory. The concepts and principles of climatic adaptability relevant to the evaluation of land resources have been described in Kassam *et al.* (1977) and FAO (1978).

Crops have climatic requirements for photosynthesis and for development. Rates of crop photosynthesis, growth and yield are related to the assimilation pathway, and its response to temperature and radiation. However, the phenological climatic requirements which must be met are not specific to a photosynthetic pathway. Crops may be classified into climatic adaptability groups according to fairly distinct photosynthesis characteristics. Each group then comprises crops of 'equal ability' in relation to potential productivity, and the differences between groups in the response of photosynthesis to temperature and radiation determine productivity, provided phenological requirements are met. Crop adaptability groups and their characteristic average photosynthesis response to temperature and radiation are presented in Table 3.

The time required to form yield depends on the phenological constraints on the use of time available in the growing period; the location of yield in the plant (e.g. seed, leaf, stem, root) also has an important influence. Temperature controls rates of growth and development in general and, within the optimum temperature range for growth, it may influence the growth of specific parts and the accumulation of yield.

Table 3. Average photosynthesis response of individual leaves of four groups of crops to radiation and temperature

Characteristics	Crop adaptability group*			
	I	II	III	IV
Photosynthesis pathway	C_3	C_3	C_4	C_4
Rate of photosynthesis at light saturation at optimum temperature ($g\ CO_2\ m^{-2}\ h^{-1}$)	2–3	4–5	>7	>7
Optimum temperature (°C) for maximum photosynthesis rate	15–20	25–30	30–35	20–30
Irradiance for maximum photosynthesis rate (Wm^{-2} total radiation)	150–400	200–500	>1000	>1000
Crops	Wheat Barley Triticale Lentil Broadbean Chickpea White potato	Rice Groundnut Pigeonpea Sweet potato Cassava Yam	Pearl millet Lowland sorghum Lowland maize	Highland sorghum Highland maize

* For further information on crop adaptability groups see Tables 1.1 to 1.5 in Kassam *et al.* (1977) or Tables 3.1 to 3.5 in FAO (1978).

For example, in hardy winter wheat and barley, chilling is required for flower initiation, but the optimum temperatures at the time of flowering (anthesis) and subsequent yield formation are much higher. Similarly, optimum temperatures for growth in sugarcane are greater than 20 °C but during the ripening period and because the yield is located in the cane, a lower temperature in the range 10 to 20 °C is desirable for the concentration of sugar of the right kind. On the other hand, optimum temperatures for growth, development and yield formation in rice are higher than 20 °C, and lower temperatures lead to head sterility.

Winter wheat (C_3-species, Group I) is an annual with a determinate growth habit. It requires 120 to 150 days of active growth. Its yield is located in the terminal inflorescence in seeds and the yield formation period is the last one-third of its active growth cycle. The period from flowering to maturity must be frost-free. Its climatic adaptability qualifies it to be considered for further matching in areas where mean daily temperatures are less than 20 °C. In the Mediterranean environment and in the 'low' altitude areas, these temperatures are experienced between October and May. In the 'high' altitude areas, they are experienced through most or all of the year.

In different parts of the globe, temperature and the availability of water from rainfall exert different constraints on crop production. In warm tropical regions, the major constraint limiting the time available for crop production is availability of water. In subtropical regions with winter rainfall, low temperatures and radiation during the winter period may limit crop growth when water is available. During the summer period in such areas, water availability may limit crop growth despite a favourable temperature and radiation climate.

When rain is abundant, temperature regime in one region may permit growth of a crop from Groups II and III only, whereas the conditions in another region may permit growth of a crop from Groups I and IV only. These situations occur particularly when temperature changes occur spatially e.g. a decrease in temperature due to increase in altitude or latitude. However, when the period is long and temperature changes are seasonal in nature, part of the period may be suitable for growth of a crop from Groups II and III only, while another part of the period may be suitable for growth of a crop from Groups I and IV e.g. in the tropics in areas away from the Equator and in the subtropics.

The growing period has been used as a basis for the assessment of climatic resources, defined as the period in which water availability and temperature permit crop growth. An inventory of climatic resources has been prepared to allow:

i   differentiation of the region into major climatic divisions reflecting changes in the geographical and seasonal distribution of the crop;
ii quantification of the period during which rainfed production is possible;
iii calculation of yield that can be attained under conditions which are free from constraints;
iv partial quantification of yield losses related to agro-climatic constraints.

Major climatic divisions

To take into account temperature requirements that limit the distribution of a crop on a regional scale, prevailing temperature regimes were categorized by identification of major divisions. Mean daily temperatures of less than 20 °C during the growing period were regarded as being suitable for consideration of crops from Groups I and IV.

Table 4. Characteristics and areas of major climates of the region during the growing period

| Basic thermal climate | Major climates during growing period | | Mean daily temperature (°C) regime during the growing period | Suitable for consideration during the growing period for crop group | Area (10⁶ ha) | | |
	No.	Descriptive name			North Africa	West Asia	Total (per cent)
Tropics	1	Warm tropics	More than 20	II and III	250	130	380 (25.3)
	2	Cool tropics	15–20 5–20	IV I	4	17	21 (1.4)
	3	Cold tropics	Less than 5	Not suitable	–	–	–
Subtropics	4	Warm subtropics (summer rainfall)	More than 20	II and III	121	–	121 (8.1)
	5	Cool subtropics (summer rainfall)	15–20 5–20	IV I	2	–	2 (0.2)
	6	Cool subtropics (summer rainfall)	Less than 5	Not suitable	–	–	–
	7	Cool subtropics (winter rainfall)	5–20	I	441	414	855 (56.9)
	8	Cold subtropics (winter rainfall)	Less than 5	Not suitable	6	100	106 (7.1)
Temperate	9	Cool temperate	5–20	I	–	4	4 (0.3)
	10	Cold temperate	Less than 5	Not suitable	–	14	14 (0.9)
				TOTAL	824	679	1503

To arrive at the major climatic divisions for a region, the effect of latitude on mean temperature was first taken into account and thermal tropics, subtropics and temperate regions were distinguished. The subtropics were then separated into areas where the rainfall was in the cooler part of the year, i.e. *subtropics with winter rainfall* and where it was in the warmer part of the year, i.e. *subtropics with summer rainfall*.

To take into account the effect of altitude on mean temperature during the growing period, the tropics and the subtropics with summer rainfall were each divided into three major climates, while the subtropics with winter rainfall and the temperate areas were divided into two each. This procedure defined 10 major climates for the region, listed in Tables 4 and 5.

Table 5. Areas of major climates by countries in North Africa and West Asia (10^6 ha)

Country	Major climate								Total
	1	2	4	5	7	8	9	10	
Morocco	—	—	—	—	38.9	5.7	—	—	44.6
Algeria	4.0	—	57.1	2.3	174.5	0.3	—	—	238.2
Tunisia	—	—	—	—	16.4	—	—	—	16.4
Libya	—	—	34.2	0.1	141.7	—	—	—	176.0
Sudan	243.2	4.5	2.9	—	—	—	—	—	250.6
Egypt	2.9	—	26.9	—	69.3	—	—	—	99.1
North Africa	250.1	4.5	121.1	2.4	440.8	6.0	—	—	824.9
Jordan	—	—	—	—	9.8	—	—	—	9.8
Saudi Arabia	59.6	6.1	—	—	149.3	—	—	—	215.0
PDR Yemen	27.4	1.3	—	—	—	—	—	—	28.1
Yemen AR	10.2	9.3	—	—	—	—	—	—	19.5
Oman	21.2	—	—	—	—	—	—	—	21.2
UAE	7.6	—	—	—	0.8	—	—	—	8.4
Qatar	—	—	—	—	2.2	—	—	—	2.2
Bahrain	—	—	—	—	0.1	—	—	—	0.1
Lebanon	—	—	—	—	1.0	0.1	—	—	1.1
Israel	—	—	—	—	2.1	—	—	—	2.1
Syria	—	—	—	—	18.3	0.2	—	—	18.5
Turkey	—	—	—	—	53.3	9.8	2.4	11.4	76.9
Iraq	—	—	—	—	43.4	0.1	—	—	43.5
Kuwait	—	—	—	—	1.6	—	—	—	1.6
Iran	4.1	—	—	—	100.3	56.9	1.4	2.1	164.8
Afghanistan	—	—	—	—	31.8	33.0	—	—	64.8
West Asia	130.1	16.7	—	—	414.0	100.1	3.8	13.5	678.2
Region	380.2	21.2	121.1	2.4	854.8	106.1	3.8	13.5	1503.1

12

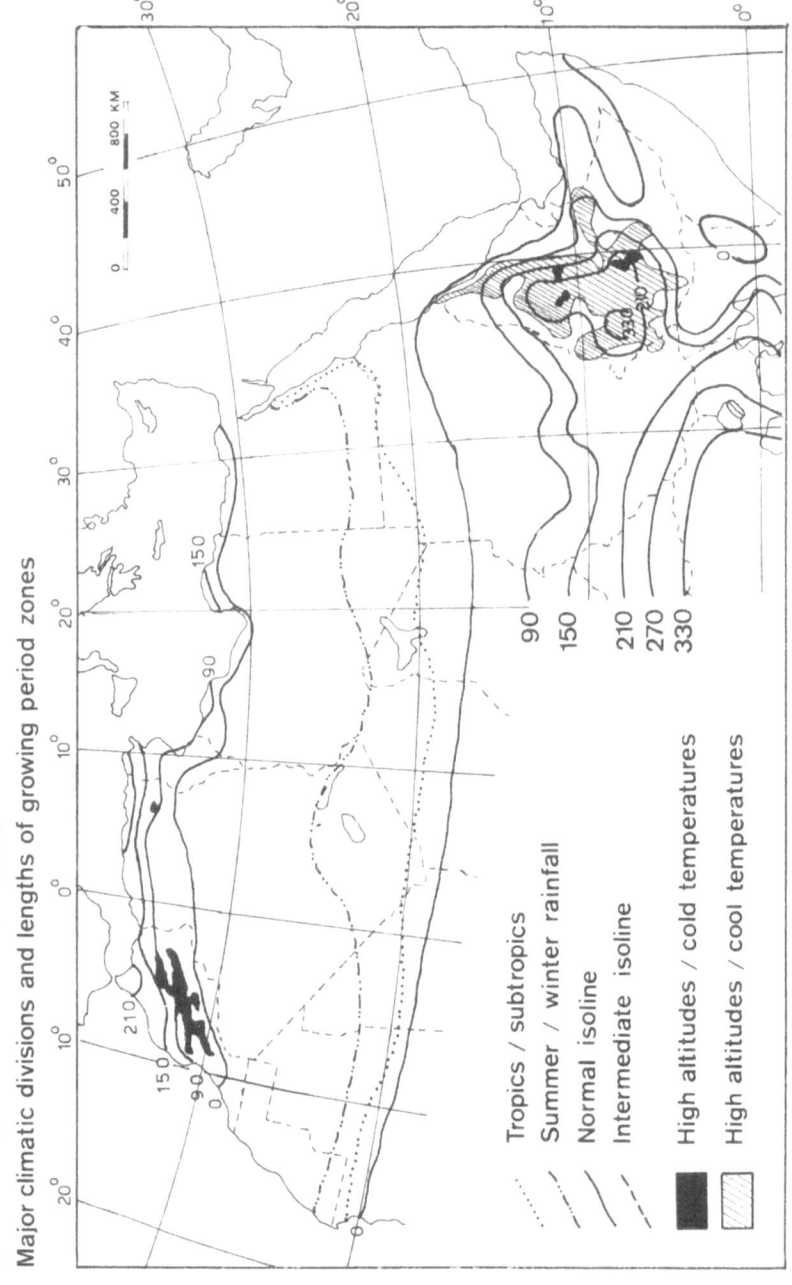

GENERALIZED CLIMATIC INVENTORY
Major climatic divisions and lengths of growing period zones

90
150
210
270
330

Tropics / subtropics
Summer / winter rainfall
Normal isoline
Intermediate isoline
High altitudes / cold temperatures
High altitudes / cool temperatures

Fig. 2.   North Africa.

13

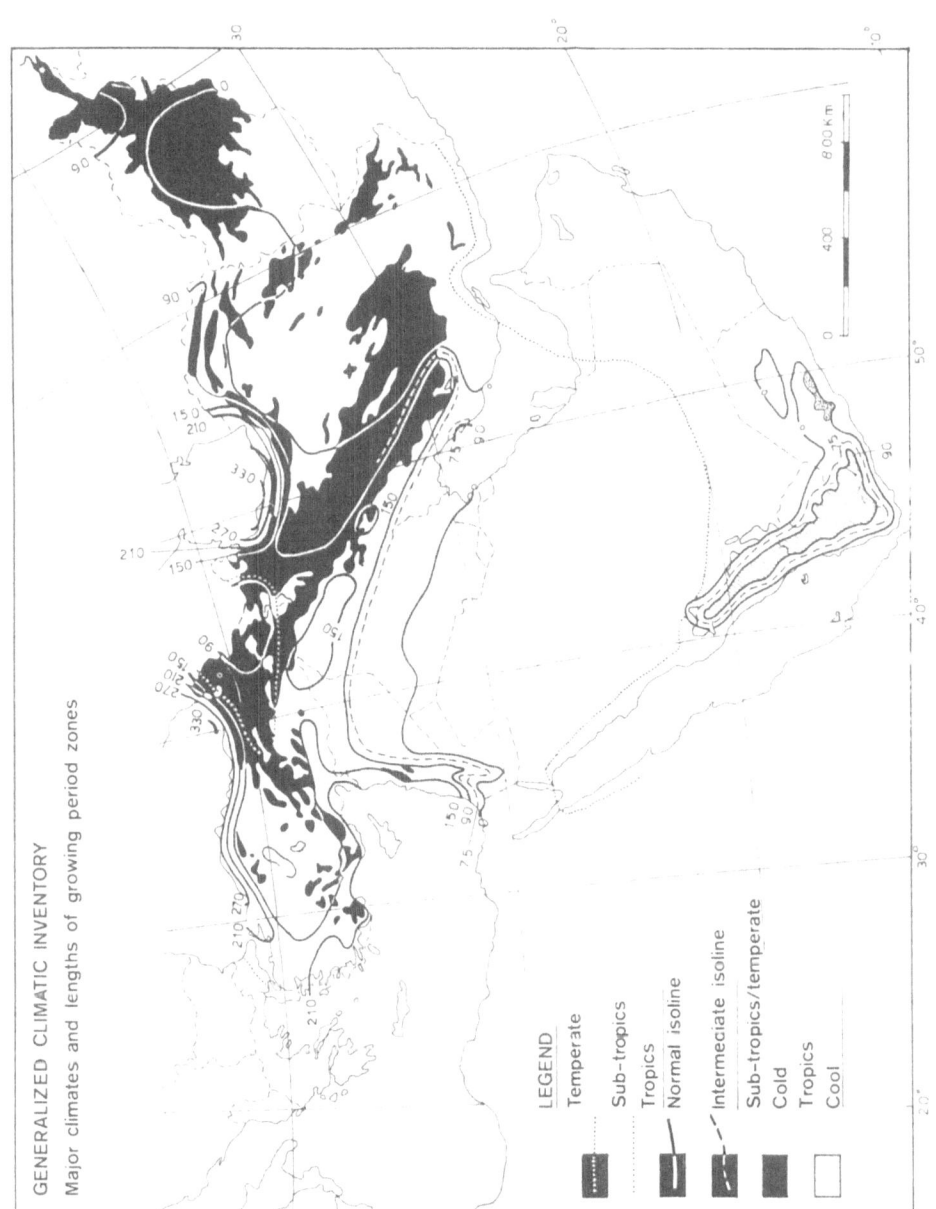

Fig. 3. West Asia.

Growing period

The following working definition of 'growing period' has been used. The growing period is the period during the year when temperatures are higher than 5 °C and when precipitation exceeds 0.5 potential evapotranspiration-PET (including a period required to evaporate a nominal 100 mm water stored in the soil from excess precipitation, or less if not available).

A growing period with a humid period, i.e. a period with an excess of precipitation over PET, is treated as a 'normal' growing period; a growing period with no humid period as an 'intermediate' growing period. Areas where rainfall does not exceed 0.5 PET are treated as 'dry', i.e. no growing period, while areas where temperatures during the period of water availability are continuously less than 5 °C are treated as 'cold', i.e. the areas in major climates 8 and 10.

In the inventory, growing period zones were delineated by isolines of growing periods with values of 0, 75, 90, 120, 180, 210, 240, 270, 300 and 330 days.

Table 6. Areas of growing period zones by major climates in North Africa and West Asia (10^6 ha)

Growing period zones (days)	Major climate								Total	
	1	2	4	5	7	8	9	10	Area	%
330–364 N	–	–	–	–	0.3	–	–	–	0.3	0.02
300–329 N	–	–	–	–	0.3	–	–	–	0.3	0.02
270–299 N	2.5	–	–	–	1.7	–	–	–	4.2	0.3
240–269 N	8.5	0.2	–	–	2.8	–	–	–	11.5	0.8
210–239 N	8.5	0.2	–	–	14.4	–	–	–	23.1	1.5
180–209 N	18.2	0.1	–	–	17.8	–	–	–	36.1	2.4
150–179 N	33.3	–	–	–	19.4	–	1.6	–	54.3	3.6
120–149 N	14.0	4.3	–	–	23.0	–	–	–	41.3	2.7
90–119 N	15.2	6.6	–	–	42.9	–	0.7	–	65.4	4.3
75– 89 N	11.2	–	–	–	3.9	–	1.4	–	16.5	1.1
1– 74 N	79.1	0.1	–	–	28.4	–	–	–	107.6	7.2
0 Dry	168.5	5.5	121.0	2.4	620.0	–	–	–	917.4	61.0
1– 74 I	15.3	1.4	–	–	65.9	–	–	–	82.6	5.5
75– 89 I	6.1	2.8	–	–	12.3	–	–	–	21.2	1.4
90–119 I	–	–	–	–	1.5	–	–	–	1.5	0.01
0 Cold	–	–	–	–	–	106.2	–	13.5	119.7	8.0
Total:										
Area	380.3	21.2	121.0	2.4	854.6	106.2	3.8	13.5	1502.9	100
Per cent	25.3	1.4	8.1	0.2	56.9	7.1	0.3	0.9		

N = Normal zone; I = Intermediate zone.

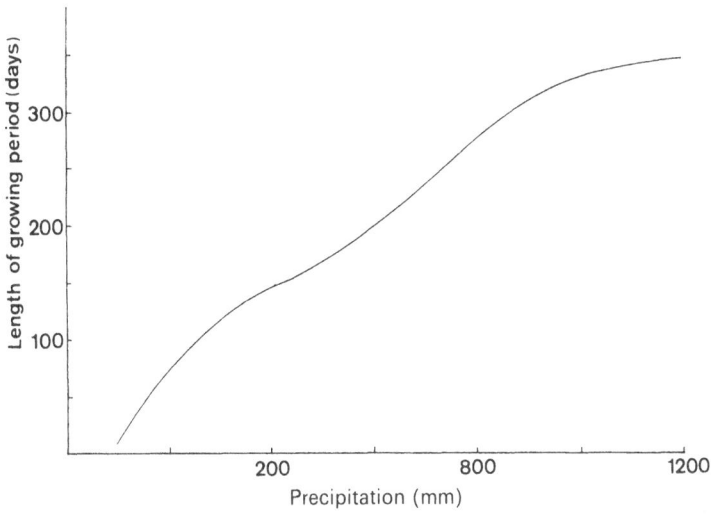

Fig. 4. Relationship between length of growing period and precipitation in major climates 7 and 9.

A generalized version of the inventory for North Africa and West Asia is shown in Figs. 2 and 3 respectively. The distribution of normal and intermediate growing period zones in different major climates for North Africa and West Asia is given in Table 6. Further, a general relationship between precipitation and length of growing period is given in Fig. 4.

The climatic inventory shows that 275.6×10^6 ha (18.3% of the regional total) have a growing period more than 75 days while 190.2×10^6 ha (12.7%) have a growing period of less than 75 days. Areas that have no growing period (i.e. dry) comprise 917.4×10^6 ha (61.0%). Taking the Mediterranean-type environments separately, (i.e. areas in major climates 7, 8, 9 and 10), the inventory shows that 144.0×10^6 ha (14.7%) have a growing period of more than 75 days while 94.3×10^6 ha (9.6%) have a growing period of less than 75 days. Areas than have no growing period (i.e. dry) comprise 620.0×10^6 ha (63.4%) while areas that are above 1500 m in altitude and designated as 'cold' comprise 119.7×10^6 ha (12.2%).

ICARDA focuses research activities in the areas with 200 to 600 mm rainfall. This rainfall range is roughly equivalent to a growing period range of 75 to 210 days (Fig. 4). These areas comprise 125×10^6 ha or 86 per cent of the total area that have a growing period of more than 75 days. Precipitation (P) and potential evapotranspiration (PET) during the growing period in such areas shown in Fig. 5. On average, there is little surplus rainfall in the growing periods of less than 150 days, equivalent to about 400 mm total precipitation. For a 210 days growing period, the surplus amounts to 150 mm. Therefore, in general the depth of wetting of the soil profile does not exceed the depth which can be effectively explored by roots.

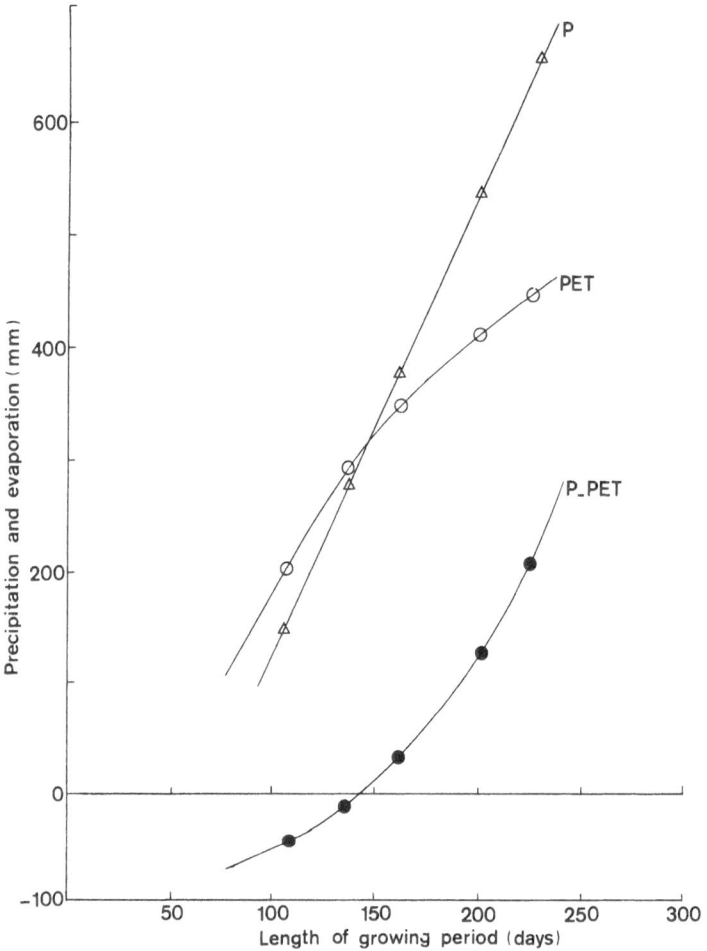

Fig. 5. Relationship between precipitation (P), potential evapotranspiration (PET) and surplus or deficit (P-PET) as functions of the length of growing period.

Mean daily temperatures during the growing period were below 20 °C and seasonal mean values were in the range 12 to 14 °C. Seasonal mean radiation values were in the range 12 to 14 MJ m^{-2} day^{-1} (Sarraf, 1977).

SOIL INVENTORY

The 1:5 million FAO/Unesco Soil Map of the World was the source of information for the soil inventory used in the assessment. The soil mapping units are associations of individual soil units, and there are 106 different soil units making up the 26 major soil groups. The complete definitions of the soil units are given in Volume I (Legend) of the FAO/Unesco Soil Map.

In addition to providing data on the composition and extent of each soil mapping unit, the Map also provides information on the texture class of the dominant soil and the main slope class in the mapping unit. Three texture classes (coarse, medium and fine) and three slope classes (0 to 8 per cent, 8 to 30 per cent and greater than 30 per cent) are recognised. Additionally, the Map provides information on land characteristics which are not reflected by the composition of the mapping units but which are significant for the use or management of the land. This information is presented as a phase overprint on the mapping units. A total of 12 phases are recognised (e.g. stony, lithic, petric, petrocalcic, petrogypsic, saline, sodic).

A summary of the extents of soil groups, combined for all slopes, textures and phases, by growing period zones, is given in Table 7 for North Africa, and Table 8 for West Asia. Sudan has no areas in the subtropics with winter precipitation (i.e.

Table 7. Areas of soil groups by growing period zones in North Africa* (10^6 ha)

Soil group	>210	75–210	1–75	0 Dry	0 Cold	Total	Sudan (all zones)	Region (all zones)
Acrisols	–	–	–	–	–	–	3.4	3.4
Cambisols	2.0	7.0	1.1	0.1	0.5	10.7	3.9	14.5
Rendzinas	0.2	0.9	0.5	<0.1	0.5	2.1	–	2.1
Ferralsols	–	–	–	–	–	–	5.6	5.6
Gleysols	<0.1	0.1	0.1	0.9	–	1.2	12.9	14.0
Phaeozems	–	0.1	<0.1	<0.1	–	0.1	–	0.1
Lithosols	0.7	5.4	8.9	95.4	2.5	112.8	3.2	116.0
Fluvisols	0.5	2.1	2.7	22.1	0.1	27.5	12.6	40.2
Kastanozems	0.7	1.3	<0.1	–	0.2	2.2	–	2.2
Luvisols	2.3	3.8	0.7	0.1	0.8	7.7	11.1	18.8
Nitosols	<0.1	<0.1	–	–	–	0.1	0.8	0.9
Histosols	–	–	–	–	–	–	3.3	3.3
Arenosols	–	0.1	<0.1	0.1	–	0.2	21.7	21.9
Regosols	0.2	2.4	1.6	56.2	0.4	60.9	43.2	104.2
Solonetz	<0.1	0.1	0.1	<0.1	–	0.3	1.0	1.3
Andosols	–	0.1	–	–	<0.1	0.1	0.2	0.2
Vertisols	0.7	1.2	0.1	0.1	–	2.1	42.2	44.3
Planosols	–	0.2	–	–	–	0.2	0.1	0.2
Xerosols	–	5.6	5.2	3.2	0.6	14.7	16.4	31.2
Yermosols	–	3.1	12.0	180.7	0.3	196.1	35.9	232.0
Solonchaks	0.2	1.3	2.5	13.7	0.1	17.7	1.5	19.2
Dunes & salt flats						117.6	3.1	120.7
Total	7.7	34.7	35.7	373.7	6.1	574.3	250.6	824.9

* Morocco, Algeria, Tunisia, Libya, Egypt.

Table 8. Areas of soil groups by growing period zones in West Asia for major climates 7, 8, 9 and 10 and for the region (10^6 ha)

Soil units	Major climates 7, 8, 9 and 10						Region (all zones)
	>210	75–210	1–75	0 Dry	0 Cold	Total	
Acrisols	1.2	0.8	–	–	0.1	2.1	2.1
Cambisols	3.8	10.2	<0.1	<0.1	6.2	20.2	20.2
Rendzinas	0.5	2.7	<0.1	–	0.7	3.9	3.9
Gleysols	0.9	1.0	0.1	1.3	0.5	2.6	2.6
Phaeozems	0.1	0.8	–	–	0.1	1.0	1.0
Lithosols	1.8	20.6	7.8	60.3	51.9	142.5	174.4
Fluvisols	0.8	1.9	4.2	4.3	0.6	11.7	15.8
Kastanozems	<0.1	1.2	–	–	3.2	4.4	4.4
Luvisols	1.3	7.7	<0.1	<0.1	1.1	10.2	10.2
Arenosols	<0.1	<0.1	1.4	25.4	0.3	27.0	37.0
Regosols	0.3	7.9	6.5	39.2	15.1	69.0	89.9
Solonetz	–	0.1	0.9	2.8	0.3	4.1	4.2
Andosols	–	0.1	–	–	<0.1	0.1	0.1
Rankers	0.3	0.2	<0.1	–	2.5	3.0	3.0
Vertisols	0.1	4.1	1.3	<0.1	0.1	5.8	5.8
Xerosols	0.1	22.0	3.2	0.6	14.7	40.7	40.7
Yermosols	0.1	6.3	19.1	98.8	12.3	136.5	190.2
Solonchaks	0.4	1.9	10.9	22.4	3.4	39.0	49.0
Dunes, salt flats & ice caps		0.4	0.6	5.9	0.5	7.4	23.6
Total	11.8	89.8	55.9	259.9	113.7	531.2	678.0

major climate 7). The extent of soil groups for Sudan is therefore presented separately from the other countries in North Africa.

The inventories show that the area covered by Morocco, Algeria, Tunisia, Libya and Egypt is dominated by Yermosols (34.1 per cent), Lithosols (19.6 per cent) and Regosols (10.6 per cent) which together make up 64.3 per cent of the total extent of 574 × 10^6 ha. If the growing period zones 75 to 210 days (the zones of primary interest to ICARDA) are considered separately, then this area is dominated by Cambisols (20.1 per cent), Lithosols (15.5 per cent), Luvisols (11.0 per cent) and Xerosols (16.2 per cent) which together make up 62.8 per cent of the total extent.

In West Asia, the region is dominated by Lithosols and Calcic Yermosols which occupy 25.6 per cent and 17.2 per cent of the total area of 678 × 10^6 ha respectively.

The areas with winter rainfall (i.e. major climates 7, 8, 9 and 10) are dominated by soil groups Lithosols (26.8 per cent), Yermosols (25.7 per cent), Regosols (13.0 per cent) which make up 65.5 per cent of the total extent of 531×10^6 ha.

If the growing period zones 75 to 210 days are considered separately, then ten soil units account for 76.4 per cent of the total area of 90×10^6 ha. These soils are Lithosols (28.8 per cent), Calcic Xerosols (13.3 per cent), Calcaric Regosols (7.1 per cent), Calcic Yermosols (5.2 per cent), Chromic Luvisols (4.7 per cent), Luvic Xerosols (4.0 per cent), Eutric Cambisols (3.7 per cent), Gelic Cambisols (3.7 per cent), Chromic Vertisols (3.4 per cent) and Rendzinas (2.5 per cent).

MATCHING SOIL REQUIREMENTS OF WHEAT WITH SOIL INVENTORY

The suitability of the soils for wheat production was assessed by matching the crop's soil requirements with the properties of the soil units. The rating of soil units for the production of wheat at two levels of inputs is presented in Table 9. The ratings are based on how far the conditions of a soil unit meet the crop requirements under a specified level of inputs. If the soil unit largely meets the crop's requirements, it is adjudged S_1 i.e. the soil conditions do not affect the yield potential set by the climate.

If the soil unit only partly meets the crop's requirements, it is adjudged S_2, i.e. the soil does not allow the full climatic yield potential of the crop to be attained. Failure to meet the crop's minimum soil requirements, outside the range of properties necessary for growth, results in a grading of N, meaning that the soil cannot adequately support production of the crop. Where combination ratings, e.g. S_1/S_2 are given for a soil unit, it is considered that half of the area occupied be of one rating, and the remaining half the other.

In the matching exercise, Fluvisols are treated separately from other soil units because crops are most frequently produced on moisture remaining after rainy season flooding. Therefore the inventory of climatic conditions of the growing period is not totally applicable to the period when crop growth is possible on these soils. In the suitability assessment, areas of Fluvisols (excluding Thionic Fluvisols) were classified 55 per cent suitable for wheat production at both input levels in all growing period zones except zones with 0, 1 to 74I, 240 to 269, 270 to 299, 300 to 329, and 330 to 364 days which are classified as not suitable because of either excessive aridity: salinity or excessive wetness: flooding.

Modifications to the soil unit ratings were made according to significant limitations imposed by slope, texture, and phase conditions. Areas with slopes of 8 to 30 per cent, i.e. 'b' slopes, are modified as follows: for low level inputs, i.e. hand cultivation, one-third of the area remains unchanged, one-third is decreased by one class and the remaining one-third is downgraded to N. For high level inputs,

Table 9. Soil unit ratings for winter wheat

Soil Unit	Input		Soil Unit	Input		Soil Unit	Input	
	Low	High		Low	High		Low	High
Acrisols			*Kastanozems*	S1	S1	*Andosols*		
Orthic	S2	S1				Ochric	S2	S1
Ferric	S2	S2	*Luvisols*			Mollic	S1	S1
Humic	S2	S1	Orthic	S1	S1	Humic	S1	S1
Plinthic	S2/N	S2/N	Chromic	S1	S1	Vitric	N	N
Gleyic	N	N	Calcic	S1	S1			
			Vertic	S2	S1	*Rankers*	N	N
Cambisols			Ferric	S2	S1/S2			
Eutric	S1	S1	Albic	S2	S1	*Vertisols*		
Dystric	S2	S1	Plinthic	S2/N	S2/N	Pellic	S2/N	S1
Humic	S2	S1	Gleyic	N	N	Chromic	S2/N	S1
Gleyic	S2	S2						
Gelic	N	N	*Nitosols*			*Planosols*		
Calcic	S1	S1	Eutric	S1	S1	Eutric	N	S2
Chromic	S1	S1	Dustric	S2	S1	Dystric	N	S2
Vertic	S2	S1	Humic	S2	S1	Mollic	N	S2
Ferralic	S2	S1/S2				Humic	N	S2
			Histosols	N	N	Solodic	N	N
Rendzinas	S2/N	S2/N				Gelic	N	N
			Arenosols					
Ferralsols	N	N	Cambic	N	S2	*Xerosols*		
			Luvic	N	S2	Haplic	S1	S1
Gleysols	N	N	Ferralic	N	N	Calcic	S2	S2
			Albic	N	N	Gypsic	N	N
Phaeozems						Luvic	S1	S1
Haptic	S1	S1	*Regosols*					
Calcaric	S1	S1	Eutric	S1	S1	*Yermosols*	N	N
Luvic	S1	S1	Calcaric	S1	S1			
Gleyic	S2	S2	Dystric	S2	S1	*Solonchaks*	N	N
			Gelic	N	N			
Lithosols	N	N						
			Solonetz					
Fluvisols	See text		Orthic	N	S2			
			Mollic	S2	S2			
			Gleyic	N	N			

one-third of the extent of 'b' slope land remains unchanged, but the remaining two-thirds are downgraded in entirety to N, as mechanized cultivation is normally not possible on some two-thirds of these slopes. All extents of 'c' slopes i.e.

steeper than 30 per cent, are considered as 'b' slopes and the modifications for that slope class apply.

Texture modifications are governed by the following rules: all areas of group 1 textures (coarse) are decreased by one class, except for soil units where light texture limitations have already been applied in the soil unit gradings, e.g. Arenosols. All textures in groups 2 (medium) and 3 (heavy-fine) remain unchanged, no modifications being necessary as limitations imposed by heavy textures are dealt with in the soil unit gradings.

Modifications applied to the soil unit gradings to take into account limitations imposed by phase conditions, are given in FAO (1978). They include, for example, downgrading of soil unit ratings for soils with (a) stony and lithic phases because of difficulties with mechanized cultivation, (b) hardpans at shallow depths, (c) petrocalcic phase because of high calcium carbonate content, (d) saline and sodic phases.

These gradings of soil units were used to modify the agro-climatic suitability assessment described in the next section.

AGRO-CLIMATIC SUITABILITY ASSESSMENT

The agro-climatic suitability classification was derived in two steps, namely:
i calculation of constraint-free potential yields of the crop in suitable major climates distinguished by growing period;
ii calculation of agronomically expected yields, by amending the constraint-free yields through an application of factors that reflect losses likely to occur due to such constraints as water stress, pests and diseases, workability, according to their severity in each growing period zone.

Details of the methodology for the calculation of constraint-free yields are given in Kassam (1977), FAO (1978) and Doorenbos and Kassam (1979). In the model, which uses seasonal mean temperature and radiation data, it was assumed that the growth cycle of a crop was 120 to 150 days, its maximum leaf area index was 5 and its harvest index 0.4. Constraint-free yields were calculated for the high input conditions. The potential yields for the low input conditions were assumed to be 25 per cent of those under the high input conditions (Oram, 1978).

Expected yields were calculated from the constraint-free yields by assuming the following constraints and yield losses:
(a) water stress would decrease potential yields by 50 per cent in the normal zones 75 to 89 days, and 90 to 119 days, and by 25 per cent in the zones 120 to 149 days, and 150 to 179 days.
(b) pests and disease constraints would decrease potential yields by 25 per cent in zones 240 to 269 days and 270 to 299 days.

Table 10. Anticipated yields (with constraints) in t ha^{-1} dry weight and agro-climatic suitability classification for winter wheat by length of growing period in North Africa and West Asia (from Kassam, 1978)

Crop	Input Level		Growing period (days)									
			75–89	90–119	120–149	150–179	180–209	210–239	240–269	270–299	300–329	330–364
W W H I E N A T E T R	High	Yield*	0.5–0.6	1.0–1.4	2.0–2.5	2.9–3.8	3.6–4.9	3.6–4.9	2.7–3.6	2.7–3.6	1.4–1.8	1.0–1.4
		% of Max.	9 13	20 28	38 52	77	100	100	75	75	40	28
		Suit.**	NS	MS		S		VS		S		MS
	Low	Yield	0.1–0.2	0.3–0.4	0.5–0.6	0.7–1.0	0.9–1.2	0.9–1.2	0.7–0.9	0.4–0.9	0.4–0.5	0.2–0.3
		% of Max.	9 13	20 28	38 52	77	100	100	75	75	40	28
		Suit.	NS	MS		S		VS		S		MS

* Values refer to yields on an annual basis. In the growing period zones 75–89 days, 90–119 days, and 120–149 days, the land is cropped 1 year in 3, 1 in 2, and 2 in 3 respectively; and the anticipated yields on a when actually cropped basis are 1.4–1.9, 2.0–2.8, and 2.9–3.8 t ha^{-1} respectively for the high input level and 0.4–0.5, 0.5–0.7 and 0.7–1.0 t ha^{-1} respectively for the low input level.

** VS – very suitable; S – suitable; MS – marginally suitable; NS – not suitable.

(c) workability constraints, in addition to pest and disease constraints would decrease potential yields by 50 per cent in zones 300 to 329 days, and 330 to 364 days.

Data on expected yields for high and low input conditions are presented in Table 10. Based on the work of Janssen (1972), it has been assumed that the land in the growing period zones 75 to 89 days, 90 to 119 days and 120 to 149 days is cropped one year in three years (i.e. one crop year followed by two years of fallow to conserve moisture), one in two and two in three respectively. Only when the growing period is longer than 150 days, is the land assumed to be cropped without fallow. Therefore the expected yield values in Table 10 for growing period zones of less than 150 days are the annual equivalents of the yields obtained when the land is cropped in the assumed crop: fallow cycles.

Two additional sets of data are also presented in Table 10. The first data set shows the expected yield data from each length of growing period zone as a percentage of the maximum attainable, e.g. at the high input level, the maximum expected yield attainable is 4.9 t ha^{-1} in the 180 to 239 days zones. Therefore the yield of 1.4 t ha^{-1} from the 90 to 119 days zone is, as shown, 28 per cent of the maximum. When a length of growing period zone completely accommodates the longest growth cycle, i.e. 150 days, the upper yield figure reflects the full potential of that zone, and therefore only one corresponding percentage of maximum yield figure is presented.

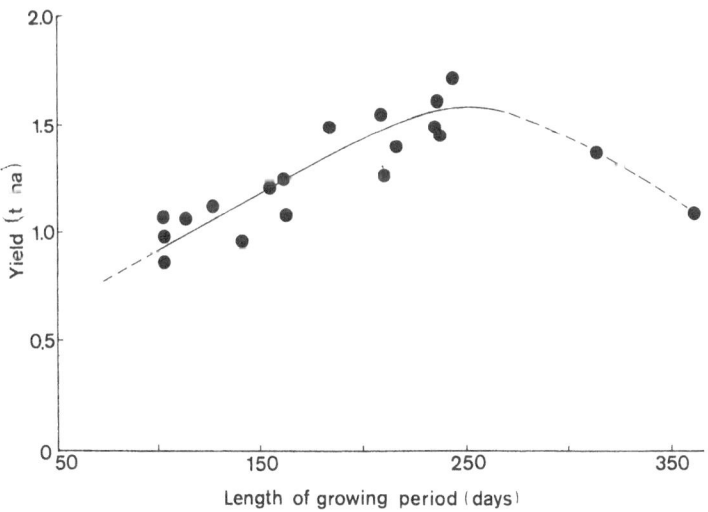

Fig. 6.   Relationship between average winter wheat yield and length of growing period in Turkey.

The second data set contains agro-climatic suitability classes. Suitability was assessed by considering all the agronomically possible yield ranges and classifying each individual growing period zone into one of four classes, defined in terms of a percentage range of the maximum attainable without constraints. Growing period zones capable of yielding less than 20 per cent of the maximum yield were classified as agro-climatically not suitable (NS); zones yielding 20 per cent to less than 40 per cent as marginally suitable (MS); zones yielding 40 per cent to less than 80 per cent as suitable (S); and zones yielding 80 per cent or more as very suitable (VS).

Values in Table 10 postulate that production potential increases with increase in the length of growing period up to about 240 days and then decreases with further increase. This trend is also observed in reality and is seen in the relationship between length of growing period and average (17 years) winter wheat production from 19 agricultural regions in Turkey (Fig. 6).

LAND SUITABILITY ASSESSMENT

In the final assessment of potential land suitability, the soil assessment was superimposed on the agro-climatic assessment. In the case of areas of soils judged S_1, no change was made in the agro-climatic suitability classification. Areas of soils adjudged S_2 had their agro-climatic suitability downgraded by one land suitability class. For soils judged as N the land is not suitable for cropping. Subsequently, the ratings for different soil phases, soil slope classes and soil texture classes were consecutively applied to arrive at the final land suitability assessment.

Four land suitability classes are therefore employed as for the agro-climatic suitability classes, and each land class is linked to expected yields from the two levels of inputs considered.

The above classification refers to areas in the normal growing period zones. In the intermediate growing period zones (i.e. 75 to 89 and 90 to 119 days), there is an additional constraint of 'continuous' water stress. Accordingly, the land suitability class of an intermediate zone was taken as one class lower than for the normal zone of similar duration.

The division between 'not suitable' and 'marginally suitable' is related to the break-even point of the value of the produce in comparison to production cost. Economic data from the region indicate that yields of 150 to 250 kg ha^{-1} with low inputs are not profit-making. This yield corresponds with the dividing line between 'marginally suitable' and 'not suitable' when fallow period requirements are included in the calculations of yields on an annual basis for interzonal suitability comparisons.

Table 11. Areas of land suited to the production of winter wheat in major climates 7 and 9 in North Africa and West Asia at high and low levels of inputs (10⁶ ha)

Length of growing period (days)	Very suitable		Suitable		Margin. suitable		Not suitable		Total
	High	Low	High	Low	High	Low	High	Low	
330–364 N			<0.1	<0.1	<0.1	<0.1	0.3	0.3	0.3
300–329 N			0.3	0.2	<0.1	<0.1	0.3	0.9	0.3
270–299 N			0.4	0.6	<0.1	0.2	1.4	1.3	1.7
240–269 N	0.3	0.3	2.0	2.8	<0.1	0.7	2.1	2.8	2.8
210–239 N	2.5	1.6	1.4	3.4	<0.1	1.1	10.0	8.9	14.4
180–209 N	4.2	3.0	5.9	3.1	0.1	1.0	12.2	11.2	17.8
150–179 N			3.4	2.8	1.4	3.6	13.7	14.1	21.0
120–149 N			0.5*	0.5*	3.2	4.5	16.4	15.7	23.0
90–119 N			0.1*	0.1*	6.1	5.6	37.1	36.8	43.6
75– 89 N							5.2	5.2	5.3
1– 74 N			0.9*	0.9*			27.5	27.5	28.4
0 Dry							620.0	620.0	620.0
1– 74 I							65.9	65.9	65.9
75– 89 I			0.4*	0.4*			11.8	11.8	12.3
90–119 I			0.2*	0.2*			1.3	1.3	1.5
Total: (0–364)									
Area	7.0	4.8	15.4	14.9	10.9	16.7	825.1	821.9	858.3
Percent	0.08	0.06	1.8	1.7	1.3	1.9	96.1	95.8	100
Total: (75–210)									
Area	4.2	3.0	11.8	10.3	10.7		97.7	94.8	124.5
Percent	3.3	2.4	9.5	8.2	8.6		78.5	76.2	100
Expected yield range (t ha⁻¹)	4.9–3.9	1.2–1.0	3.9–2.0	1.0–0.5	2.0–1.0	0.5–0.3	1.0–0	0.3–0	

* Fluvisols.

The extent of the land variously suited to the production of winter wheat in major climates 7 and 9, i.e. 'lowland' winter rainfall areas, in North Africa and West Asia is given in Table 11. The proportion of the total area in the different land suitability classes is broadly similar for both input levels. About 90 per cent of the total extent of 858×10^6 ha of the 'lowland' areas with winter rainfall emerges as being 'not suitable'. Less than one per cent of the total area is 'very suitable' while 'suitable' and 'marginally suitable' classes each comprise less than 2 per cent of the total.

The area with growing periods of 75 to 209 days comprises 125×10^6 ha (14.5 per cent of the total area). Only 20 to 23 per cent of this area of primary interest to ICARDA appears as suitable for rainfed winter wheat production in the land utilization type considered.

It may be argued that some of the land classified as unsuitable in growing period zones 75 to 89I, 75 to 89N and 90 to 119I days, is cultivable, and is indeed cultivated particularly for barley production. The total extent of such land is between 17 and 18×10^6 ha (14 to 15 per cent of the total area in 75 to 209 days zones). If no distinctions were made in suitability classes between the land in the intermediate zones and that in the equivalent normal zones, then the zone 90 to 119I days would be classed as agro-climatically marginally suitable on an annual basis or suitable on a when-cropped basis, i.e. one year out of two years.

Similarly, zones 75 to 89I and 75 to 89N days would be classed as agro-climatically not suitable on an annual basis but marginally suitable one year out of three years. Assuming that half the areas of these zones have no soil limitations, then the zone 90 to 119I would provide 0.65 (0.32 annually) $\times$ 10^6 ha; zone 75 to 89I would provide 5.9 (1.96 annually) $\times$ 10^6 ha, and zone 75 to 89N would provide between 1.92 (0.63 annually) $\times$ 10^6 ha at low input level, and 2.6 (0.78 annually $\times$ 10^6 ha at high input level. This would add a total of 8.8 (2.91 annually) $\times$ 10^6 ha and 9.2 (3.1 annually) $\times$ 10^6 ha at low and high input levels respectively, corresponding to 6.6 and 7.2 per cent of the total extents of 75 to 209 days zones.

In the zone 120 to 149 days, the land is assumed to be cropped two years out of three years so that 66 per cent of the suitable land is available annually in this zone. The total annually available suitable land (not including 'suitable' areas from zones 75 to 89I, 75 to 89N and 90 to 119I) is equal to 25.3×10^6 ha (19.8 per cent of the total area in 75 to 209 days zones) under high input conditions and 25.6×10^6 ha (20.9 per cent) under low input conditions. If 'suitable' land from zones 75 to 89I, 75 to 89N and 90 to 119I is included, then the total annually available land would be 27.4×10^6 ha (22.3 per cent) and 28.6×10^6 ha (23.2 per cent) under high and low input conditions respectively.

These figures highlight a widely known regional characteristic: suitable land is a scarce agricultural resource in the region. However, on an annual basis, these land resources of 27.4 and 28.6 $\times$ 10^6 ha in the zones 75 to 210 days are capable of offering an average rainfed wheat production of 69 $\times$ 10^6 tonnes, up to 87 $\times$ 10^6 t, with high inputs and 17 $\times$ 10^6 t, up to 22 $\times$ 10^6 t, with low inputs. It is interesting to set these figures against the present cereal deficit of 20 to 25 $\times$ 10^6 t in the region with a total cereal production (rainfed and irrigated) of 45 to 50 $\times$ 10^6 t per annum.

DISCUSSION

I am keenly aware of the general and broad nature of this regional resource analysis, and the whole exercise needs to be repeated at a more appropriate scale using better models, both static and dynamic. I am also conscious of omissions, most notably that of assessment of lands above 1500 m altitude.

Expansion of production through research, increased inputs and investment should be planned and achieved in the context of a comprehensive inventory of land and its production potential. Land resource surveys will need to specifically categorize variables of importance to crop growth, combining climate, soil and management factors which affect the optimum use of water and nitrogen.

Several recent advances in the quantitative analysis of crop production in relation to nitrogen cycling and water use in the Mediterranean environments are pertinent in the present context (e.g. Harpaz, 1975; van Keulen, 1975; Noy-Meir and Harpaz, 1977). These studies show that about 50 kg ha^{-1} of plant biomass can be produced from each mm of transpired water but to support this conversion, in a winter cereal crop, about 0.6 kg ha^{-1} of nitrogen is needed.

In the natural state, the main nitrogen gains are in rain, and fixation by non-symbiotic microorganisms; the amount of both will depend on precipitation and its seasonal distribution. The organic matter content of the soil is low and the amount of mineralized nitrogen released from it each year depends on rainfall, its distribution and cultivation practice.

In the absence of nitrogen fertilizer, the potential contribution of symbiotic nitrogen fixation is large compared to other likely sources of gains. However, the realization of this potential depends on the introduction of suitable legumes and rhizobium strains on the one hand and suitable soil and climatic conditions on the other. Increase in the nitrogen supply to the crop through regular additions of nitrogenous fertilizer generally results in considerable increases (two to eight fold) in yield, except in the very dry areas where water limits the increase of yield and the natural source of nitrogen is adequate to meet the crop's demand for nitrogen.

In comparison with the characteristic features of physical environment of the seasonally arid tropics (Kowal and Kassam, 1975) many of the characteristic features of the Mediterranean-type environments in relation to nitrogen and water tend naturally towards efficient utilization. In particular, the seasonal nature of the thermal and water regimes, the moderately cool temperatures during the growing period and the moderate depth of wetting of soil by rainwater all help to retain most of both the stable and mineralized reserves of nitrogen within the root zone.

Regardless of the method used to increase the supply of nitrogen to crops, the efficiency of utilization is likely to be high because most of the unused or surplus nitrogen will be used in subsequent years. Further, because Mediterranean-type environments are transitional in nature between the tropics and temperate areas, the possible choice of crops, cropping patterns and rotations, and farming systems is wider.

Research into ways of increasing the efficiency of utilization of water and nitrogen will have to take account of the fact that the processes involved have different time spans. Further, the processes of accumulation, conservation and utilization may all operate concurrently on a piece of land in one rotational system whereas they may occur in different pieces of land and in different years in another system but all contributing towards the overall efficiency of utilization for the rotation. This implies that, for a rotation which involves a fallow year to conserve moisture, the efficiency of utilization of water and nitrogen for the fallow in the fallow year would be zero but the efficiency of accumulation would be greater than zero. Efficiencies of accumulation, conservation and utilization may all have to be assessed for each crop season within each experimental rotation system.

When processes of utilization alone are operating, the efficiency of utilization may be assessed in terms of yield per unit of transpired water per unit of nitrogen taken up by crop, while aiming to obtain the largest possible production. However, when the processes of accumulation and conservation are operating in conjunction with the processes of utilization, account must be taken of the net acquisition of the 'new stock' and its retention in the 'stock room'.

Yield, transpiration and nitrogen uptake, are linked to each other through crop: environment interactions. These interactions are now well recognised, and in many cases they can be quantified. Research efforts in management, agronomy and plant breeding towards better utilization of land resources of water and nitrogen need to take proper account of this ecophysiological 'chain' because improvement in efficiency invariably implies improvement through or in one or more of its 'links'.

Lastly, recommended farming systems for increasing the efficiency of utilization of land resources need to be compatible with prevailing economic, sociocultural and political conditions. However, a rational assessment of compatibility of new technologies requires that the research includes variables and levels that operate outside the prevailing absorption capacity of the farming communities being served.

REFERENCES

Beek, K. J. 1978 Land evaluation for agricultural development. International Institute of Land Reclamation and Improvement (ILRI). Publication 23, Wageningen.

Darling, H. S. 1979 International Research in Agriculture. 5. ICARDA. *Span* 22, 55–57.

Doorenbos, J. and Kassam, A. H. 1979 Yield response to water. Irrigation and Drainage Paper 33, FAO, Rome.

FAO 1976 A Framework for Land Evaluation. Soil Bulletin 32, FAO, Rome.

FAO 1978 Report on the agro-ecological zones project Vol. I. Methodology and results for Africa. World Soil Resources Report 48/1. FAO, Rome.

FAO 1979 Report on the agro-ecological zones project Vol. 2. Results for South West Asia. World Soil Resources Report 48/2. FAO, Rome.

Harpaz, Y. 1975 Simulation of the nitrogen balance in semi-arid regions. Ph.D. Thesis, Hebrew University, Jerusalem.

Janssen, B. H. 1972 The significance of the fallow year in the dry-farming system of the Great Konya Basin, Turkey. Neth. J. Agric. Sci. 20, 247–260.

Kassam, A. H. 1977 Net biomass and yield of crops with provisional results for tropical Africa. Consultant's Report. Agroecological Zones Project, AGLS, FAO, Rome.

Kassam, A. H. 1978 Agro-climatic suitability assessment for rainfed wheat in the Near and Middle East by growing period zones. Consultant's Report. Agro-ecological zones project, AGLS, FAO, Rome.

Keulen, van H. 1975 Simulation of water and herbage growth in arid, regions, simulation monograph, Pudoc, Wageningen.

Keulen, van H. 1975 Simulation of water and herbage growth in arid, regions, simulation monograph, PUDOC, Wageningen.

Kowal, J. and Kassam, A. H. 1975 Agricultural ecology of Savanna. Clarendon Press, Oxford.

Noy-Meir, I. and Harpaz, Y. 1977 Agro-ecosystems in Israel. Agro-Ecosystems 4, 143–167.

Oram, P. 1979 Crop production systems in the arid and semi arid warm temperate and Mediterranean zones. In: Soil, Water and Crop Production, pp. 193–228, Thorne, D. W. and Thorne, M. D. (eds.). AVI Publishing Co., Westport, Connecticut.

Sarraf, S. 1977 Determination de zonas agroecologiques de l'Afrique du Nord et du Moyen-Orient. Consultant's Report. Agro-ecological zones project, AGLS, FAO, Rome.

2. Environmental resources and restraints to agricultural production in a mediterranean-type environment

RICHARD C. G. SMITH* and HAZEL C. HARRIS

Dept. of Agronomy and Soil Science, University of New England, Armidale, Australia

Analysis of the agro-climatology of the semi-arid and arid zones of the Near East has been undertaken by de Brichambaut and Wallén (1963), and Kassam (1981) reviewed the land resources and agricultural potential of the region. To complement these broad studies, this paper attempts a more detailed analysis for a few locations.

The agricultural potential of Mediterranean areas is strongly determined by the incidence of winter rainfall, modified by radiation and temperature. Low temperature limitations are important in the plateau regions of Turkey, Iran and north-east Iraq, but less significant in the lowland areas surrounding the Mediterranean Basin. Temperature regimes are moderately predictable and the local crops and agronomy are probably well adapted to this factor.

The main uncertainties in the thermal regime are the potential for damage by radiation frosts (Single, 1975) and high temperatures during grainfilling. Frost effects on crop yield are hard to predict from general temperature records as considerable spatial variation exists in the temperature inversions that develop and are the principle cause of frost (Rosenberg, 1974). At best, only broad analyses of the data of the last frost can be attempted at present (Harris, 1978).

Similar to temperature, rainfall has a distinct seasonal pattern, but here the similarity ends. Rainfall is sporadic and highly variable. Some rain may be lost to the system when intensities exceed the infiltration capacity of the soil. The effectiveness of infiltrated rain depends on the prevailing evaporative demand, plant cover and the ability of soil to store water for periods of insufficiency. Fortunately, the winter rainfall pattern coincides with a period of low evaporative demand, allowing crop production to occur on small amounts of rainfall. The year to year variability in rainfall implies a corresponding wide range in the production potential of crops. The challenge to research, given this inherent variability of a scarce resource, is to devise systems that will maximize the efficiency of water use in the face of rainfall uncertainty and variability.

The objective of this review is to examine the implications of rainfall uncertainty in crop production systems in a Mediterranean-type environment. How best to deal with the implication of this uncertainty will be left to later articles in this volume by Turner (1981) and Fischer (1981). We have chosen to emphasize soil water.

Solar radiation and temperature are important, but the period of the year when a crop can utilize solar radiation in rainfed agriculture is dependent on the growing season which sets the limits to the period when radiation can be captured. During winter, solar radiation and temperature may be limiting crop growth but, at the same time, they are major determinants of evaporative demand which influences the effectiveness with which rainfall is stored. The interactions are complex, and the challenge is to move from the traditional descriptive approach to a prescriptive analysis of climatic data that can define more accurately avenues for improving productivity.

THE PRESCRIPTIVE APPROACH

Insight into the nature and uncertainty of the water resource requires an analysis of historical rainfall data. We propose that the most useful means of relating such an analysis to crop growth and yield, is through an appropriate model based on a background of experimentation to ensure its validity. Such a model should be dynamic and describe the processes necessary for an adequate definition of the system.

Farmers are typically averse to risk (Anderson, 1074), and since climatically induced risk is part of the system, it should be quantified. New cultivars or 'improved' systems need to be risk efficient and evaluated by risk analysis (Anderson, 1974; Smith et al., 1978b).

Dynamic models of crop growth have beeen increasingly used in climatic analysis (Fitzpatrick and Nix, 1970; Smith et al. 1978a; Nix, 1975). These models have been limited by their simplicity. More comprehensive crop models such as that of de Wit et al. (1978) have, after simplification, found application in predicting potential crop production under water limiting conditions (van Keulen, 1975; Slabbers et al., 1979).

OUR MODEL

To assess the relative importance of the environmental constraints, we have developed a dynamic model of a wheat crop to analyse climatic data from some sites in the lowland region of the Near East, relevant to ICARDA (Fig. 1).

The major environmental variables, the sub-systems in the overall crop system and basic processes included in the model, are detailed in Fig. 2.

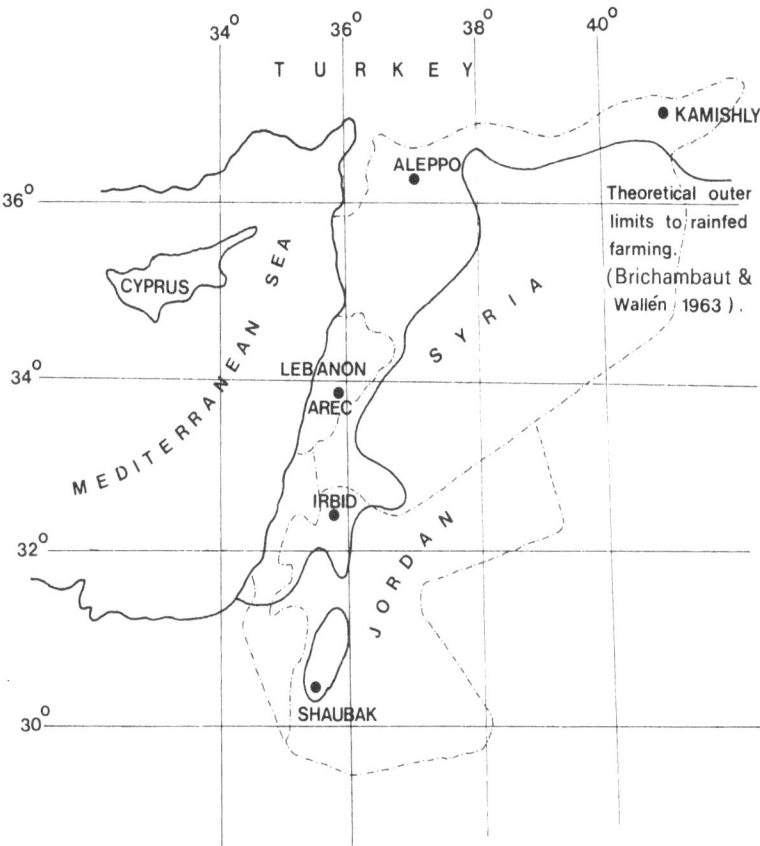

Fig. 1. The lowland area of crop production in the Near East and typical sites (●) selected for climatic analysis.

Phenological development was predicted using the equations of French *et al.* (1979) for a spring wheat (cv. Halberd) which are based on seeding date and temperature. The events simulated after sowing were floral initiation, mid-flowering and soft dough (physiological maturity). The cultivar Halberd is grown widely in South Australia in Mediterranean environments with rainfall varying from 250 to 350 mm during the growing season.

Crop photosynthesis was predicted using the simplification by Goudriaan and van Laar (1978) of de Wit's (1965) model. Respiration was predicted according to McCree (1974) with coefficients for wheat from Hodges and Kanemasu (1977). Net photosynthesis was assumed to be converted to plant biomass with 70 per cent efficiency; partitioning to roots declined linearly from 40 per cent at the beginning of growth to zero at the end, and the Photosynthetic Area Index (PAI)

34

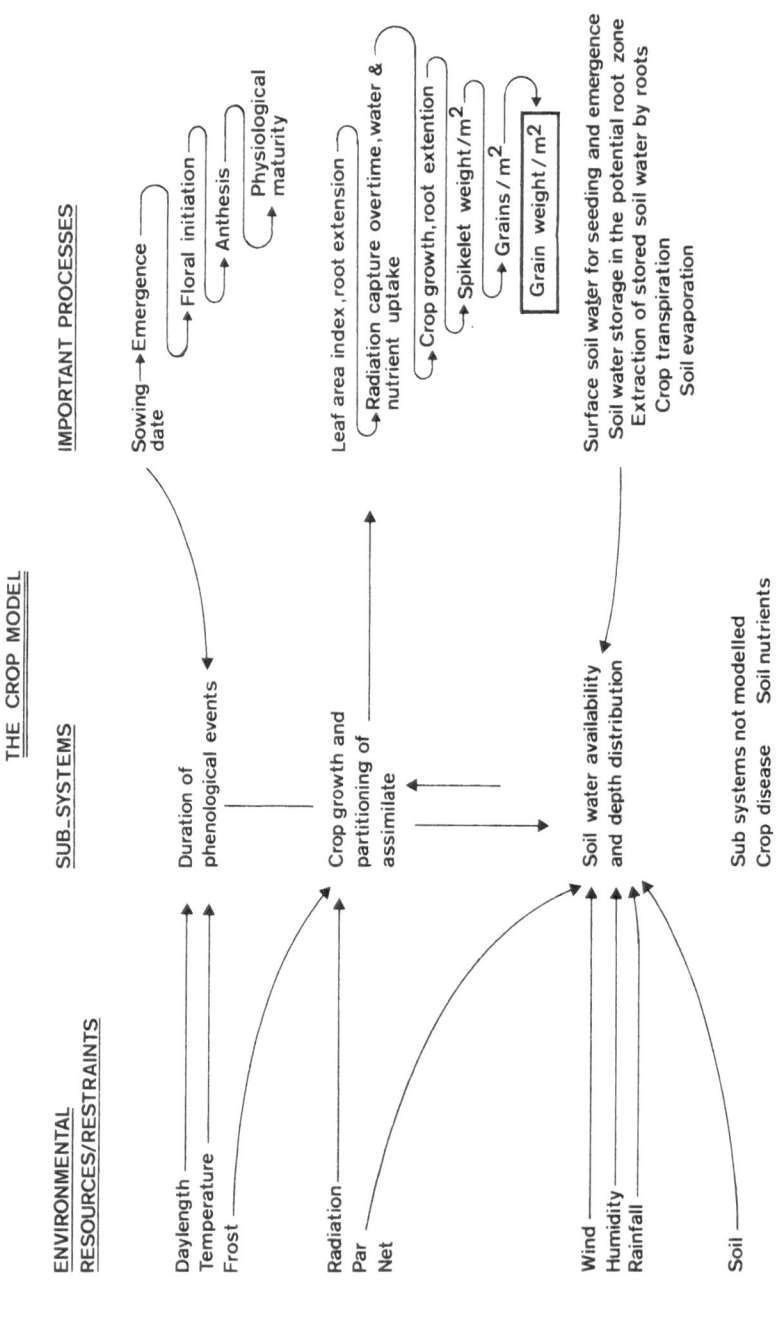

Fig. 2. The outline of the environmental resources/restraints of a Mediterranean environment and a crop model to analyse their effect on crop yield.

declined from 0.025 PAI per g m^{-2} of dry matter to zero at the end of growth. Senescence was simulated after anthesis to match the decline measured by Connor (1975) for a spring wheat in Victoria, Australia (cv. Sherpa). Kernel number m^{-2} was simulated as a function of total dry matter at flowering, using a relationship derived for the wheat cultivar, Heron (Fischer, 1979). This cultivar was considered to be the most typical of the more traditional Near Eastern lowland wheats.

Fischer (1979) also gives an equation for the dwarf Mexican wheats which produce a much higher number of kernels per unit weight of dry matter at anthesis. Kernels were assumed to have an upper size limit of 50 mg so that kernel number and size set an upper sink limitation to yield. A major determinant of this sink limitation was the time spent by the crop between each of the phenological events. Kernel weight was predicted by assuming that 80 per cent of the post-anthesis dry matter production went into kernel, filling at a maximum grain growth rate of 2 mg/kernel/day.

Frost was assumed to occur when the minimum screen temperature was 0°C, assuming that, with a temperature inversion, ground level ambient temperatures may be within the range to damage sensitive tissue. Kernel number was reduced by 50 per cent on days when the minimum temperature was $\leq 0\,°C$, but was allowed to increase at the 'normal' rate on subsequent days. However the sensitivity of the ear to frost damage depends on the degree of protection of the ear by surrounding tissue and previous cold hardening. Therefore greatest sensitivity occurs as the ear emerges and when there are high daytime temperatures promoting active growth (Single, 1975). We believe that prediction of frost damage in the plant is possible only in very general terms, and that more specific predictions of yield reduction are probably impossible on present knowledge.

Similar uncertainty surrounds the effect of high temperatures during the grain-filling stage. Hot spells at this period are frequently associated with high pressure cells causing circulation of air from desert regions into the more humid zone (Fisher, 1978). This results in the winds known variously as the 'sirocco' or 'hamseem' responsible for the advection of energy into the cropping zones which may severely restrict the yield of crops. The prediction of yield loss depends on the availability of data for the frequency of occurrence of these events, and on a clearer definition of the effect of exposure to high mid-day temperatures ($> 33\,°C$) on photosynthetic capability and grain development in crops.

Evapotranspiration and soil water balance were predicted using the model of Ritchie (1972, 1974), modified by Ritchie (pers. comm.) to simulate soil water distribution by depth, using the tipping bucket concept for infiltration and the reverse concept for water extraction. The model separates soil and plant evaporation which is important in annual crops with incomplete canopy cover for a significant period of the season.

We are currently working with ICARDA to both validate and improve this model to enable us to have greater confidence in its use for climatic analysis.

ANALYSIS OF CLIMATIC DATA

Primary data

Climate data from a typical lowland site at Aleppo, Syria, where sowing commonly occurs from mid-November to mid-December is considered (Fig. 3). Radiation declines to a minimum of 6.3 MJ m^{-2} day^{-1} (150 cal cm^{-2} day^{-1}) associated with mean temperatures of about 6 °C (Fig. 4). Under these conditions, growth and development will be slow but should not cease completely (Friend, 1966). In the high plateau country where mean mid-winter temperatures are below 0 °C, growth will cease for extended periods.

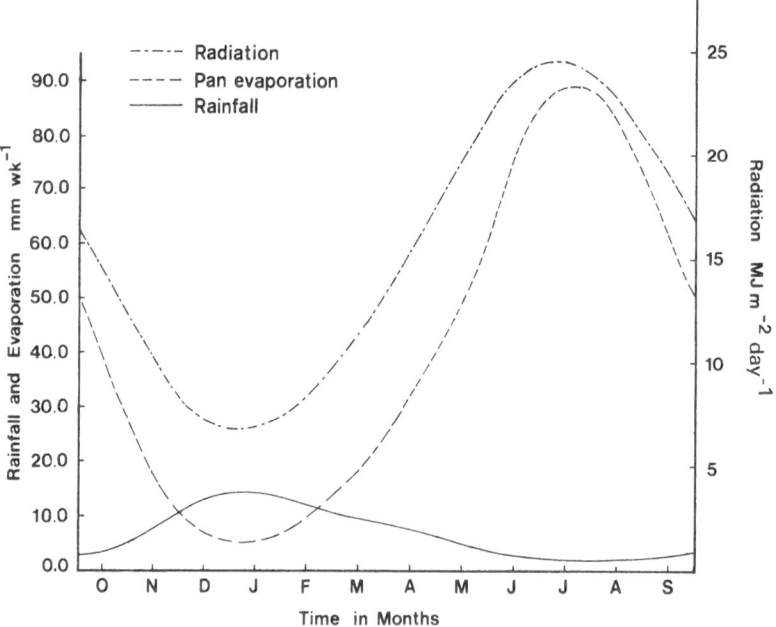

Fig. 3.   Mean weekly radiation, pan evaporation (Lambrecht) and rainfall for Aleppo, Syria.

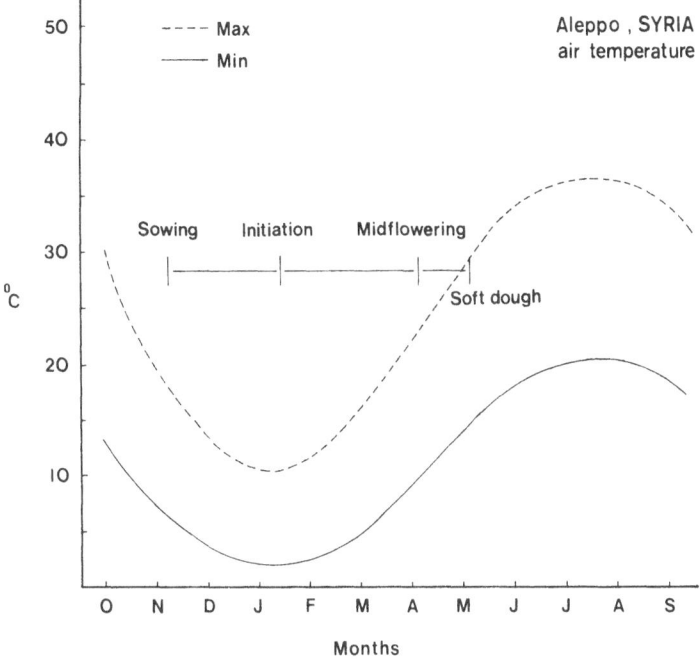

Fig. 4. Mean maximum and minimum screen temperatures for Aleppo, Syria with approximate phenological events for wheat. From French *et al.* (1979).

It is more significant that low radiation and temperature, in association with high humidity and low windspeed (Fig. 5), restrict the evaporative demand conditions, and with increasing rainfall, soil water content increases. Crop growth is possible in this midwinter climate but conditions are also favourable for crop disease in some leguminous crops (Hawtin, 1979). After floral initiation, the crop experiences stronger sunshine and higher temperature through and beyond physiological maturity. Frosts cease by the beginning of April, several weeks before mid-flowering (Harris, 1978). The final grainfilling period occurs with mean temperatures rising from 15 to 20 °C from the end of April until the middle of May.

In general, kernel numbers per ear and final grain size are inversely related to temperatures above 10 to 15 °C (Wardlaw, 1975). Therefore the temperature regime between floral initiation and physiological maturity may have a profound effect on yield (Warrington *et al.* 1977). These effects of temperature may be important in determining the yield potential of an area for wheat and should be included in a full prescriptive climatic analysis. The effects are included in the model.

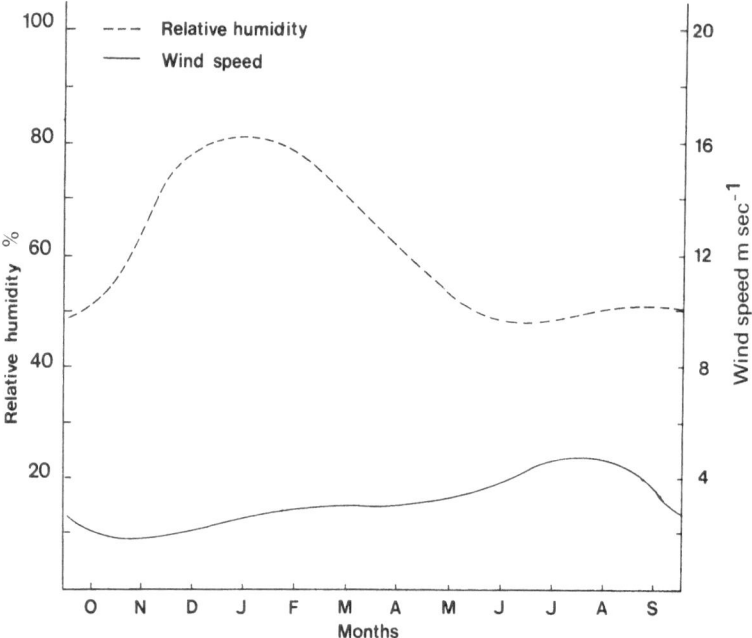

Fig. 5. Mean relative humidity and windspeed for Aleppo, Syria.

Growing season length at five lowland locations

The production of crop dry matter depends primarily on the amount of solar radiation intercepted by active photosynthetic tissue over time. The period of the year in which interception by photosynthetic tissue can occur is limited by the period when soil water is sufficient to sustain photosynthetic tissue. We define the potential growing season (PGS) from the time when continuous growth begins until the time it must cease because of the shortage of soil water. The actual growing season (AGS) realized in practice will be somewhat shorter, due to a lag after the break of season to permit soil water to accumulate for safe establishment, seedbed preparation and weed control.

At the end of the season, crop growth may cease due to the natural processes of maturation of an annual crop. In this case, growth may cease before the end of the potential season. Alternatively, the cessation of growth or, if the crop has reached the grainfilling stage, the hastening of maturity, may be caused by declining soil water and rising evaporative demand.

Table 1. Soil water profile characteristics assumed for a good wheat soil based on the difference between the maximum and minimum observed soil water contents (Gibbon pers. comm.) giving the extractable soil water (ESW).

Depth m	ESW per cent	Total ESW mm
0 –0.2	20	40
0.2–0.4	18	36
0.4–0.6	15	30
0.6–0.8	12	24
0.8–1.0	10	20
		150

The PGS is determined by the interaction between climate, crop water use and stored soil water. The model was used to determine the PGS from long-term daily rainfall data for five locations (Fig. 1). Sowing occurred on the first day after the beginning of November when the moisture content exceeded the value of 60 per cent of the maximum extractable water. The assumed soil water characteristics are set out in Table 1, based on data supplied by Gibbon (pers. comm.).

The simplifying assumption, made in the PGS analysis, is that each growing season is independent of its predecessor; in particular there is no carryover of soil water from one season to the next. This assumption allows the potential variability in length of growing season to be defined, and from this basis the consequence of practices that result in soil water carryover can be quantified in terms of the reduction in this variability.

For the determination of the PGS, to give effect to this assumption the day-degree requirement of the crop was doubled to ensure that crop growth and hence water use would continue until terminated by a lack of soil water in the root zone. This was assumed to occur when the ESW in the root zone fell below 2 per cent of the total. The lower limit for potential evaporation (Ritchie et al., 1972) was assumed to occur when 70 per cent of the ESW in the root zone had been extracted. The ratio of actual to potential evapotranspiration then declined to zero at zero ESW.

The results are presented as a frequency analysis (Fig. 6 and 7). Mean PGS varied from 138 days at Shaubak to 200 days at Kamishly. The values are 10 to 20 days higher than the values quoted by Kassam (1980) where the growing season is based on the period when rainfall exceeds 50 per cent of the potential evapotranspiration. Growth can probably begin before this point where soil evaporation predominates and rapidly passes from Stage I evaporation to the falling rate of

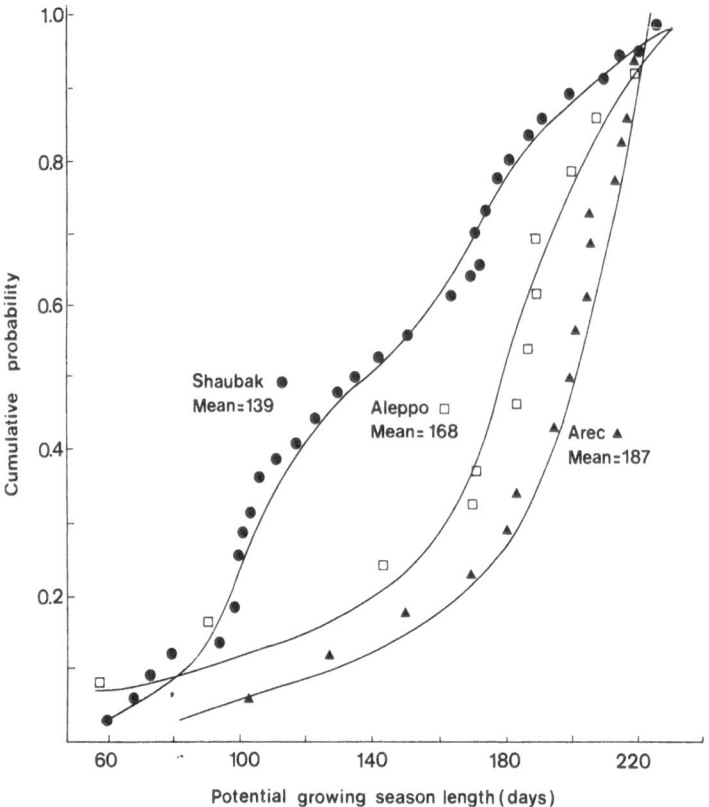

Fig. 6.   The cumulative probability distribution for potential length of growing season.

Stage II evaporation, in the days after rainfall. Similarly, at the termination of the season, crop growth can continue for several weeks on stored soil water. Brichambaut and Wallén (1963) quote lengths of rainy season (e.g. Aleppo 180 days) which agree with our predictions. For modelling purposes, we need a more precise definition for the beginning and end of the growing season.

A significant feature of the analysis is that the variation in PGS within sites is greater than the mean difference between sites. For example, at Aleppo, the PGS can vary from 60 to 220 days, but the probability distribution is positively skewed with a greater number of seasons falling above than below the mean.

This variability poses problems for the development of systems and crop varieties to make efficient use of the environmental resource. However, this variability should dominate our thinking because it is under these conditions that the farmer has to produce his crops and survive. Being averse to risk, he

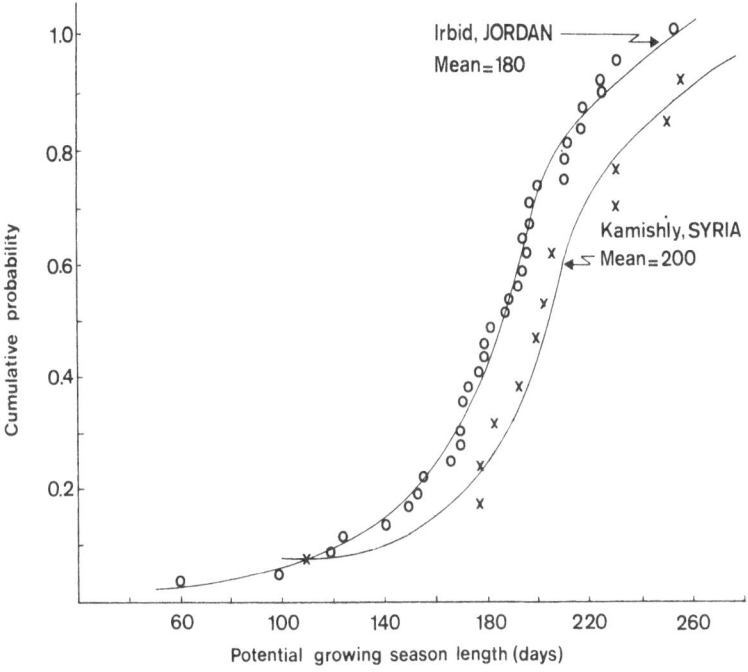

Fig. 7. The cumulative probability distribution for potential length of growing season.

wishes to minimize failure but, being a profit taker, he wishes to maximize yield. To what extent are these two objectives compatible? Within the realm of biological possibilities we need to strive to make them so.

To increase our insight into this year-to-year variability, we have pursued the analysis further with the Aleppo data. Because the variation in PGS at Aleppo is equal to or greater than that between sites, conclusions derived from Aleppo, encompassing a large number of years, can be extrapolated to sites of different rainfall and mean PGS if they have a similar thermal and radiation regime.

Relationship of PGS with other environmental variables at Aleppo

Daily rainfall data at Aleppo was analysed for the years from 1961 to 1974. Annual rainfall during the growing season in this period varied from 113 mm to 453 mm and was positively correlated with PGS ($r^2 = 0.86$), which was the highest correlation of the five stations. The correlations at the other stations were

Kamishly	0.56
Shaubak	0.42
Irbid	0.05
AREC*	0.35

This indicates that the distribution of rainfall was often as important as the amount of rainfall.

The PGS at Aleppo was negatively correlated ($r^2 = -0.76$) with the predicted germination date (break of season) (Fig. 8), but this relationship was largely determined by a single point when a severe drought curtailed the growing season. Analysis of the correlations between the beginning and end of the growing season revealed no significant relationship.

Variation in PGS length is closely associated with the heat sum available for phenological development (Fig. 9). The wheat cultivar Halberd, referred to previously (French *et al.* 1979), has an average day-degree requirement based on mean ambient temperature of 1940 to soft dough (Physiological Maturity), so would probably be able to complete its development to this stage in 60 per cent of all years. In 40 per cent of all years, its development would probably be prematurely terminated by soil water stress. This variation in PGS is also closely associated with the sum of total solar radiation received during the growing

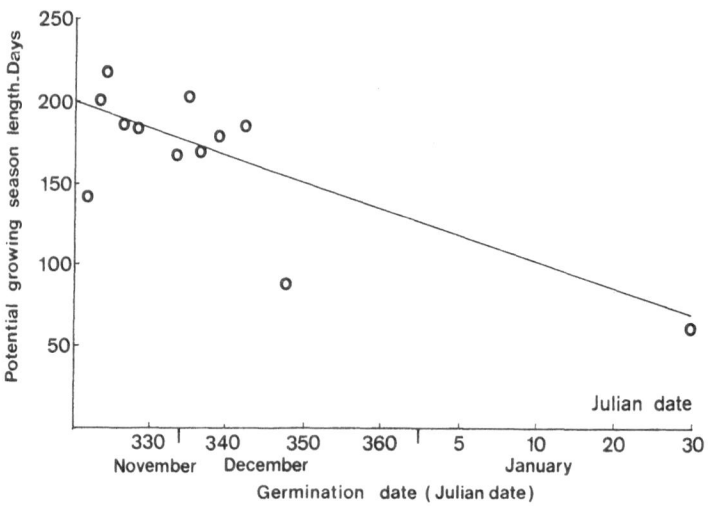

Fig. 8.   Relationship between length of growing season and the break of season.

* The Agricultural Research and Education Center of the American University of Beirut.

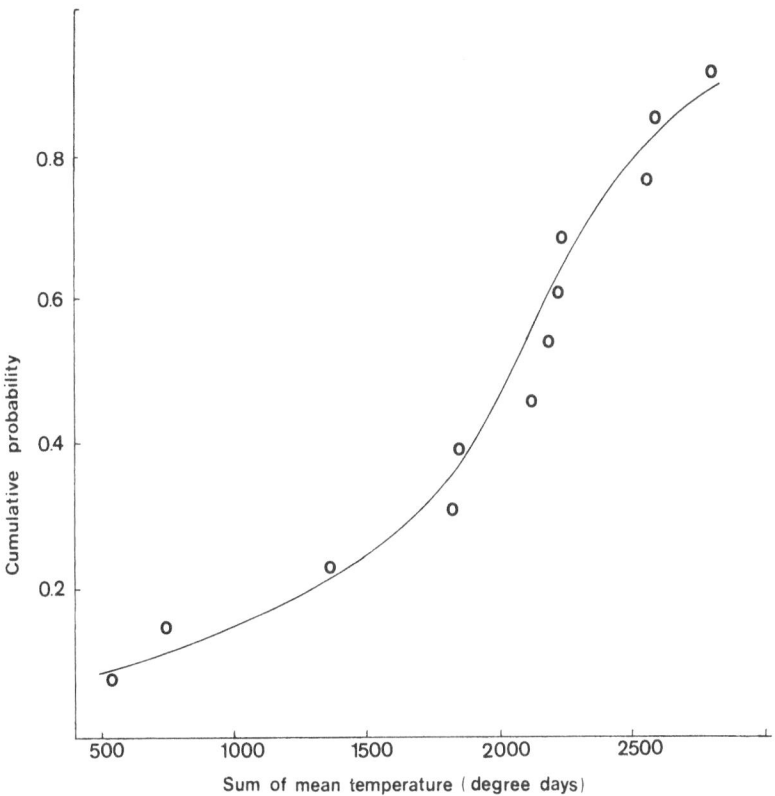

Fig. 9. The variations in the summation of mean temperature during the PGS at Aleppo from the analysis of climatic data from 1962 to 1974.

season (Fig. 10), which implies that a similar order of variation should exist in potential dry matter production and grain yield. We estimate potential above ground dry matter production, taking account of the normal pattern of leaf area development, to vary from 1 t to 12 t ha^{-1}, implying a potential grain yield variation from 0 to 4.6 t ha^{-1}. These estimates are dependent on enough soil water being available to utilize the large variations in total solar radiation interception. From our model, it appears that evaporative demand from emergence to physiological maturity for only 1 of the 12 years studied was less than rainfall during the growing season (Table 2); suggesting that water stress during crop growth should not be a major factor limiting crop yield except at the end of the season. Therefore, within the growing season, soil water levels will generally be adequate to realize the potential growth rate determined by the other limiting factors such as radiation, temperature and nutrition.

44

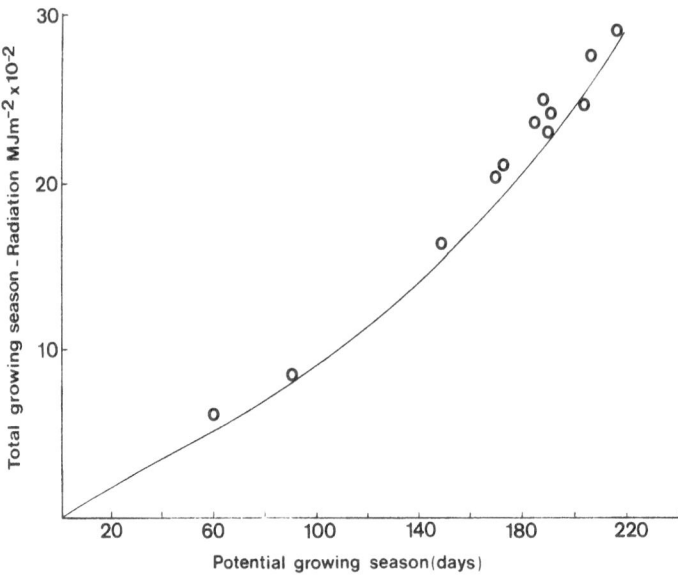

Fig. 10. Relationship between the sum of solar radiation and potential length of growing season.

Table 2. Rain before and during the period of potential crop growth with an ESW = 150 mm, compared with predicted evapotranspiration and evaporative demand.

Growing season	Rain during the growing season mm	*EA from crop emergence mm	**EO from crop emergence mm	Rain from crop emergence mm
1962–63	398	337	388	338
1963–64	302	267	285	215
1964–65	372	293	340	312
1965–66	154	98	104	66
1966–67	468	374	466	374
1967–68	407	317	438	341
1968–69	450	317	363	351
1969–70	197	164	173	142
1970–71	303	273	283	226
1971–72	453	388	498	362
1972–73	113	77	99	53
1973–74	457	288	337	370
Mean	339	274	315	263

*EA – estimated actual evapotranspiration
**EO – estimated potential evaporation.

The length of growing season also tends to restrict the length of the frost-free period for flowering. In the shorter seasons (120 days), this period appears to be restricted to about 20 to 30 days, whereas in the longer seasons (180 days) the frost-free period is correspondingly longer.

Relationship of PGS with soil water store at Aleppo

The major factor which affects the storage of water is the depth and texture of the soil and hence the maximum amount of stored water the soil can hold. The effect of variations, which can be considered as reflecting differences in soil depth, was tested with the model (Fig. 11).

The effect of maximum soil water store depended on rainfall. In dry years, the maximum water-holding capacity of the soil was not utilized therefore there was

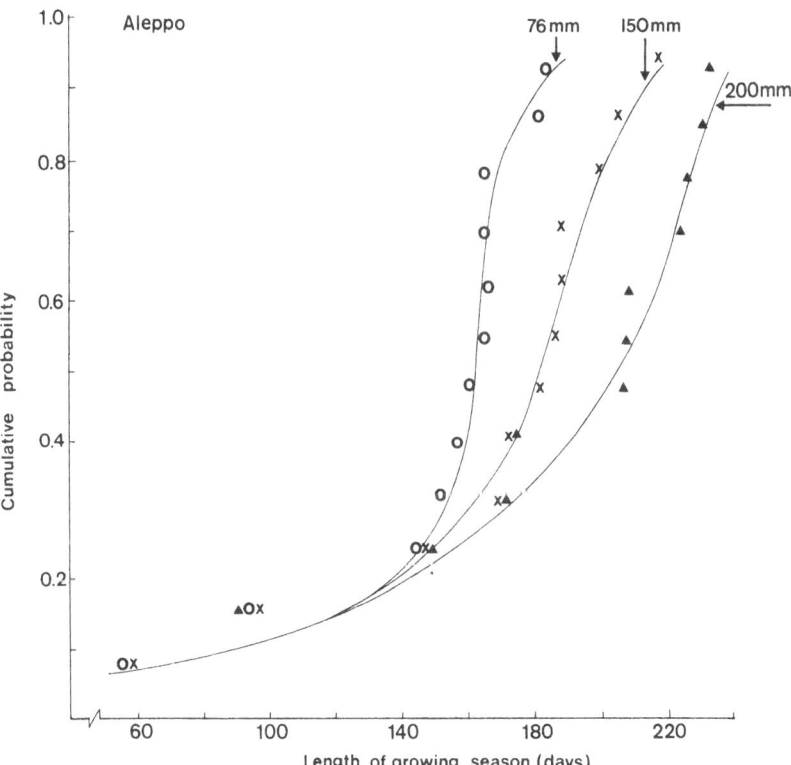

Fig. 11.   Effect of the soil water storage capacity of the soil on the length of growing season. Changes in soil water storage were made to reflect possible changes in the depth of soil.

little impact of this parameter on the length of the growing season. In the wetter years, a greater proportion of the rainfall percolated to depth, and was stored and used later to extend the final stages of crop growth. In shallow soils, greater deep drainage or runoff will occur and there is little benefit from wetter years.

An approximate doubling of the maximum soil water store, from 76 mm to 150 mm, extended the PGS from an average of 151 days to 168 days. This additional 17 days represents a significant resource for crop growth, in particular grain-filling, but only in the wetter years. It occurs at the end of the season when irradiance and temperature are rising. The additional water for crop growth will occur at depth, and crops would need a deep rooting habit (Hurd, 1974) to capitalize on this additional resource.

Distribution of soil water with depth and over time

The pattern of soil water over time with depth and its year to year variation was analyzed by assuming the following total extractable soil water profile (ESW): –

Depth (m)	ESW ($cm^3 cm^{-3}$)	Total ESW mm
0 –0.2	0.20	40
0.2–0.4	0.19	38
0.4–0.6	0.17	34
0.6–0.8	0.15	30
0.8–1.0	0.13	26
1.0–1.2	0.11	22
1.2–1.4	0.05	10
		Total 200

The decrease with depth may reflect the decline in root length density that normally occurs, and the associated reduction in ability to extract the soil water.

Year to year variability was analysed statistically by simulating the daily soil water balance for 12 years (1962–74), taking the mean value of extractable water in the whole profile for each week and then deriving the relative frequency of different levels for each week. For graphical presentation, we have chosen three points from the cumulative frequency distribution; the median value exceeded by 50 per cent of all values; the lower quartile exceeded by 75 per cent of all values; and the upper quartile exceeded by 25 per cent of all values. Fifty per cent of all values fall within the interquartile range and 50 per cent outside. The method of calculating these values is given by Smith *et al.* (1978b).

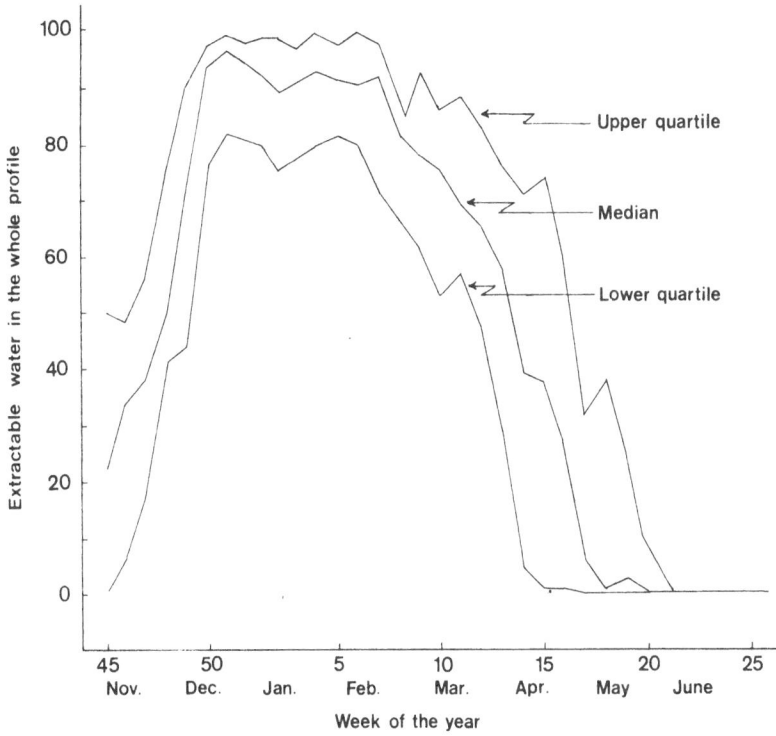

Fig. 12.   Per cent ESW in the whole soil profile at Aleppo, Syria , 1962 to 1974.

The general pattern of soil water averaged over the whole profile is typical of a Mediterranean-type climate, increasing gradually after the break of the season until February-March, then declining to the end of the season (Fig. 12). The variation, measured by the interquartile range, is high early in the season, decreases to a minimum through the winter months, then increases to maximum levels from about mid-March. Of more relevance is the soil water level in the root zone (Fig. 13). The wetting up of the upper soil layers is much more rapid; leading to the break of season during the later part of November. The main source of variation occurs at the onset of the decline in soil water from January onwards, culminating in the termination of the growing season.

Analysis of depth of wetting (Fig. 14) shows that, in the drier years, the soil water is largely in the surface layers of the soil, not penetrating beyond 0.4 m in 25 per cent of the years. In 50 per cent of all years, the water did not penetrate beyond 1 m and only in the wetter years did the water penetrate beyond 1.4 m. Therefore, in the drier years, the surface soil layers still have moderately high soil water

48

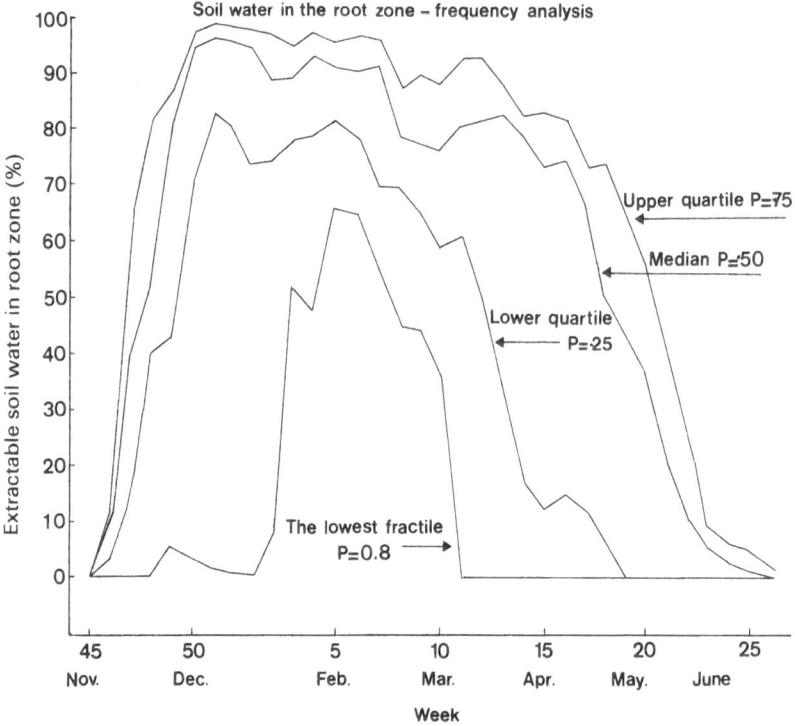

Fig. 13. Per cent ESW in the predicted root zone at Aleppo, Syria.

contents but dry out more rapidly at the end of the season. In the wetter years, significant amounts of water are stored at depth and then extracted towards the end of the season. Exploitation of this water is obviously important for extension of the growing season. Also the significance of these wetting patterns will be related to the rainfall distribution. In those years when rainfall is evenly distributed through the season, the accumulation of moisture at depth will be less significant than when rainfall occurs more sporadically.

This general analysis indicates that water stress should not be a significant factor in yield until the later stages of crop growth, i.e. from anthesis onwards, depending on the maturity type. But when this stress begins, it is likely to be terminal. This conclusion would need to be confirmed for other locations.

Examination of individual years confirms this general trend. Fig. 15a shows a typical pattern, whereas Fig. 15b shows a pattern where a more severe stress developed in April, but rains came and the season extended into May.

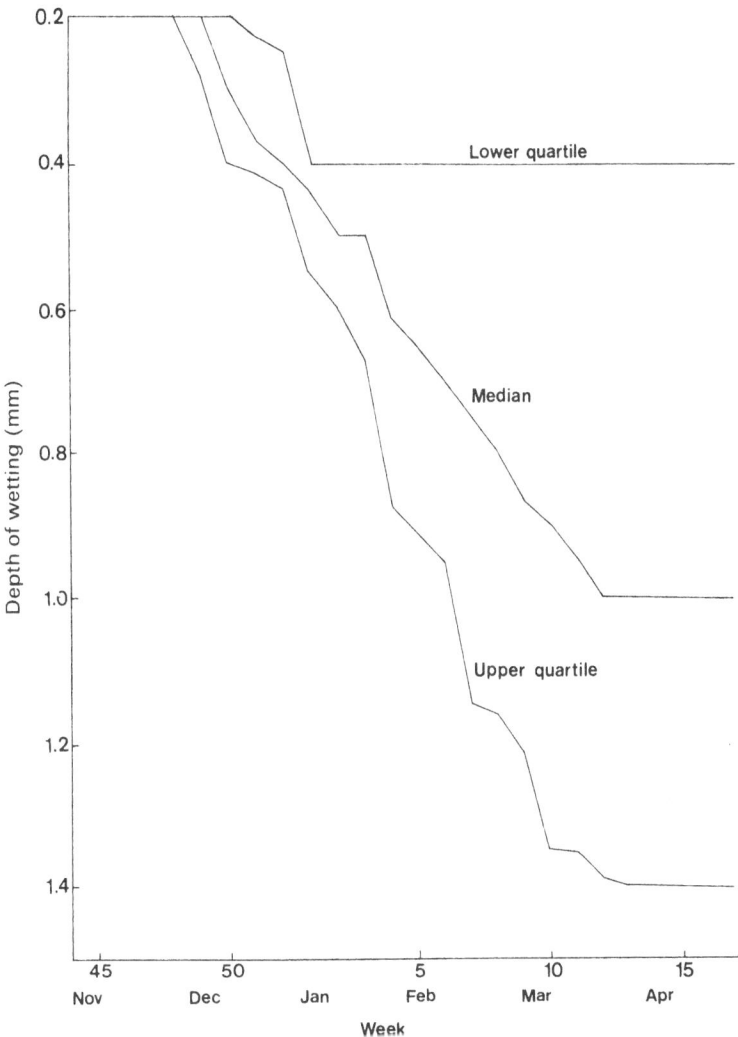

Fig. 14. Predicted depth of wetting at Aleppo, Syria.

Fallowing and crop rotation

In the PGS analysis, we have assumed that no extractable water was carried over from one season to the next, but this is obviously an oversimplification where fallowing occurs or shallow rooted or early maturing crops are grown in the preceding season. Therefore in the next stage of our climatic analysis, we will

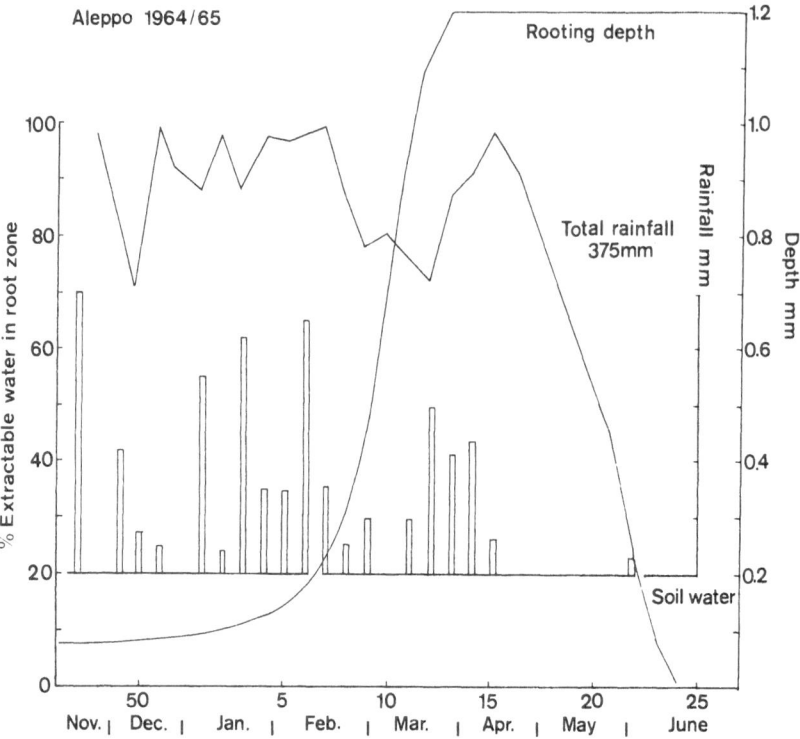

Fig. 15a. Simulated root zone soil water for the crop season 1964/65, Aleppo, Syria.

need to look at cropping or fallowing sequences which are more realistic than those considered here. A preliminary analysis of the climatic data for potential growing season length, assuming a carryover of soil water from the previous season, confirms this need (Fig. 16).

In this analysis, which predicts an increase in growing season of 22 days, the carryover was set at an arbitrary 75 mm of extractable water, over and above current rainfall. This increased the growing season, but, over a sequence of years, total rainfall will be the same and therefore total and mean PGS should be largely unaltered. Therefore the main effect of interseasonal carryover, as revealed in Fig. 17, will be a reduction in the year-to-year variability in growing season. This effect will be most apparent when a dry year follows a wet year. The effect will probably be less or negligible when a wet year follows a dry year because of an absence of carryover. It is therefore apparent that the future development of the PGS type analysis will depend on the simulation of realistic crop sequences over a longer run of years.

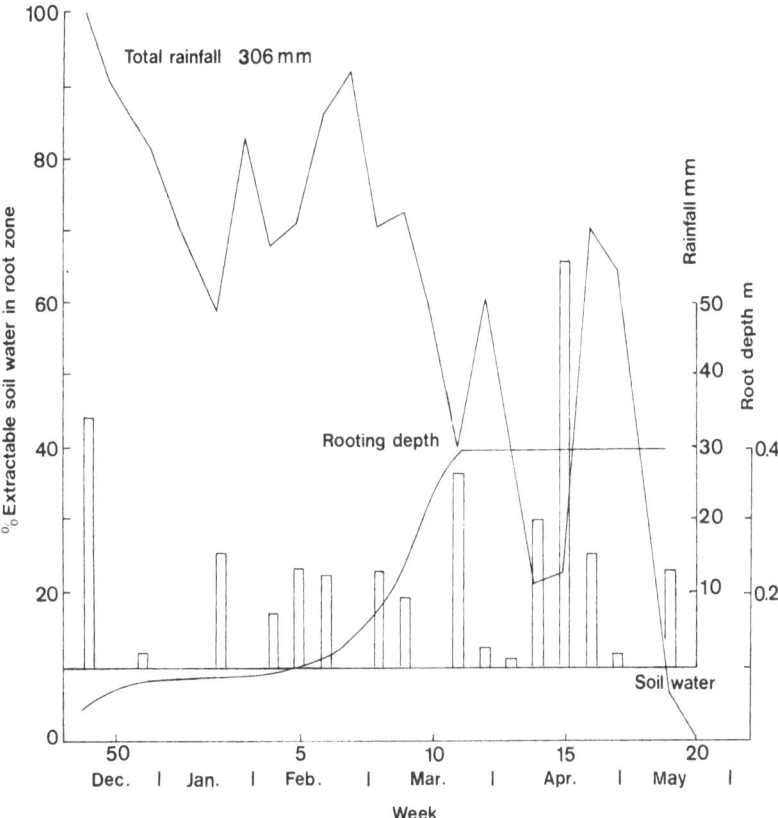

Fig. 15b. Simulated root-zone soil-water for the crop season 1970/71 at Aleppo, Syria.

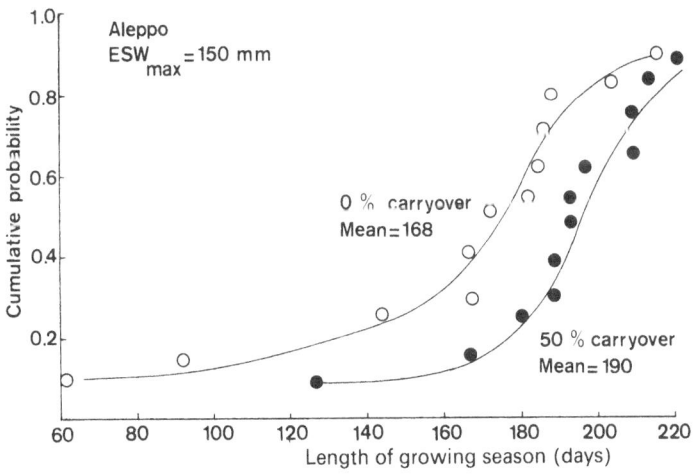

Fig. 16. Effect of carryover of 50 per cent of the ESW from one season to the next, due to either fallowing or growing a short season shallow rooted crop.

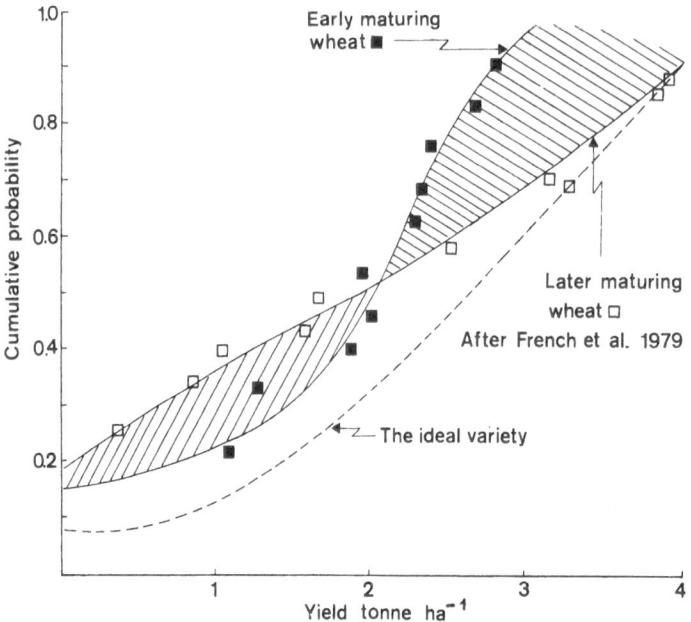

Fig. 17. Predicted relationship between wheat yields and the maturity of the wheat variety grown for Aleppo, Syria.

Our current water balance model was found to be inadequate to predict redistribution of soil water during the dry summer period, therefore we shall need to adopt a more sophisticated model based on diffusion theory and soil hydraulic characteristics (Van Keulen, 1975).

Nitrogen-water interaction

Previous analyses of the environmental limitations to plant growth in Mediter-ranean-type environments indicate that nitrogen rather than water may be the most limiting factor (van Keulen, 1975). Noy-Meir and Harpaz (1978) quote the average concentration of nitrogen in rainwater in semi-arid climates to be about 2 ppm which, with an annual rainfall of 200 to 400 mm, gives an annual input of 4 to 8 kg N ha^{-1}. Reported values of non-symbiotic nitrogen fixation by micro-organisms average about 5 kg ha^{-1} (Noy-Meir and Harpaz, 1978). To this must be added nitrogen from mineralization of soil organic matter which contributes about 2 per cent of its nitrogen content per year; typically yielding from 3 to 16 kg ha^{-1} depending on the humus content of the soil. In addition some nitrogen becomes available from the mineralization of residues. Noy-Meir and Harpaz (1978) estimate a possible annual contribution from this source of from 19 to 56 kg N ha^{-1}.

We calculated total annual biomass yields of from 1.2 to 15 ha^{-1}. At a nitrogen content of one per cent, these yields imply annual nitrogen requirements of from 12 to 150 kg N ha^{-1}. In a non-fertilized soil, and in drier years, water will probably be the major limiting factor, whereas in wetter years, nitrogen will be. For a particular soil, there will be a level of rainfall at which both factors are of equal importance.

Only in about 30 per cent of all years is rainfall at Aleppo likely to be sufficient for leaching beyond the root zone to occur from the end of March onwards in deep soils (Fig. 14). By this stage, the crop will have had ample opportunity to explore the root zone and to take up most of the available nitrogen. For this reason, the system is likely to operate with a high level of efficiency, the main loss being the transport of grain products out of the system. With grain at 1.5 per cent nitrogen this is likely to vary from 15 to 45 kg N ha^{-1} yr^{-1}. Therefore, for a cereal crop, some additional input of nitrogen may be required to achieve potential productivity.

The wide year-to-year variability in potential productivity makes it hard to predict the optimum rate of fertilizer nitrogen to apply. However, because of the absence of leaching and of denitrification of nitrogen in drier years there should be a significant carryover of unused nitrogen from one year to the next.

CONCLUSIONS

In natural systems, nitrogen may be the major resource which is limiting production. However by use of artificial fertilizers or legumes a shortage of nitrogen can be overcome more easily than a shortage of water.

Our analysis of climatic data suggests that the major water-related limitation to yield will be the length of the growing season. Rainfall usually appears to be adequate for growth to occur following the break of season, which itself is variable, until the terminal end of the season. Closely associated with the length of growing season is the amount of incoming radiation and hence potential production of dry matter. Realization of the yield potential of this environment will depend on developing crop varieties and associated farming systems which are able to adapt to and exploit the variability of the growing season.

The variability in the growing seasons is 'managed' by using fallowing to even out the availability of soil water from year to year (Janssen, 1972). The effectiveness of rotations which include a fallow can be evaluated by simulation. Similarly, any change in the crops used in the rotations should be preceded by evaluation of their water use and of the longer term effects of any increase in water use on the soil water carryover and variability of the system. Despite management to remove some of the variability in the length of growing season a considerable amount will still exist (Fig. 16), and selection of crop species or varieties which are able to adapt to this variability will be important for improving productivity.

The phenotypic model used in this study (French et al., 1979) implies a fairly rigid and uniform developmental pattern within the crop giving a high level of uniformity, the objective of much modern plant breeding (Matheson, 1975). Comparison of the predicted yield of this maturity type with one where the day degree requirement was reduced by an arbitrary 20 per cent for each stage (Fig. 17) indicated that the later maturing type would yield more in the wetter years but less in the drier years. Overall it had a higher predicted mean yield ($1.90 \, t \, ha^{-1}$ v. $1.76 \, t \, ha^{-1}$) but a greater variability in yield. The climatic analysis suggests that the ideal variety would combine the attributes of both types, giving modest yields in dry years but being yield-responsive in the wetter years. An 'ideotype' (Donald, 1968) emerges of a variety that distributes the setting of its seed over time. If the season is short, it will produce some seed before the termination of the season, but if good soil water conditions continue then it will be able to continue setting grain.

Perhaps the concept of multi-lines (Jensen, 1952) is relevant to crop adaptation in a variable climate, or a plant that produces several generations of tillers in discrete bursts, filling the grain of one set before beginning the grainfilling of the next. Non-determinate crops such as cotton and soybeans tend to exhibit this type of plasticity. This need for plasticity has been recognized by plant breeders (Matheson, 1975) but is not clearly defined.

Two factors which would be likely to limit the contribution to production of early flowering plants or tillers would be frost or the sprouting of the grain in the head.

Agronomic research for improving resource use should be concerned with maximizing the use of solar radiation during the growing season. Seeding rate should ensure a large initial leaf area index before the onset of winter and early seeding to maximize the duration of energy capture. However, this may lead to an earlier termination of the growing season as more water is used in the early growth stages. In addition, an understanding of the nitrogen cycle and the interaction between water and fertilizer nitrogen is important in achieving an efficient use of water. Other factors such as disease and pest management, and weed control are important in maximizing the use of solar radiation and soil water.

Development of modelling may have value for handling the problem of longer-term climatic variability, in evaluating results of research or breeding, and as a basis for general recommendations. To use this approach, a better understanding of the system is required. Empirical models of phenological development such as that developed by French *et al.* (1979) are needed for the major cultivars. Also the relationship between phenological development and yield needs to be understood better. The accurate modelling of the dynamics of soil water is important. In particular, a knowledge of soil types and their extractable soil water content under different crops is needed.

Modelling allows another dimension to be added to agronomic research, a dimension of time and space that is often difficult to encompass adequately; using field experimentation. It represents a sophistication of existing approaches but does not replace them. However, the ultimate limitation is our understanding and ability to pose the right questions for research that will lead to a better use of our environmental resources.

Our ability to predict changes in soil water is generally recognized to be good, due to advances in crop physiology and soil and environmental physics. Analysis of temperature limitations to crop yield in different locations is needed, but poses greater problems because of the complexity of the response to temperature of the source/sink processes of crop yield. However as our understanding increases, such analyses will become both feasible and meaningful. At present both are feasible but their validity may be in doubt.

ACKNOWLEDGEMENTS

We are grateful to Dr. Joe T. Ritchie for allowing us to use a Fortran copy of his wheat model as a basis for our own model. Dr. David Gibbon kindly gave us access to his unpublished soil water and crop yield data. ICARDA provided a research grant to support the collection and analysis of climatic data from the Near East.

REFERENCES

Anderson, J. R. 1974. Risk efficiency in the interpretation of agricultural production research. Rev. Marketing Agric. Econ. 42, 131–184.
de Brichambaut and Wallen, C. C. 1963 A study of Agro-climatology in semi-arid and arid zones of the Near East. World Meteorological Organization Tech. Note No. 56.
Connor, D. J. 1975 Growth water relations and yield of wheat. Aust. J. Plant Physiol. 2, 353–366.
Donald, C. M. 1968 The breeding of crop ideotypes. Euphytica 17, 385.
Fisher. W. B. 1978 The Middle East. Methuen, London.
Fischer, R. A. 1979 Dryland wheat yield in Australia. Growth and water limitation: a physiological framework. J. Aust. Inst. Agric. Sci. 45, 83–94.
Fischer, R. A. 1981 This volume, p. 249–278.

Fitzpatrick, E. A. and Nix, H. A. 1969 A model for simulating soil water regime in alternate fallow-crop systems. Agric. Meteorol. 6, 303–319.

Fitzpatrick, E. A. and Nix, H. A. 1970 The climatic factor. In: Australian Grasslands. Moore, R. M. (ed.), A.N.U. Press, Canberra.

French, R. J., Schultz, J. E. and Rudd, C. L. 1979 Effect of time of sowing on wheat phenology in South Australia. Aust. J. Exp. Agric. Anim. Husb. 19, 89–96.

Friend, D.J.C. 1966 The effects of light and temperature on the growth of cereals. In: The Growth of Cereals and Grasses. Milthorpe, F. J. and Ivins, J. D. (eds.) Butterworths, London.

Goudriaan, J. and van Laar, H. H. 1978 Calculations of daily totals of the gross CO_2 assimilation of leaf canopies. Neth. J. Agric. Sci. 26, 373–382.

Harris, H. C. 1978 Some aspects of the Agro-climatology of West Asia and North Africa. In Food Legume Improvement and Development. Hawtin, G. C., Chancellor, G. J. (eds.), ICARDA, Aleppo, Syria and IRDC, Canada.

Hawtin, G. C. 1979 Strategies for the genetic improvement of lentils, broad beans, and chick-peas, with special reference to research at ICARDA. In Food Legume Improvement and Development Hawtin, G. C. and Chancellor, G. J. (eds.), ICARDA, Aleppo, Syria and IRDC, Canada.

Hodges, T. and Kanemasu, E. T. 1977 Modelling daily dry matter production of winter wheat. Agron. J. 69, 974–978.

Hurd, E. A. 1974 Phenotype and drought tolerance in wheat. Ag. Meteor. 14, 39–55.

Janssen, B. H. 1972 The significance of the fallow year in the dry farming system of the Great Konya Basin, Turkey. Neth. J. Agric. Sci. 20, 247–260.

Jensen, N. F. 1952 Intra-varietal diversification in oat breeding. Agron. J. 44, 30–34.

Kassam, A. H. 1981 This volume, p. 1–29.

Keulen van, H. 1975 Simulation of water use and herbage growth in arid regions. Simulation Monographs, Pudoc, Wageningen.

McCree, K. 1974 Equations for the rate of dark respiration of white clover and grain sorghum, as functions of dry weight, photosynthetic rate and temperature. Crop Sci. 14, 509–514.

Matheson, E. M. 1975 Exploration and exploitation of genetic variation. In: Australian Field Crops Vol. 1: Wheat and other temperate cereals. Lazenby, Alec and Matheson, E. M. (eds.), Angus & Robertson, Sydney.

Nix, H. 1975 The Australian climate and its effect on grain yield and quality. In: Australian Field Crops, Vol. 1; Wheat and other temperature cereals. Lazenby, Alec., Matheson, E. M. (eds.) Angus & Robertson, Sydney.

Noy-Meir, I. and Harpaz, Y. 1978 Agro-Ecosystems in Israel. In: Cycling of mineral nutrients in agricultural ecosystems. Frissel, M. J. (ed.), Elsevier, Amsterdam.

Ritchie, J. T., Burnett, E. and Henderson, R. C. 1972 Dryland evaporative flux in a subhumid climate: III. Soil water influence. Agron. J. 64, 168–173.

Ritchie, J. T. 1972 Model for predicting evaporation from a row crop with incomplete cover. Water Resources Res. 8, 1204–1213.

Ritchie, J. T. 1974 Evaluating irrigation needs for southeastern U.S.A. In: Contribution of irrigation and drainage to world food supply. Symposium of Am. Soc. C. Eng., August 1974, Biloxi, Mississippi.

Rosenberg, N. J. 1974 Microclimate, the biological environment. Wiley, New York, pp. 315.

Single, W. V. 1975 Frost injury. In Australian Field Crops, Vol. 1: Wheat and other temperate cereals. Lazenby Alec, Matheson, E. M. (eds.), Angus & Robertson, Sydney.

Slabbers, P. J., Horrendorf, V. S. and Stapper, M. 1979 Evaluation of simplified Water – Crop Yield Models. Agric. Water Management 2, 95–129.

Smith, R. C. G., Anderson, W. K. and Harris, Hazel C. 1978 A systems approach to the adaptation of sunflower to new environments. III. Growth and yield predictions for continental Australia. Field Crops Res. 1, 215–228.

Smith, R. C. G., English, S. D. and Harris, Hazel C. 1978 A systems approach to the adaptation of sunflower to new environments. IV. Yield variability and optimum cropping strategies. Field Crop Res. 1, 229–242.

Turner, N. C. and Begg, J. E. 1981 This volume, p. 97–131.

Wardlaw, I. F. 1975 Physiology and development of temperate cereals. In Australian Field Crops, Vol. 1: Wheat and other temperate cereals. Lazenby, Alec and Matheson, E. M. (eds.), Angus & Robertson, Sydney.

de Wit, C. T. 1965 Photosynthesis of leaf canopies, Pudoc, Wageningen.

de Wit, C. T. 1978 Simulation of assimilation, respiration and transpiration of crops. Simulation Monographs, Pudoc, Wageningen.

Warrington, I. J., Dunstone, R. L. and Green, L. M. 1977 Temperature effects at three developmental stages on the yield of the wheat ear. Aust. J. Agric. Res. 28, 11–27.

3. Rainfed farming systems in the mediterranean region

DAVID GIBBON

School of Development Studies, University of East Anglia, Norwich, England

Mediterranean-type environments* are found in the Mediterranean Basin, Central and Southern California, Central Chile, South-West Cape Province, Southwest of Western Australia and in the southern part of South Australia. They support a wide range of farming systems, but nearly all contain, or have contained, common elements: cereals, small livestock, olives, vines, fruit trees and vegetables. The major influences that have given rise to differences in presentday farming systems are the history of population growth and settlement; political developments and the type of relationship with metropolitan centres; forms of land tenure and the role of farmers in the social and political structure; and the growth of urban demand and the degree of commercialisation and specialisation in production.

The Mediterranean Basin itself has been important since settled agriculture began, being a centre of origin of many of the major cereal and legume crops and of the early domestication of sheep and goats. It is also the area where dry farming techniques of growing cereal crops were first developed (White, 1963, 1970) and was a focal point for the introduction and spread of new crops and the development of intensified agricultural systems during the spread of Islam. (Watson, 1974; Grigg, 1974).

The region experienced two periods in which it probably contained the most highly organised and productive agricultural systems in the world, during the Roman administration, and later during the period of Arab domination. However, the more recent colonial experiences of the 19th and 20th centuries resulted in a period of stagnation and decline of productivity, particularly of the main food crops, as the colonising powers, supporting the settlers and the wealthy landowning classes, concentrated on high value export crops, often grown on the

---

* The geographic, topographic, climatic, edaphic and vegetation features of Mediterranean environments have been adequately described elsewhere (de Brichambaut and Wallen, 1963: Emberger, 1977: Hills. 1966: Matthews, 1924; Meigs, 1964; New'begin, 1929; Papadakis, 1973; UNESCO/FAO, 1963; Whittlesey, 1963). The principal land use systems have also been discussed (Duckham and Masefield, 1970; Grigg, 1974; Stamp, 1961; UNESCO, 1964) and a number of writers have considered present farming systems in the Mediterranean basin in comparison with the historical development of agriculture in similar regions elsewhere (Aschmann, 1977; Grigg, 1974; Oram, 1979).

most productive land. This perpetuated the under-development of large areas of semi-arid lands and marginalised the peasants (Amin, 1976). Since independence, governments have placed increasing emphasis on the production of food crops, redistributed many of the former, large land ownerships among the rural poor, and provided greater support for poorer farmers through price control, credit facilities and cooperatives.

At the same time, governments have a major control over the resources and inputs necessary for production, and several favour the development of large scale highly capitalised with a small labour input farming systems as a means of supplying national food and export requirements. Knowledge of what has been achieved in similar environments elsewhere, and advice from aid agencies serve to reinforce this view. Consequently, large amounts of capital for development are channelled into large scale, predominantly irrigated, production projects.

During the past 30 years, the population of the region has increased rapidly and mobility has increased among both rural and urban populations. Migration of labour is an important feature of some of the farming systems that are discussed in this chapter. Migration occurs to the wealthy countries of the region and to northern Europe, and the returns are often a significant component of household, and indirectly of national, income.

Present day farming systems in the Mediterranean Region are therefore influenced by a range of circumstances related to their past and recent history, the extent of intervention by governments, forms of land tenure, and ownership and inheritance, all of which are unlike those found in similar climatic regions elsewhere. Hence environmental similarities between Mediterranean regions are not necessarily relevant to the potential for change.

The links that exist between the main types of farming systems make it very difficult, and in some ways misleading, to classify systems into distinct categories. Indeed, this type of crude classification may have created some false perceptions of farming systems.

The categories that follow merely identify dominant trees, crops or livestock within farming systems. The whole farming system generally centres on the needs of the household for food, clothing, housing and capital. The choice and balance of enterprises, both on an off farm, is the result of many influences, including these requirements, social obligations, and the whole structural context of productive activities.

- *Steppe-based nomadic or semi-nomadic pastoral systems*, primarily with sheep and goats. Animals are generally based in the drier areas during the spring and early summer, moving into the higher rainfall, arable areas after harvest and increasingly, during recent years, relying on supplementary feeding during the autumn and winter.

- *Rainfed cereal production systems.* These are predominantly based on a cereal: fallow rotation with wheat the main cereal in higher rainfall areas, and barley in areas below 300 mm and on shallower soils in the wetter areas. This system is found throughout the low-altitude regions of the Near East and North Africa and in the high-altitude areas of Spain, Turkey, Iran, and Afghanistan.

  Where there are deep (> 100 cm) soils in areas of more than 300 mm rainfall, many more crops are grown, including the grain legumes, but also a wide range of summer crops are grown using stored moisture after a winter fallow. In these areas, more complex rotations have developed, but none is longer than four years. In several countries, all the main operations of these systems have been mechanised.

- *Rainfed mixed tree and arable crop systems,* often with dairy cattle and some goats. This type of system is found in the wetter areas (higher than 600 mm annual rainfall) of the region and tree crops (olives, figs, fruit crops, vines), cereals and legumes are grown; usually in combination.

- *Irrigated farming.* Within this category it is possible to identify a range of systems, from those based entirely on the supply of water from a dam or river, to those where enough water is available only for the supplementary watering of summer crops. It is also important to distinguish large irrigation schemes run by government agencies, from those with individual ownership of land and water rights. Each may give rise to very different farming systems, but these are not relevant to this chapter.

It is important to recognise that there are transition zones between these categories and also areas where significant interrelationships occur. At the wetter end of the predominantly rainfed cereal areas (400–600 mm annual rainfall), tree crops, principally olives, are an increasingly important element in the farming system, and tend to replace cereals as the main source of income.

One conceptualisation of a trend in present day farming systems, constantly referred to in the literature, is the growing polarisation of production systems, particularly between livestock and crop systems. This concept has presumably arisen from a limited view of existing farming patterns in some countries and the writers are not aware of the complex interrelationships that can exist between livestock owners and arable farmers (who may be the same family) in ensuring that livestock are provided with a year-round supply of adequate feed. Arable owners may get some return from renting crops for grazing in dry years, and the soil many benefit from organic residues.

In the drier, rainfed cereal areas (< 250 mm annual rainfall), sheep production is an increasingly important part of the farming system. In many areas in this zone there is a cereal: sheep system with sheep based in the arable area for some of the year, and the young stock moving to the steppe for grazing during the spring.

Throughout the rainfed arable areas, the presence of ground water, even in limited quantities, makes irrigation possible, and the range of alternative crops and systems is increased considerably where it has been developed. The inter-relationships between rainfed and irrigated systems are often important here, particularly with respect to allocation of scarce resources between systems.

Other significant links between systems include the seasonal migration of people from the drier areas after their cereal harvest, to higher rainfall or irrigated areas for the cotton and olive harvests.

STABILITY OF PRODUCTION SYSTEMS

The main factor affecting stability is rainfall. This has been discussed in physical terms in previous papers. One of our main concerns is the ability of farmers to develop strategies to overcome the inherent variability in rainfall. We may consider three types of problems

(i) Short-term seasonal lack of moisture, late arrival of rains, low annual rainfall or an early end to the rainy season. All of these may occur in areas of low annual rainfall, and generally result in farmers deciding to let their sheep graze their crops. It may also necessitate the migration of animals to wetter areas, or an increase in supplementary feeding. In the year after a dry season, the cereal may be planted again on the same land and output is usually lower than it would have been, had it followed a fallow;

(ii) A succession of dry years in areas of low average annual rainfall, tends to result in an increased reliance on income from non-agricultural work through migration to urban centres or outside the country, or from agricultural labouring in higher rainfall areas. In such areas, income generated from outside the village can become the major source of income; and

(iii) In some areas, major shifts in rainfall patterns have occurred. Grigg (1970) quotes an example from Tunisia, where the 200 mm isohyet shifted 200 km north between 1931–34 and 1944–47. This type of shift will have a profound effect on the whole nature of farming and the possibilities for future development of stable systems.

The annual variation in production of cereals is influenced by the distribution and amount of rain falling. The area planted in one season appears to vary with the previous season's annual rainfall: poor rains in one season result in a reduction in the area planted in the following season (ICARDA, 1979). The area harvested is also highly variable, as crops sown on shallow soils and in drier areas may not survive and certainly will not be harvested in dry years. Yield potential varies considerably with soil type; crops growing on the shallow soils in the

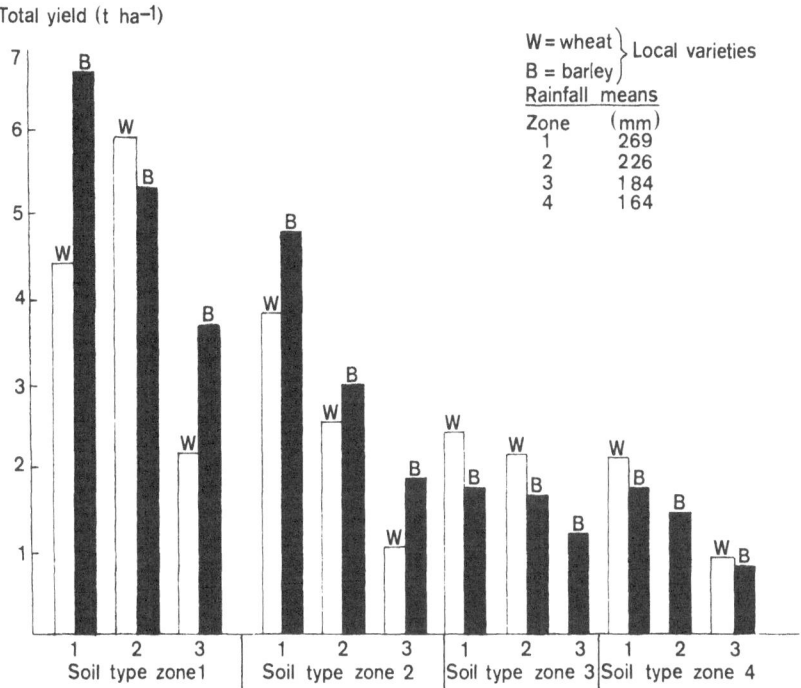

Fig. 1a. Wheat and barley total yield by soil type and rainfall zone 1978–79.

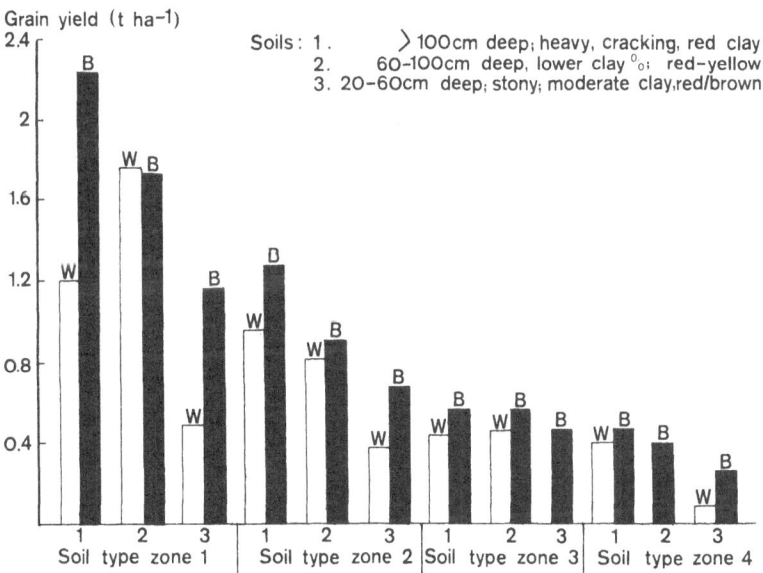

Fig. 1b. Wheat and barley grain yield by soil type and rainfall zone 1978–79.

higher rainfall areas producing comparable yields to those on much deeper soils in drier areas. Some data from farmers' fields illustrate this. (Fig. 1).

The adherence to a cereal: fallow rotation has provided an element of stability, with the fallow giving the opportunity to control weeds, contribute a small amount of moisture, and allow the mineralization of some nutrients for the succeeding crop. In the drier areas, abandonment of the fallow in favour of continuous cereals, may create instability in the system by increasing indebtedness and dependency.

A number of other important factors affect stability of production. The first, is the extent of government intervention in input and output channels and in controlling prices of commodities. Access to inputs at the right time and in the right quantities, the marketing and payment arrangements and the timing of announcement of prices, in addition to their relative levels, all can have an effect on decision making with respect to the allocation of resources and time, and thus have repercussions throughout any farming system.

PRODUCTIVITY OF FARMING SYSTEMS

As some of the factors affecting primary productivity in rainfed systems, namely availability of moisture and soil nutrients, and the efficiency with which they are utilized by crops, are dealt with in other papers I will consider a number of other factors that may influence productivity in farming systems, but are rarely considered in technical discussions. When we compare existing productivity with potential productivity, we frequently make comparisons between average farm yields and the maximum potential yield achieved in experimental trials or on demonstration plots. (IRRI, 1974; Oram, 1977). Emphasis is then placed on identifying the principal physical and technical constraints in existing production systems and in devising methods of closing this 'yield gap'.

This can be a very misleading approach, particularly when average farm yields are compared with experimental station yields or even demonstration plot yields, and also where the yield of certain crops is considered in isolation from the output of other crop or livestock enterprises in the system. In all crop production systems, the output from one crop is strongly influenced by previous cropping or previous land management, the sequence of crops, the input from animals, the allocation of resources between crops and livestock, competing demand for labour use, the timing of cultivations, weeding and fertilizer use. Other demands on labour of a social and political nature, problems of infrastructure and the longer term interests and objectives of farming families may also influence output. Some of these interrelationships are indicated in Fig. 2.

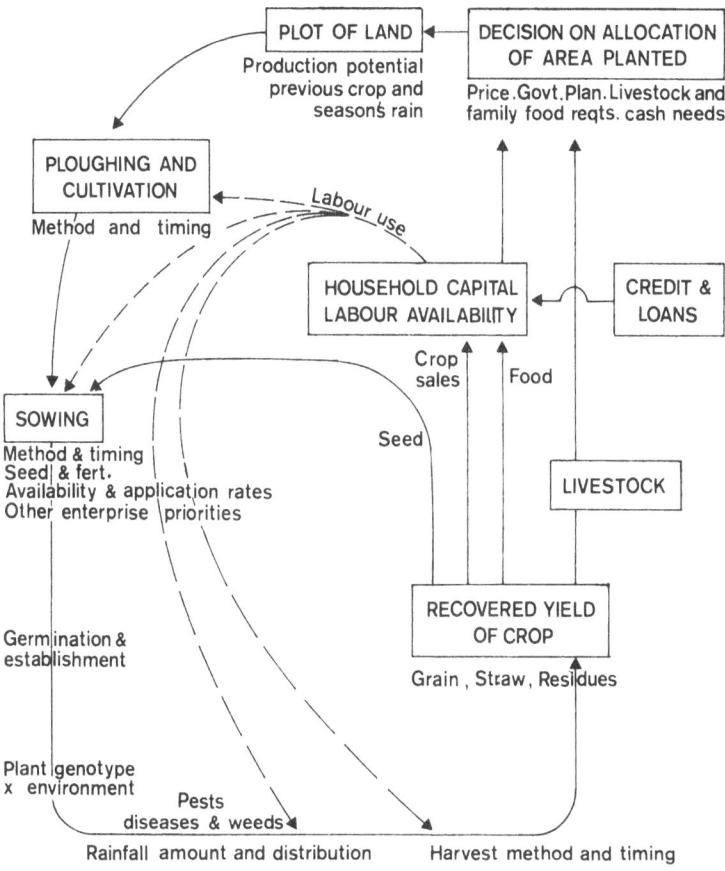

Fig. 2.   Some factors affecting production and returns of a crop within a farming system.

Estimates of productivity that take account of the range of actual total biological and 'economic' yields and of potential productivity under conditions that are fully specified, may give a more realistic picture of the potential for yield improvement (Fig. 1). Yields in Zones 1 and 2 compare favourably with yields in so called 'high technology' areas (French, 1978; Russell, 1967).

Small land-holding size and small plot size are sometimes regarded as being important factors which limit productivity and modernisation of farming (Grigg, 1974; Oram, 1979). So far, we have no evidence to support this contention from work done in northern Syria. Practically all farmers have access to a tractor and cultivation equipment, and cultivation, planting and harvest of all the main crops are done without any major delays from small sized plots or small holdings. Fig. 3

66

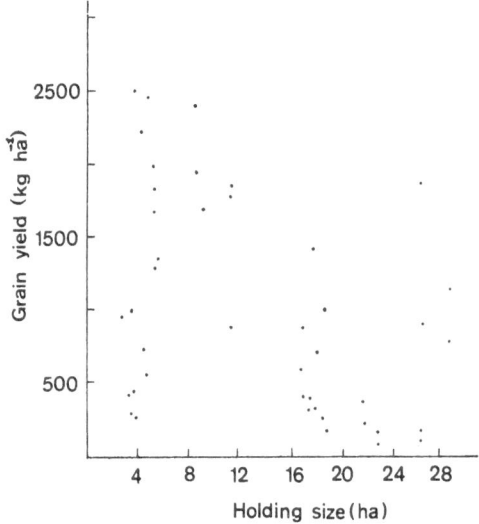

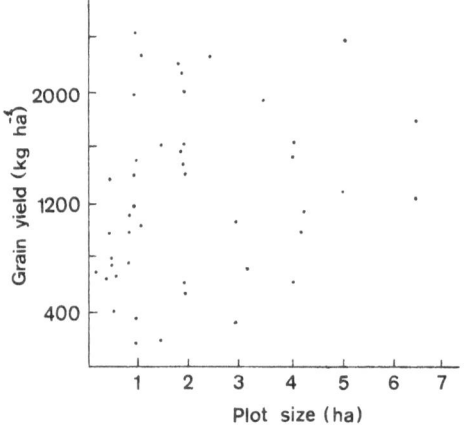

Fig. 3. Grain yield x holding size and plot size. Data from farmers' plot records, Aleppo Province, 1978-79.

illustrates yields from different-sized holdings and plots in Aleppo Province. In no case is there any significant correlation. The assumption by some people (e.g. Carter, 1977) that 'modernisation' of agriculture can come about only through the use of large machinery on large areas needs to be questioned.

Two examples from studies in Aleppo Province may illustrate some of the characteristics of existing systems in relation to the links between cropping and

Table 1. Farm family sample Aleppo Province 1977–78

	Village 2A/06	Village 4/04
Number of families in sample	8	10
Areas (ha)		
average cultivated	14.75	20.8
total cultivated	117.9	208.1
wheat	28.9	37.0
barley	27.6	136.6
lentil	19.0	—
vetch	6.2	—
Yields (kg/ha)		
wheat	651	235
barley	824	307
lentil	838	—
vetch	379	—
cereal straw	112	110
legume straw	798	—
Rainfall (mm)		
average annual	351.0	222.0
1977/78	316.8	240.5*
People in sample families	·52	67
Land/person ratio (ha: people in family)	2.27	3.11
Wheat consumption (kg per person per day)	0.61	0.77
Number of milking ewes and goats	102	95
Average yield of milk (kg/head)	55	60
Consumption of metabolizable energy (GJ per productive ewe)	5.5	5.4

* 34 mm of this fell in three showers during mid-October, and the next significant rain fell on December 4.

livestock systems, and the flows of human food and livestock feed energy within village farming systems. In each of the following studies, the combined information from a 25 per cent sample of farming families is used to illustrate these features and the sample profile characteristics are given in Table 1.

68

Case 1 Village 2A/06

This village has 283 people. Total land area is 500 ha with three main soil types ranging from deep, red cracking clays to shallow, stony soils. Average family size is 7.2 people (range, 2–13) and average land holding is 12.9 ha (range, 1.5–40 ha). Average annual rainfall is 331 mm. There are 714 sheep plus goats in the village and eight tractors. Common rotations include cereals, legumes and summer crops in two, three·or four yeár rotations. The summer crop provides an opportunity for weed control and also provides a useful cash income in wet years. Cereals usually follow summer crops which leave some moisture in the soil at planting (about 30 mm per m depth more than the profiles left after cereals and lentils). This may be significant in drier than average years (Fig. 4).

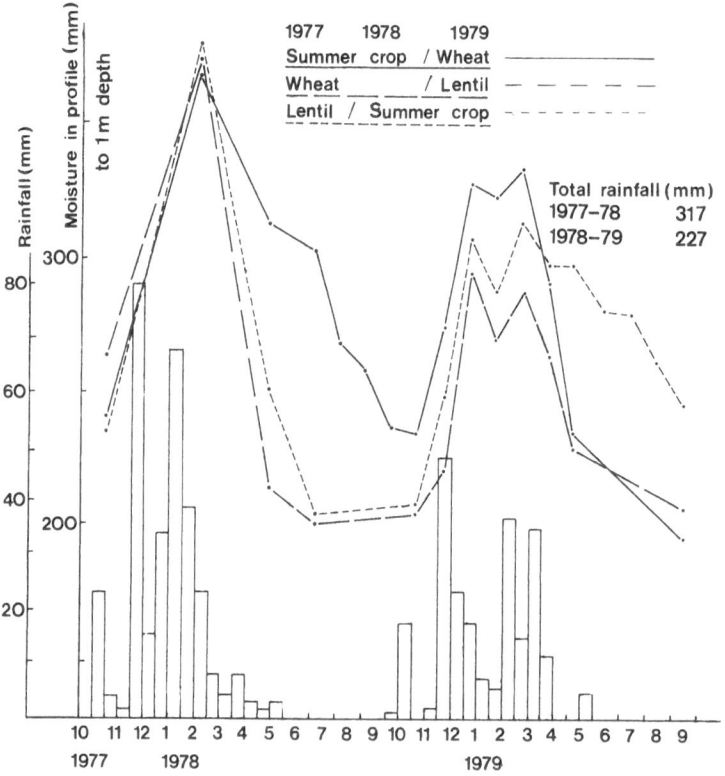

Fig. 4.   Village 2A/06. Changes in soil moisture under three crops over two seasons.

69

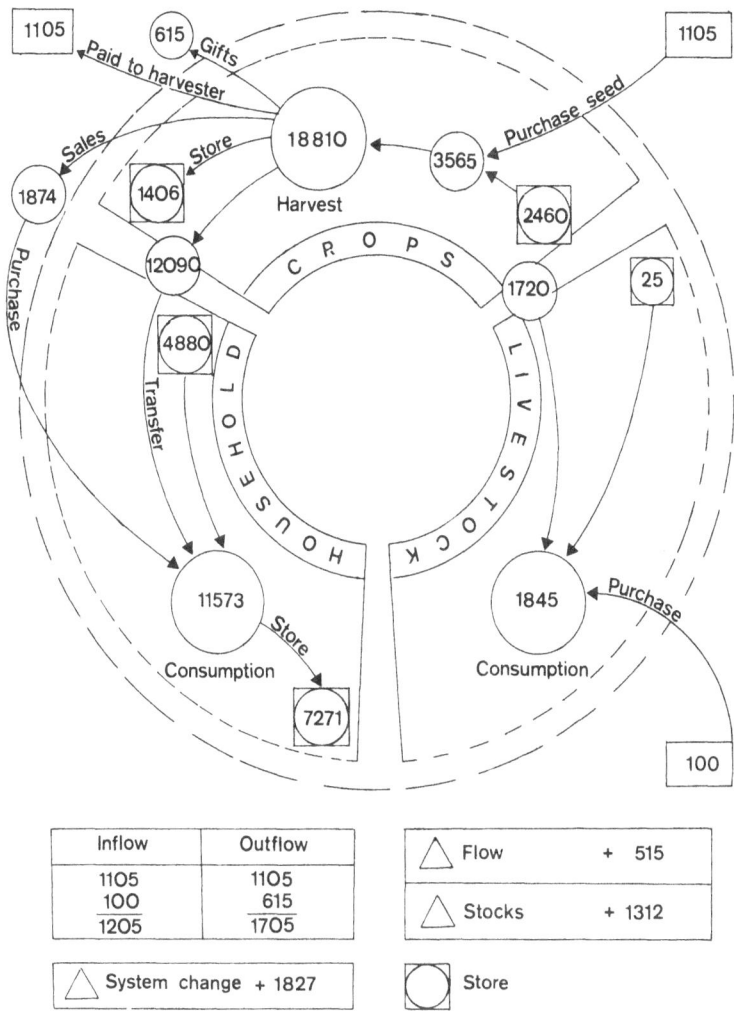

Fig. 5.　Village 2A/06: Wheat flows (kg) November 1977-October 1978.

For village 2A/06, families generally aim to achieve self sufficiency in wheat in most years even though barley normally outyields wheat by 50 per cent in average rainfall years. Fig. 5 shows that for 1977–8 a moderate surplus of wheat was produced with a harvest: seed ratio of 5.28:1. This supplied all of the household needs, and the straw contributed to livestock feed requirements.

70

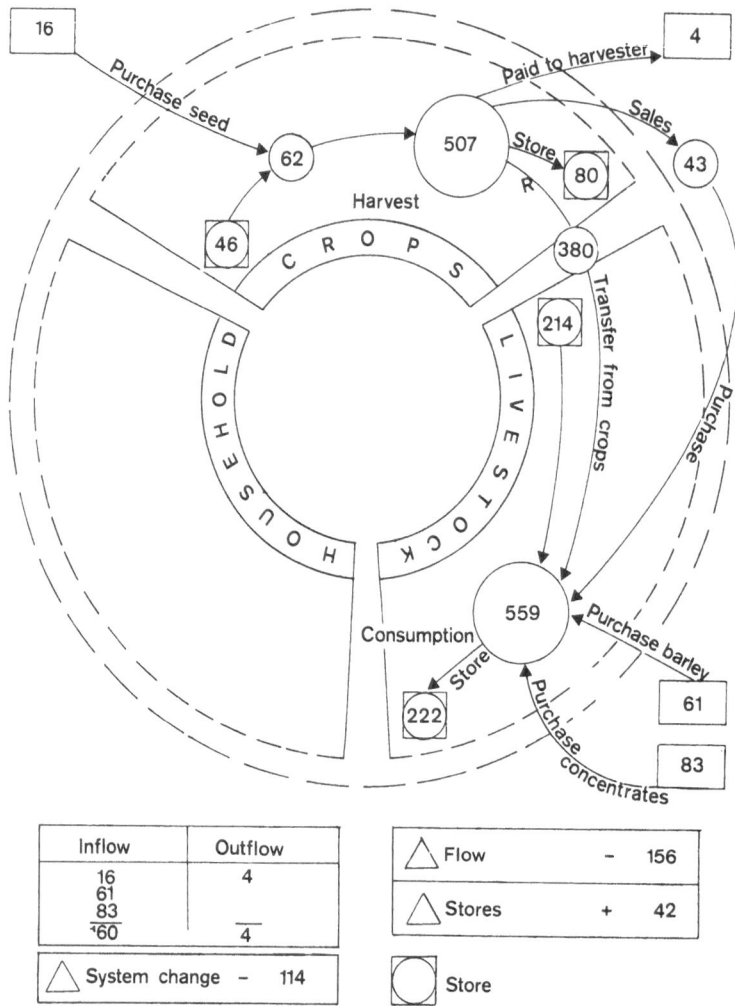

Inflow	Outflow
16	4
61	
83	
⁺60	
	4

△ Flow	− 156
△ Stores	+ 42

△ System change − 114

◯ Store

75 per cent of metabolizable energy input to livestock generated within
the system

Fig. 6. Village 2A/06. Animal feed flows (gigajoules of metabolizable energy) November 1977-
October 1978.

In considering animal feed flows* within the system (Fig. 6), all feeds have been converted to gigajoules of metabolizable energy (1 gigajoule = 10^9 joule). Although the feed system shows a deficit of 114 GJ, farmers are reluctant to make this up from home grown crops as this would upset the winter crop: summer crop balance. The deficit could, for example, have been made up by an extra 11.5 ha of barley (at 1977–8 production levels) but this would have to have come from summer crop land (29.5 ha available) or some other crop. In 1977–8 the winter crop: summer crop ratio was 2.76: 1; to substitute barley for summer crops would change this ratio to 5: 1. The same argument applies to increasing the legume area, plus the fact that lentils and vetches are too labour intensive to increase beyond present levels.

So long as the sales of the surplus crops (lentils and summer crops) can offset the energy deficit, the system 'balances'. The 83 GJ of purchased concentrates supplies proteins, minerals and vitamins which are not produced in the farm feed, and the present flows in the system are further supported by the availability of cheap feeds from the General Organisation for Feeds.

Case 2 Village 4/04

This village has 192 people. Total land area is 501 ha with three main soil types which have a lower clay content and a poorer structure than those of village 2A/06. Average family size is 6.4 people (range 1–11) and average land holding is 19.7 ha (range 17–27 ha). There are 328 sheep plus goats in the village and one tractor. The cereal crops alternate with a fluctuating proportion of fallow and in the recent succession of drier-than-normal years, barley crops are being grazed and the proportion of land fallowed has declined from 45 per cent to 17 per cent. The whole season fallow contributes very little to the store of moisture available for the succeeding cereal crop. (Fig. 7). In this situation, the problems of livestock feed supply are closely linked to the deterioration of the range through overgrazing and an increase in cereal cultivation in marginal areas. Reliance on supplementary feeding is common

In this village, farmers also plan to grow enough wheat to supply their family needs, but in 1977–8 only 33 per cent of the wheat supply of the household was produced within the system (Fig. 8).

---

* In the cropping sector, the calculation includes apportioned allowances for seed plus stores of wheat and lentils as their contribution to feed. Thus 8 per cent of wheat harvest went to animals, so 8 per cent of wheat in store/purchased was 'chargeable to animal feeds'. The equivalent figures for barley, lentil and vetch are 100 per cent, 14 per cent and 100 per cent respectively. Of the metabolizable energy input from outside the system, 61 GJ was for barley plus straw and 83 GJ was for cotton seed cake, cotton hulls and other products.

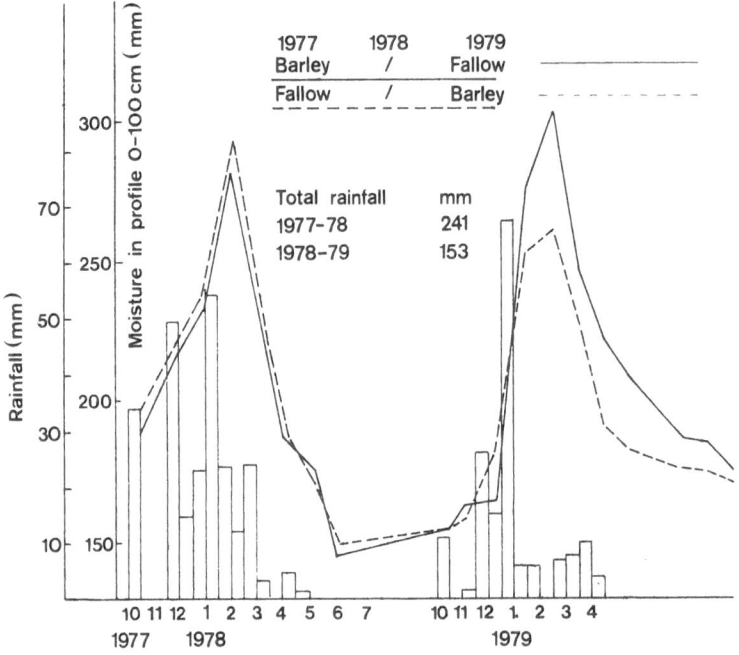

Fig. 7. Village 4/04. Changes in soil-moisture under cereal/fallow over two seasons.

In order to balance, the yield would have to be 595 kg/ha which is generally expected in 'average to good' rainfall years. Alternatively, an extra 56.6 ha of wheat would have been necessary to ensure self-sufficiency in 1977–8 leaving only 80 ha available for barley. However, this would have increased the feed deficit by 244 GJ giving a total deficit of 270 GJ on feeds. Wheat and barley prices and costs of production are broadly similar, but barley is expected to give a higher yield, so the bias towards barley in the area planted seems sensible.

With regard to the feed balance (Fig. 9) the system deficit is 46 GJ which is equivalent to 11 ha of barley or 4182 kg of barley grain. In a more 'normal' rainfall year, the additional 31 kg/ha needed to cover this deficit would be met.

So, although in a 'normal' year the system would balance satisfactorily, the last three poor years have created a very difficult situation, particularly as a portion of the metabolizable energy that could be supplied by the system for feeds is sold cheaply and bought back dearly.

These examples give only a partial picture of flows and links in farming systems but they illustrate the close interdependence of crop output, livestock needs and household food requirements. It would be even more informative to examine

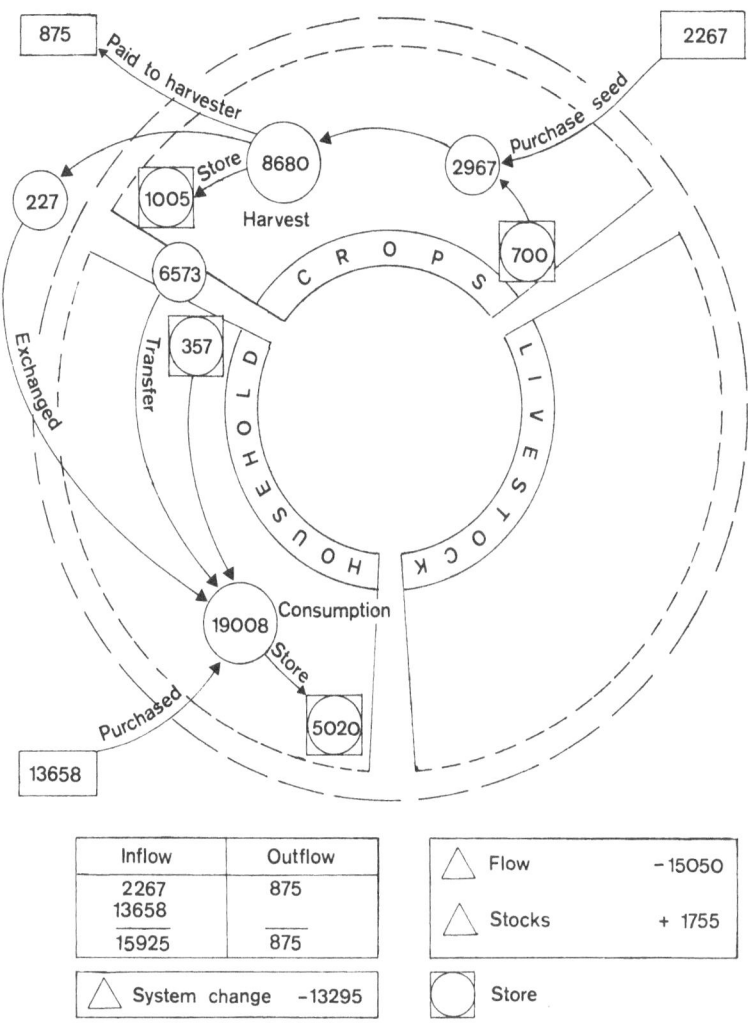

Fig. 8.   Village 4/04. Wheat flows (kg) October 1977-September 1978.

how these relationships change over time and, in particular, how the systems are able to cope with continuing adverse conditions in village 4/04. The scope for alternative crops in village 2A/06, such as forage, appears to be limited, given current relative prices and in view of the repercussions which such a change would have throughout the system.

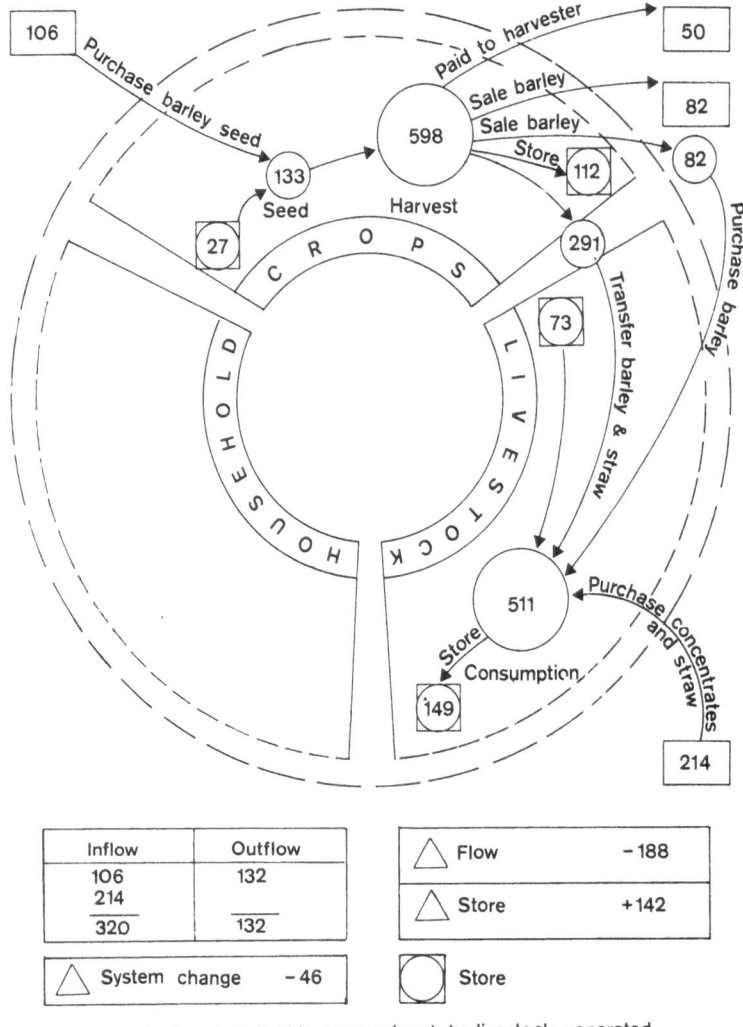

Fig. 9. Village 4.04. Animal feed flows (gigajoules of metabolizable energy, sheep) October 1977-September 1978.

Loans, credit and income from work off the 'home' farm help to sustain a farming system in many ways. Most of the farmers in our study area are obtaining money or credit from a variety of sources and its role is indicated in Fig. 10 (based on earlier figure, Gibbon and Martin, 1978) which combines some physical flows

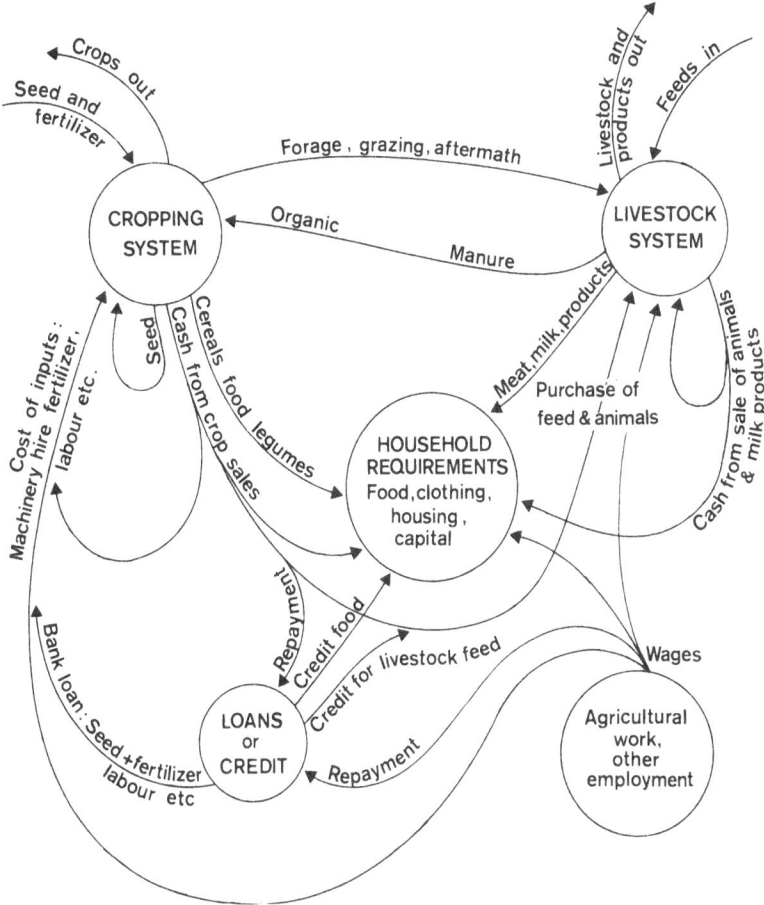

Fig. 10.   Some farm system interrelationships.

with cash flows. This figure, along with the previous flow diagrams, highlights the importance of developing an understanding of household needs as these constitute the focal point of the farming system.

THE POTENTIAL FOR IMPROVEMENT

Attention is repeatedly drawn to the relatively slow improvement in agricultural production in the Mediterranean Basin over the past 30 years, in contrast to the spectacular developments that have occurred in California, in relation to irrigated agriculture, and in South Australia, with respect to rainfed agriculture (Arnon, 1971; Oram, 1977).

In discussing the possibilities and potential for improvement, many writers have considered that a series of relatively simple technical changes in existing systems will transform agriculture, making it more stable and much more productive. (Arnon, 1971, 1979; Bolton, 1979; Carter, 1975; Leeuwrick, 1975; Oram, 1956, 1979.) While there is no doubt that more productive systems and mixed crop: livestock systems, proven elsewhere under totally different political, social and economic environments, can be shown to be technically feasible in parts of the Mediterranean (Doolette, 1977; Leeuwrick, 1975; Lozoides, 1979), the fact remains that the so called 'complete systems' advocated so far have not been widely adopted, nor, indeed, are they always actively encouraged or supported by governments, except in Cyprus (Photiades, 1979).

Perhaps too much emphasis is being placed on particular types of technical change and on importing technology, while the structure of farming systems in the region and the fact that constant changes are taking place within present farming systems are imperfectly understood. It is also not generally appreciated that changes in the use of resources in farming systems are only part of a wider process of rural development, from which they cannot be isolated.

We can initiate changes that will provide more food and secure a more reliable income to many people, only if we first understand the processes and inter-relationships within existing systems as illustrated, in part, by Figs. 2, 5, 6, 8, 9, 10; and secondly if we involve rural people themselves in the development of alternative technologies. Technologies that are imported complete, or con-structed inside fenced research stations by research scientists, are likely to have minimal application as only the ultimate users of technologies can decide what is appropriate for their own circumstances (Biggs, 1978; Bunting, 1978; Taylor, 1969).

I nevertheless accept that much knowledge from elsewhere can be important at various stages in the development of our research programme. The wealth of material presented at the workshop indicates how much is already known of the basic processes operating in this and similar environments. Our primary task would seem to be to see how much of this information is relevant to our situation, how much new knowledge is needed, and what are the appropriate experimental techniques to be used.

We also need to appreciate that farming families themselves can make a major contribution to a research programme through their intimate knowledge of their environment and of soil and crop behaviour, their constant experimenting with alternative techniques and their ability to develop strategies to cope with adverse conditions. Such information should have a major influence on the way we conduct our research, but it rarely does. (Belshaw and Hall, 1972; Richards, 1979; Swift, 1979.)

RESEARCH STRATEGIES

The assembly, evaluation and use of information of the kind which I have discussed necessitates a systems approach to research and requires a team of people whose experience spans a range of disciplines. To understand how existing farming systems work and the nature of problems which farmers face, team members must be committed to a type of programme very different from that adopted by conventional research institutes in developing countries. Only by following such a programme can we hope to identify the new knowledge which is needed to explore the alternative techniques and inputs which may increase agricultural production.

A framework for a systems research programme is given in Fig. 11. This scheme merely identifies the components of the programme and indicates how they interrelate. In the actual organisation of the programme, a sequence of steps may be followed, though not in the conventional sequence: identification of constraints – experimental station trials – off-site trials – farmer testing – extension. Recent discussion and writing have improved our whole approach to this methodology (Biggs, 1978; CGIAR/TAC, 1978; Flynn, 1978; Norman, 1976).

We need alternatives to the development of an 'improved package of practices', particularly as farmers already have their own set of inputs, quite logical for their own circumstances. They usually select from a technology package made up by research scientists whatever they consider to be initially beneficial and appropriate in their circumstances (Mann, 1975; Ryan and Subrahmanyam, 1975).

The continuing study-of, and involvement-in, existing farming systems becomes the focus of the whole programme for developing an understanding of change, generating ideas and knowledge, and for evolving alternatives for making more efficient use of available resources. Thus, experiments of various kinds, ranging from the testing of a new crop variety under farmer management, to site-replicated factorial trials, under alternative management systems, are also important components of the programme.

At the same time, studies should continue on such topics as the effect of government intervention, the utilization of labour, the interrelationships between livestock and arable systems, the effect of power structure on the distribution and access to resources, and studies of the methods of exchanging information with farming communities.

78

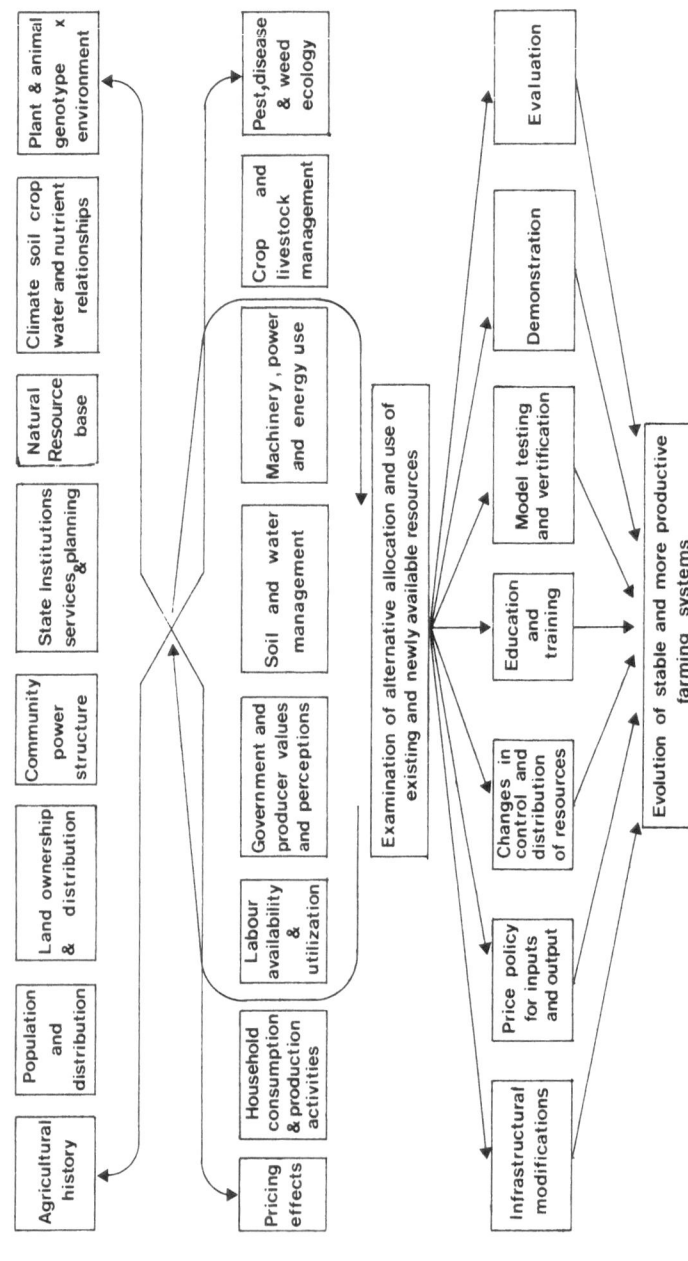

Fig. 11. Framework for a systems research program.

ACKNOWLEDGEMENTS

I am grateful to all members of the ICARDA Farming Systems Research Team who have collected the original data and contributed towards the development of the approach presented in this paper. Particular thanks are due to James Harvey for the assembly and presentation of the Figures 5, 6, 8, and 9, and for the following discussion, and also to Adrienne Martin and James Harvey for critical comments on earlier versions of the paper.

REFERENCES

Amin, S. 1976 Unequal development. Monthly Review Press, London.
Arnon, I. 1970 Crop Production in Dry Regions, Vol. I. Leonard Hill, London.
Arnon, I. 1979 Optimising yields and water use in Mediterranean agriculture in soils. In: Mediterranean type climates and their yield potential, 14th Colloq. of the Int. Potash Institute Sevilla. Spain, May 1979.
Aschmann, H. 1977 Historical development of agriculture in semi-arid regions of winter precipitation. Proc. Int. Symp. Rainfed Agric. in Semi-Arid Regions, Riverside, Calif. April 1977.
Belshaw, D. G. R. and Hall, M. 1972 The analysis and use of agricultural experimental data in tropical Africa, East African J. Rural Development 5.
Biggs, S. 1978 On farm and village level research: An approach to the development of agricultural and rural technologies. Asian Report No. 6, ICRISAT Office, New Delhi, India.
Bolton, F. E. 1979 Agronomic yield constraints in rainfed cereal production systems. Fifth Reg. Cereals Workshop Algiers, May 1979.
Brichambaut, G. de and Wallén, G. 1963 A study of agroclimatology in the semi-arid and arid zones of the New East. Tech. note 56 W.M.O. Geneva.
Bunting, A. H. 1978 Science and technology for human needs, rural development and the relief of poverty. OECD Workshop Scientific and Technical Coop. with Dev. Countries, Paris, April 1978.
Carter, E. 1975 The potential role of integrated cereal-livestock systems from Southern Australia in increasing food production in the Near East and North African Region. FAO/UNDP, Karachi, 1975.
Carter, E. 1977 A review of the existing and potential role of legumes in farming systems of the Near East and North African Region. Intern. Report for ICARDA, October 1977.
CGIAR/TAC, 1978 Farming systems research at the international agricultural research centres. Analysis and Workshop on FSR., Nairobi, 1978.
Duckham, A. N. and Masefield, G. B. 1970 Farming systems of the world.
Emberger, L. 1955 A biogeographic classification of climates. Recl. Trav. Labs Bot. Geol. zool. Univ. Montpellier 7, 3–43.
Flynn, J. 1978 Agro-Economic Considerations in Cassava Intercropping Research. Cassava Intercropping Workshop, Triwandrum, India.
French, R. J. 1978 The effect of fallowing on yield of wheat. II. The Effect on Grain Yield. Aust. J. Agric. Res. 29, 669–84.
Gibbon, D. and Martin, A. 1978 Food legumes in the farming systems, ICARDA/IDRC.
Grigg, D. 1970 The Harsh Lands. MacMillan, London.
Grigg, D. 1974 Farming systems of the world. C.U.P. Camb.
Hills, E. S. (ed) 1966 The Arid Lands.
ICARDA 1979 Farming systems research report No. 1. Dec. 1979.
IRRI 1974 Report on research in progress, Los Banos, Philippines.
Leeuwrick, D. M. 1975 The relevance of the cereal-pasture rotation in the Middle East and North African Region. Proc. 3rd Regional Wheat Workshop, Tunis.

Lozoides, P. 1979 Crop rotations under rainfed conditions in a Mediterranean climate in relation to soil moisture and fertilizer requirements. FAO Reg. Sem., Amman, 5–10 May 1979.

Mann, C. 1977 Package of practices: A step at a time with clusters? Rockefeller Foundation. Middle East Tech. Univ. Ankara, Turkey.

Matthews, H. A. 1924 A comparison between the Mediterranean climates of Eurasia and the Americas. Scottish Geographical Magazine 40, 150.

Meigs, P. 1964 Classification and occurrence of Mediterranean-type dry climates. In: Land use in semi-arid Mediterranean climates. Arid Zones Res. 26 UNESCO, Paris 17.

Newbegin, M. 1929 The Mediterranean climatic type; Its world distribution and the human response. South African Geograph. J. 12, 14.

Norman, D. 1976 Farming systems research in the context of Mali. Middle East and African agricultural seminar, Tunis, Feb. 1977, Ford Foundation.

Oram, P. 1956 Pastures and fodder crops in rotations in Mediterranean agriculture. Agric. Dev. Paper, 57, FAO, Rome.

Oram, P. 1977 Agriculture in the semi-arid regions: Problems and opportunities. Proc. Int. Symp. Rainfed Agric. in Semi-Arid Regions. Riverside, Calif. April 1977.

Oram, P. 1979 Crop production systems. University of Utah.

Papadakis, J. 1973 The Mediterranean zone of Chile. FAO, Tech. Report AGL: SF/CHI 18, No. 3.

Photiades, T. 1979 Integration of livestock with rainfed agriculture in Cyprus. FAO Reg. Semi. Rainfed Agric. in Near East. Amman, Jordan, May 1979.

Richards, P. 1979 Community environmental knowledge in African rural development. In: Rural development: Whose knowledge counts? IDS Bull. 10, No. 2.

Russell, J. S. 1967 Nitrogen fertilizer and wheat in a semi-arid environment. 1: Effect on yield. Aust. J. Expl. Agric. An. Husb. 7, 28.

Ryan, J. and Subrahmanyam, K. V. 1975 An appraisal of the package of practices approach to the adoption of modern varieties. Occ. Paper 11, ICRISAT.

Stamp, L. D. 1961 A history of land use in the arid zone, UNESCO, Paris.

Swift, J. 1979 Notes on traditional knowledge, modern knowledge and development. In: Rural development: Whose knowledge counts? IDS Bull. 10, No. 2.

Taylor, D. C. 1969 The farmer and agricultural development. Man, Food and Agriculture in the Middle East. AUB, 1969.

UNESCO, 1964 Land use in semi-arid Mediterranean climates. Paris.

UNESCO/FAO, 1963 Bioclimatic map of the Mediterranean zone. Arid Zone Res. 21, UNESCO, Paris-Rome.

Watson, A. 1974 The Arab agricultural revolution and its diffusion. 700–1100. J. Economic History 34, No. 1.

White, H. 1963 Roman Agriculture in North Africa. Nigeria Geograph. J. 6, 39.

White, H. 1970 Fallowing, crop rotation and crop yields in Roman times. Agric. History 44, 281–290.

Whittlesey, D. 1963 Major agricultural regions of the Earth. Ann. Assoc. Amer. Geographers 26, 199–240.

4. Water dynamics in the soil-plant-atmosphere system

JOE T. RITCHIE

United States Department of Agriculture, Science and Education Administration, Agricultural Research Southern Region, Grassland, Soil & Water Research Laboratory, Temple, Texas, U.S.A.

The water problem in agriculture is related both to weather and to the reserves of water in the soil that are available to plants. Water dynamics in the soil-plant-atmosphere system concerns the capacity of the soil water reservoir, its depletion and replenishment, and its efficient management for crop production.

The concept of the soil as a reservoir for water is appealing and useful. Since only a small amount of water can be stored in crop plants relative to the rate of transpiration through them, it is the storage of water within the soil pores that permits transpiration to continue for several days without recharge by rainfall or irrigation. However, water storage in the soil is not similar to that in a bucket. Some water may drain out of the root zone, and not all water remaining in a drying soil can be taken up by the plant as rapidly as it is needed because it is held too tightly by soil particles.

Although methods of determining the capacity of the soil water reservoir available to the plant are not exact, the concept permits calculations of the soil water balance and its impact on crop production.

Water-balance calculations using computers are becoming more common. There should be more emphasis on water-balance technology in the future because it is needed for accurate estimation of crop yields, early warning about food shortages, better farm management, reliable irrigation scheduling and water-resource planning, etc. Because of these urgent needs, it is important to develop models of the water balance that are as general as possible so that local calibrations are eliminated or at least minimized. Models should also not depend on the input of weather records that are difficult to obtain.

The dynamics of the soil-water balance requires separate understandings of the atmospheric, plant and soil-water factors which affect the soil water balance. These factors are interdependent but will be discussed separately for simplicity.

ATMOSPHERIC INFLUENCE

The accuracy of estimating evaporation from soil and plants is of primary importance for reliable water-balance evaluations. Because climatic variables influence evaporation so strongly when adequate water is available, the proper combination of factors to estimate maximum evaporation (E_{max}) is important. In 1973, the American Society of Civil Engineers (ASCE) evaluated the accuracy of several equations for estimating E_{max} from a wide variety of locations. The Society tested energy balance and aerodynamic combination equations, humidity-, radiation-, and temperature-based equations and some miscellaneous equations. The well-known combination equation of Penman and two other equations, somewhat similar to it, were superior because they had small errors, but some other equations were impressive. Although the report did not evaluate daily errors, experience has shown that temperature-, humidity-, or pan evaporation-based calculations give high daily errors but tend to become more accurate when records from several days up to an entire season are included as a single comparison.

One E_{max} calculation method which the ASCE report did not discuss and which has gained popularity during recent years is an equation suggested by Priestley and Taylor (1972), based on a correlation found between what Priestley (1959) called equilibrium evaporation (E_{eq}) and E_{max}. The equation for E_{eq} is the same as the radiation term in Penman's combination equation,

$$E_{eq} = \frac{\Delta}{\Delta + \gamma} (R_n - G). \tag{1}$$

In this equation, Δ is the slope of the saturation vapor pressure curve at mean air temperature (mb/°K); γ is a psychometric constant (mb/°K); R_n is the net radiation at the canopy top (mm/day), and G is the heat-flux density at the soil surface (mm/day). Priestley and Taylor found that

$$E_{max} = \alpha E_{eq}, \tag{2}$$

where α averaged about 1.26 for climate with little advection when E_{eq} was determined from 24-hour R_n values and G was assumed to be zero.

There is little reason to use more complicated equations when the simpler ones such as equation 2 give more accurate and consistent results. However, the main disadvantage of equation 2 is that it does not account adequately for advection. Tanner and Jury (1976), Sumayao *et al.* (1978), and Meyer and Green (1980) have proposed making α variable when humidity is below or temperature is above certain threshold values. These empirical modifications provide equations that are somewhat similar to the Jensen and Haise (1963) radiation-based E_{max} equations, where temperature is used to modify the radiation term. It is important to understand that E_{max} cannot be calculated exactly, and that all equations are empirical and, therefore, need some calibration.

INFLUENCE OF PLANTS

Actual evaporation may not equal E_{max} because of an incomplete crop canopy or a deficiency of water in the root zone. Use of a locally fitted crop coefficient to express the canopy cover through a crop-growth cycle has been commonly used to reduce E_{max}, but two problems have prevented generality: (i) growing season times and durations shift because of variable weather and (ii) soil evaporation during partial plant cover varies greatly, depending on the wetness of the soil surface.

It is possible to separate soil and plant evaporation logically when we know the fraction of the energy intercepted by the plant canopy and the critical soil parameters (Ritchie, 1972; Tanner and Jury, 1976). Several attempts to use this logic have been successful, using measurements of leaf area index (LAI) to estimate the energy interception fractions (Al-Khafaf et al., 1978; Kanemasu, 1976).

Measurements of LAI are important, but they are time consuming and many people have no records available. Short-cut procedures are possible for estimating plant-leaf are based on regression of leaf area per plant with more easily measured variables such as plant height, length of a certain leaf, number of leaves, and stem diameter.

An important development in crop modelling is that it appears possible to reasonably calculate plant leaf area development. Rate of leaf appearance is closely coupled with plant temperature. When leaf sizes, numbers, and growth rates are known, it is possible to develop a logical system for LAI changes during a season, using only weather data required to calculate E_{max}. Such a system has been described for development of the grain sorghum leaf (Arkin et al., 1976).

Another possibility for gross evaluation of vegetative cover is through remote-sensing procedures. Allen and Richardson (1968) have proposed a theoretical possibility for remote sensing of LAI on the basis of differences between plant canopy reflectance and soil reflectance. Wiegand et al. (1979) demonstrated that three vegetational indexes derived from satellite data were correlated well enough with LAI to provide inputs to evapotranspiration (ET) models for LAI values > 0.3. Use of remote-sensing techniques, however, would require frequent overflights during periods of rapid increases in LAI and might not be economically feasible.

INFLUENCE OF SOIL WATER

Plant response to soil water deficits

The response of crop economic yield to water deficits is a dynamic process and general quantitative relations are difficult to establish. Plant water stress can be induced (i) by a deficiency of water supply in the root zone, and (ii) by an excessive atmospheric water demand from leaves. In many crop production systems, variations in soil water deficits are the major cause of year-to-year variation in yield. Although many laboratory experiments and theoretical evaluations have demonstrated that high evaporative demand causes plants growing in a wet soil to show symptoms of water deficiency, such results seldom extrapolate to field conditions because the response often depends on the environmental history of the crop. As a result, experimental measurements of crop canopy photosynthesis, transpiration, or leaf extension growth usually do not show mid-day depressions caused by high evaporative demand at that time, unless the soil water in the root zone is depleted to less than 50 per cent of the total soil water extractable by plants. However, it is generally thought that short periods of high evaporative demand and high temperature during the critical periods of plant pollination or formation of reproductive organs can irreversibly reduce yield regardless of the soil water status, but convincing quantitative evidence is lacking.

The influence of the soil water deficit on crop behavior in the field has been the subject of many agronomic field trials, the results of which are often specific to the location, climate, crop, and soil. It has been almost impossible to generalize about plant response to water deficits using soil measurements such as soil water potential or water content. Consequently, many soil scientists have recently used plant measurements as indicators of crop response to water deficits.

Early field work on crop water status by agronomists and ecologists centered on the idea that leaf stomatal diffusion should be a primary measurable factor influenced by water deficits since stomatal closing is directly linked to both photosynthesis and transpiration. Diffusion resistance meters were developed (Kanemasu et al., 1969), made commercially available and are now commonly used in many agronomic studies. Penman and Schofield (1951) developed an equation including vapor diffusion resistance to estimate transpiration when stomata were closed or partially closed. The main disadvantage of using stomatal resistance has centered on the difficulty of deriving an integrated value of resistance for an entire canopy that is generally suitable for use in evaporation equations. Although stomatal resistance measurements have provided quantitative descriptions for research, they are not to my knowledge being used on a widespread operational basis to estimate crop performance characteristics or to determine when to irrigate. One possible reason for the lack of applicability of stomatal resistance measurements to crop performance is that the physiological

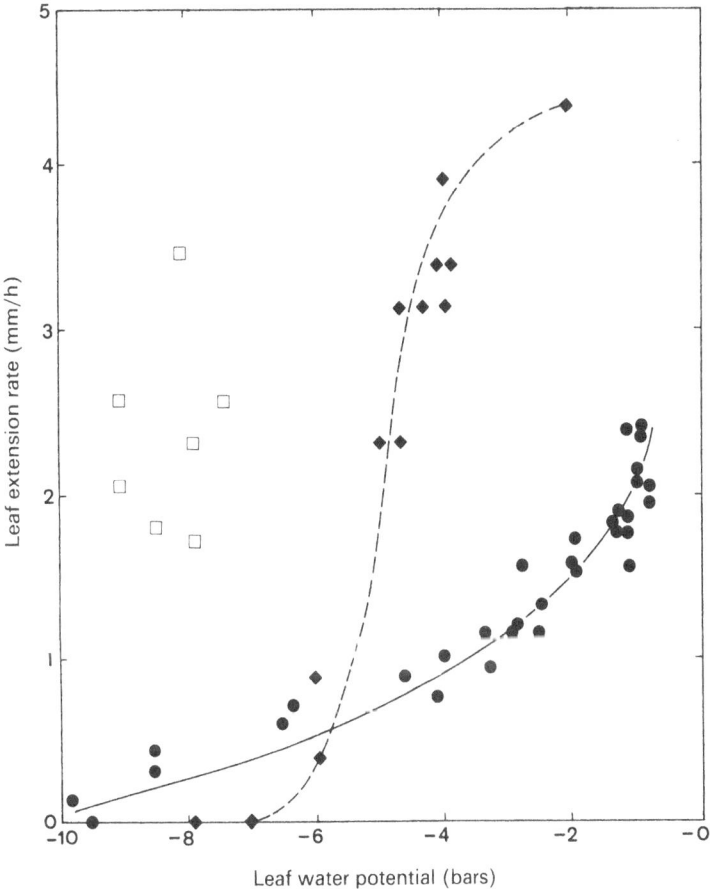

Fig. 1. Relationship between leaf extension and leaf water potential for corn growth in the field (□), or grown in controlled environment in the dark at 28 °C (●), or in the light at 30 °C (■). From Watts (1974).

processes of plant extension growth are more sensitive to plant water deficits than are the stomatal regulated processes (Hsiao, 1973) + hence, irreversible damage may have occurred in a crop before stomatal resistance measurements clearly indicate a change in plant conditions.

Another physical measurement which is sensitive to crop water deficit is the plant water potential. Like stomatal resistance, agronomists and ecologists began to make this measurement when the pressure chamber (Scholander *et al.*, 1965) was adapted for relatively simple use in the field. Problems in interpreting the results of plant water potential measurements in general, quantitative terms have almost paralleled those of stomatal diffusion. With both measurements, when atmospheric conditions are about constant day by day, little change occurs when soil water is being depleted until some threshold value is reached, following which the values usually change rapidly (van Bavel, 1967; Ritchie, 1973). Therefore, plant water potentials have limited value in estimating the onset of field water stresses for applications such as irrigation scheduling.

Evidence for the lack of a general relationship between leaf extension rate and leaf water potential in corn is shown in Figure 1 from data compiled by Watts (1974), who compared field and controlled environment data. When low night temperature does not reduce extension growth, leaf expansion continues day and night at fairly similar rates despite a usual diurnal change in leaf water potential from about − 1 to − 9 bars (Watts, 1974; McCree and Davis, 1974). Physiologists have suggested that this phenomenon is the result of a diurnal osmotic potential adjustment in the leaf, causing leaf turgor pressure to remain adequate for extension growth (Begg and Turner, 1976).

Leaf potential measurements just before sunrise appear to be generally related to daily extension growth. Pre-dawn potentials should give an idea of the integrated potential of the soil that affects plant extension growth because the plants have had long enough during the night when the transpiration is practically zero to recover to a potential in equilibrium with the root-soil system. Cutler and Rains (1977) found a fairly stable relationship between pre-dawn water potential and daily cotton leaf elongation. There was some variation in the relationship caused by levels of stress conditioning achieved by varying the frequency of irrigation during pretreatment periods.

The response of stomatal regulated processes to plant water deficits has been known for many years. However, it is generally recognized that stomata do not respond directly to leaf water potential until a critical threshold potential has been reached, after which stomata close over a narrow range of potentials and cause no further decrease in plant water deficit. Early work on these relationships implied that the critical potential threshold, when stomata would close, might provide a species-specific parameter that would be valuable in quantitative evaluation of plant water deficits. However, field work reported in the 1970's has demonstrated that there is no unique leaf water potential causing stomatal closure. Begg and Turner (1976) presented evidence that this threshold leaf potential varies with position of the leaf in the canopy, age of the plant, and growth conditions such as the number of stress cycles or whether plants are grown in the field or in controlled environments. Jordan and Ritchie (1971) found that stomata closed rapidly at -16 bar potential in cotton plants grown in a growth chamber, whereas stomata of field grown cotton plants subjected to a long water-drying cycle, did not close at the lowest measured leaf potential of -27 bar. Similar differences in response have been shown for sorghum (McCree, 1974) and vines (Kriedemann and Smart, 1971). These findings further demonstrate the limitation of leaf water potentials as indicators of plant water deficits.

Thus far, I have pointed out that expansion growth is more sensitive than stomatal processes to plant water deficits. Other important processes such as cell division, leaf wilting or rolling, tillering, leaf abscission or partial death, seed filling, pollination, seed abortion and translocation, all have different sensitivities to plant water deficits, some of which are significant only during specific phases of plant development. Differences in sensitivity of plant processes to water deficits may be related to natural survival mechanisms. Primitive progenitors of modern crop plants were subjected to a great variety of climatic conditions and their adaptive mechanisms prevented extinction. While modern crops are grown in monocultures usually at higher plant populations per unit area than their progenitors, their varying degrees of sensitivity to water deficits show that they have retained the survival mechanism.

The differences in sensitivity of plant processes to water deficits may be the primary reason for the large difference between the responses of small container-grown plants and field-grown plants to soil water deficits. When water stress develops gradually in plants, as when plants grow on stored water from deep soils, the various processes affected at different times by the stress should become evident to the close observer. Unfortunately, there have been few field studies where more than one process has been evaluated during plant water stress (Hsiao, 1973).

However, when plants are grown in containers where watering is required every one to three days to prevent water stress, it is difficult to observe the varying stress responses because of insufficient time for various regulatory mechanisms to express themselves. Consequently great caution should be used when extrapolating the response of plants in containers to field conditions. An example of this difference is the form of the relationship found between soil water and transpiration obtained experimentally in container-grown corn plants by Denmead and Shaw (1962) and that found for corn grown in a weighing lysimeter with a large soil volume (Ritchie, 1973). In the container study, transpiration was reduced under high evaporative conditions after a small fraction of the soil water was reduced, whereas in the large soil volume, transpiration was not reduced when as much as 70 per cent of the available water was extracted.

Ludlow and Ng (1976) found that the water relations of green panicum (*Panicum maximum*) grown in large pots in a growth room compared favorably with similar plants grown in similar sized pots in a field environment. The chambers were programmed to provide average values of outdoor daylength, maximum and minimum temperature, and relative humidity. Photosynthetically active radiation was 66 per cent of the outdoor value during three weeks without water. Threshold water potentials at which stomatal resistance increased and leaf elongation ceased were similar for both outdoor and growth room potted plants. Thus, it appears that growth room climates *per se* may be satisfactory for plant water deficit studies that may be extrapolated to field conditions if root volume is not restricted much more than under field conditions to ensure that the rate of onset of stress is not accelerated.

It is possible to obtain the small-container effect in the field when plants grow in shallow soil with low water-holding capacity because of the possibility of rapid stress. The rate of onset of stress is primarily affected by the water storage of the root zone; transpiration rate is secondary. To demonstrate the effect of the adaptive mechanism of osmotic adjustment, consider a crop canopy fully covering the ground growing in three soils having respectively, 1, 15, and 30 cm of extractable soil water in the root zone. Assuming a constant transpiration rate of 5 mm a day until plants had removed about 70 per cent of the extractable soil water, the water potential to which plants recover overnight (the pre-dawn value) would be expected to have relations similar to those shown in Figure 2a.

In the graph, recovery potentials of about − 5 bar represent zero to slight deficit, those between − 5 and − 15 bar, moderate deficits, and those below − 15 bar, severe deficits. The soil with 1 cm extractable water provides one day of no stress conditions, and the onset of stress is very rapid, lasting possibly only two more days until a severe stress. In this case, osmotic adjustment is minimal and plants would likely die because of the lack of time to adapt. For the soils with 15- and 30-cm extractable water, however, there would be a period of 24 and 50 days, respectively, with little stress and about 8 and 11 days with moderate stress. The stomatal-controlled functions for the three soil comparisons might provide patterns as shown in Figure 2b where the threshold leaf water potential for

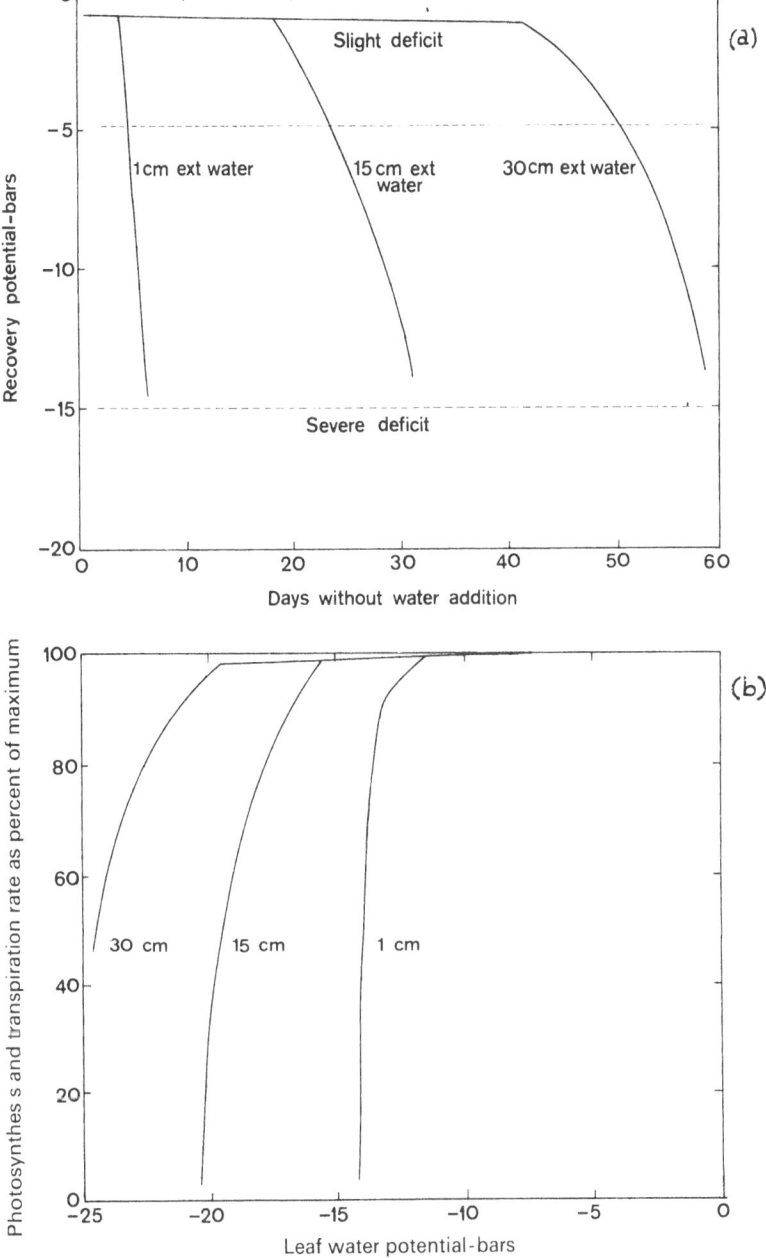

Fig. 2. Estimated plant water relations of a crop growing on stored soil water amounting to 1, 15 and 30 cm extractable water;
(a) the pre-dawn recovery potential as a function of time after water addition as related to extractable soil water and
(b) the influence of leaf water potential on relative photosynthesis and transpiration.

stomatal closure is at progressively lower values for larger soil water storage capacities. Evidence for such relationships is found in Jordan and Ritchie (1971) and Brown *et al.* (1976). Absolute threshold potential values where stomatal closure is obtained would be expected to vary with such things as species and leaf age, and the amount of adjustment should vary between drought-tolerant and drought-susceptible plants.

Because of the difficulty of using plant stomatal resistance or water potential measurements for operational purposes to determine the effect of water deficit on crop performance, empirical evaluation of various processes as related to soil water deficits should continue to be a useful option. However, recognizing that water stress causes variable responses for different physiological processes, a set of relationships needs to be established for each process for predictive purposes. Figure 3 represents a possible template for such an evaluation. The type of relationship shown in this figure is often used to estimate evapotranspiration (ET) reduction in response to soil water deficits. The general concept is that there is no reduction in the process being considered until the amount of extractable water in the entire root zone falls to some threshold value, following which the process is reduced in proportion to the extractable water.

From several analyses of the type shown in Fig. 3, it appears that threshold values for various physiological processes are about the same for many crops and soils. The diagram shows that the process of leaf elongation is more sensitive to

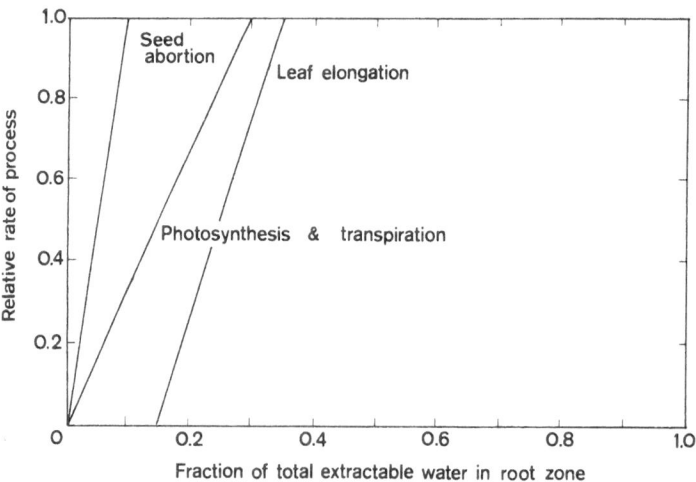

Fig. 3. Suggested possibilities for expressing the influence of extractable soil water on the relative rate of various physiological processes.

soil water deficits than other processes and that elongation stops even when some soil water remains. The sensitivity of processes regulated by stomata is less than that of elongation. Processes like seed abortion occur only under a very severe stress. Other processes such as seed filling, leaf senescence, tillering and root extension can be similarly evaluated but the rate of some processes is reduced in very wet soil when aeration is poor.

For ET estimation, all daily records of ET/E_{max} from accurate, weighing lysimeters in the field fit within the concept shown for transpiration in Fig. 3, with some deviation in the threshold soil water fraction where ET/E_{max} falls below 1 (see van Bavel, 1967; Priestley and Taylor, 1972; Ritchie *et al.*, 1972; Nkemdirim and Haley, 1973; and Meyer and Green, 1980). The exact point of the threshold water content is difficult to distinguish and is usually extrapolated from measurements taken when ET/E_{max} is clearly reduced by drought. The zero point for extractable soil water usually has to be extrapolated from measurements above that point because plants are likely to lose much of their leaf area within the extremely dry range and soil evaporation may then become a significant part of ET. In this dry range, plants undergo drastic changes in their natural adaptation for survival.

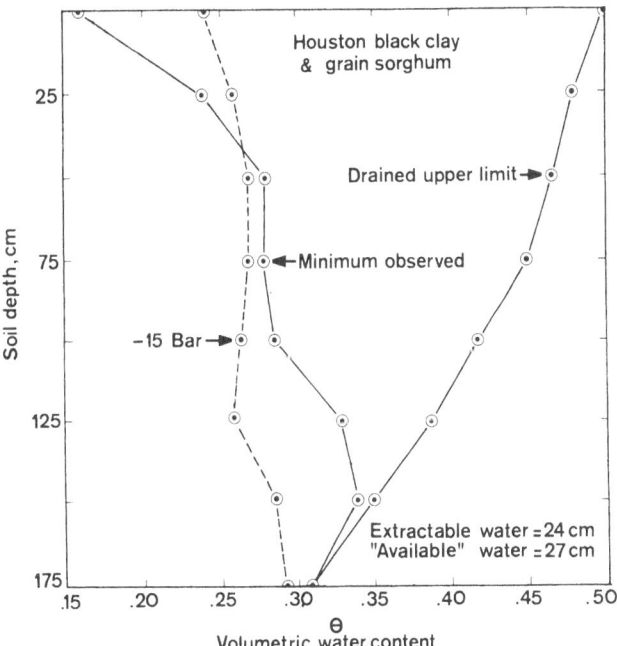

Fig. 4.   Graphic representation of the relationship between extractable soil water and the ratio of actual evapotranspiration (ET) to maximum evaporation (E_{max}).

A practical problem concerned with establishing relationships of the type shown in Fig. 3 is the evaluation of the soil extractable water. Traditionally, agronomists use water available in the root zone between field capacity and wilting point. Problems associated with this definition include determining values for root zone depth, field capacity, and wilting point, and soil bulk density for all soil depths in the root zone where physical properties change. When soil water potentials measured in the laboratory are used to define field capacity and wilting point, there are uncertainties about which potential to choose for the limit, especially for field capacity.

The concept of extractable water was introduced as a practical means of eliminating some of the problems of the available water concept. The extractable water is defined as the difference between the highest measured volumetric water content in the field (after drainage) and the lowest measured water content when plants are very dry and leaves are either dead or dormant. Extractable water thus defined is preferred to the traditional available water lower limit because it weights root distribution. This definition eliminates the need for doing soil water content/potential relationships for each soil depth where physical properties change. Field measurements of the total extractable water are often less variable spatially than available water estimated from water content-potential measurements.

Fig. 4 illustrates the distinction between extractable water and available water for Houston Black Clay. Available water integrated for the entire profile gave 3 cm more water than measured, about half of the water used during the phase of decreasing ET.

Another problem of using soil water in the root zone to estimate ET/E_{max} occurs when the soil water reservoir is partially refilled after a very dry condition and the additional water does not fill the root zone reservoir above the threshold water content where $ET/E_{max} = 1$. In this case, the added water remains near the surface where root density is greatest and provides a soil water status which is satisfactory for good plant turgor, although the entire root zone water content is low.

When $ET/E_{max} < 1$, estimating transpiration from the type of relationship shown in Fig. 3 implies that ET is a function of extractable soil water and E_{max}. In practice, E_{max} becomes less important the drier the soil, and factors affecting water transport in the soil-plant system become more important limitations on transpiration. The analysis of this dynamic problem has received much theoretical and experimental attention. None of the proposed models has been widely used and several technical problems must be clarified before general models can be developed. Some technical problems include (i) quantification or estimation of the root density in soil, (ii) accurate evaluation of water flow to roots with large soil conductivity and potential differences surrounding roots, (iii) quantification of radial and axial root conductivities, (iv) role of the gap between the soil and root surface, and (v) quantification of the water potential in the root xylem system and at the root surface.

Infiltration

Accurate water-balance modelling also requires an estimate of the amount of water infiltration into the soil from precipitation or irrigation. The amount of water that infiltrates into the soil is governed by a diversity of variables; the major ones being the amounts and rates of precipitation, soil type, amount and type of vegetative cover, land slope, surface roughness, and initial soil water content. Frozen soils add another variable to infiltration problems. The most commonly used approach to estimating infiltration is through statistical models fitted to experimental data. These models are usually developed with the sole aim of optimizing the prediction by use of appropriate regressions.

Infiltration has also been studied through physical models. Physicists usually seek to advance understanding of hydrologic processes through use of models which embody, as fully as possible, our knowledge of the physical processes. The statistical strategy may yield a useful predictive system, but such an approach ignores the physical processes and is often useless for extrapolation outside the area where experimental data have been taken. The physical strategy often leads to an intolerably complicated model with limited usefulness and high labor requirements to characterize the real system and its initial stage.

I believe that a general and useful infiltration model must take advantage of the statistical strategy by fitting 'rationally' empirical expressions to functions with physical meaning. A promising possibility for a relatively simple physically based model for infiltration is the optimal prediction of the time between initiation of rainfall and the intiation of surface runoff, or ponding time (Smith and Parlange, 1977). The technique requires inputs of soil conductivity at saturation and sorptivity.

Infiltration equations that require precipitation data for less than 24-hour periods may not be useful for many operational models because of the lack of short term rainfall data. Because of this constraint, it may not be possible to model infiltration accurately with a single general approach.

INCREASING AVAILABLE SOIL WATER THROUGH MANIPULATION OF ROOTS

Incomplete extraction of apparently available soil water can result in limitations of productivity in many rainfed agricultural regions. Enhancement of deep rooting is a possiblity for increasing the availability of soil water. Genetic variability in root growth rates and field rooting patterns has been demonstrated in several crops (Hurd, 1974; Jordan et al., 1979; Jordan and Miller, 1980; O'Brien, 1979; Taylor et al., 1978). If deep rooting genetic materials can be combined with high yielding ones, the result could have a beneficial impact on production in areas where deep soil water is available and replenishable on an almost annual basis. If additional photosynthate is necessary to form and maintain deeper root systems, the trade-off for higher yields may not be possible. Also, the water status of plants taking up water from deep in the soil may be adversely affected and thus reduce production potential.

There have been no clear field demonstrations of benefits derived from breeding plants for deeper root water extraction. Measurement of roots and receipt of rainfall during the drying cycle causes experimental difficulties. However, in my opinion, the possibilities of increasing available water through developing plants with deeper root systems deserves considerable additional research as a means of increasing production in dry regions.

CONCLUSIONS

Although precise definitions of the two concepts for the upper and lower limits of soil water availability are limited by the dynamics of the soil-plant-atmosphere system, the use of these rough limits helps greatly in evaluating the impact of soil water balance on crop production. Estimates of extractable water determined in the field overcome several problems associated with definitions of the upper and lower limit and provide a measure of the soil water reservoir which is useful in estimating the influence of soil water deficits on important processes coupled with plant growth and yield.

Physical measurements such as stomatal conductance and water potential, while they are sensitive to plant water deficits, have proven to be of limited value in operational use because they lack sensitivity under marginal conditions of stress when some growth processes are restricted.

Much is yet to be learned about the dynamics of water in the soil-plant-atmosphere system. A specific strategy to guide research to meet future production demands requires close linkage between scientists of several disciplines, especially plant breeding, plant physiology, climatology, and soil and crop management. Multidisciplinary teams will be required to meet the challenge of the future to produce optimum crop production systems that avoid or tolerate plant water stress.

REFERENCES

Al-Khafaf, S., Wierenga, P. J. and Williams, B. C. 1978 Evaporative flux from irrigated cotton as related to leaf area index, soil water, and evaporative demand. Agron. J. 70, 912–917.
Allen, W. A. and Richardson, A. J. 1968 Interaction of light with a plant canopy. J. Opt. Soc. Amer. 58, 1023–1028.
Arkin, G. F., R. L. Vanderlip, and Ritchie, J. T. 1976 A dynamic grain sorghum growth model. Transactions of the ASAE. 19, 622–624, 630.
Begg, J. E. and Turner N. C. 1976 Crop water deficits. Advances in Agronomy. In: Brady, N. C. (ed.) 28, 161–217.
Brown, K. W., Jordan, W. R. and Thomas, J. C. 1976 Stomatal response to leaf water potential. Physiol. Plant. 37, 1–5.
Cutler, J. M. and Rains, D. W. 1977 Effect of irrigation history on responses of cotton to subsequent water stress. Crop Sci. 17, 329–335.
Denmead, O. T. and Shaw, R. H. 1962 Availability of soil water to plants as affected by soil moisture content and meteorological conditions. Agron. J. 45, 385–390.
Hsiao, T. C. 1973 Plant responses to water stress. Annu. Rev. Plant Physiol. 24, 519–570.
Hurd, E. A. 1974 Phenotype and drought tolerance in wheat. Agric. Meteor. 14, 39–55.
Jensen, M. E. and Haise, H. R. 1963 Estimating evapotranspiration from solar radiation. J. Irrigation and Drain. Div. Am. Soc. Civ. Eng. 89, 15–41.
Jordan, W. R. and Miller, F. R. 1980 Genetic variability in sorghum root systems: Implications for drought tolerance. Adaptation of Plants to Water and High Temperature Stress. In: Turner, N. C. and Kramer, P. J. (eds.), Wiley-Interscience, New York (In Press).
Jordan, W. R., Miller, F. R. and Morris, D. E. 1979 Genetic variation in root and shoot growth of sorghum in hydrophonics. Crop Sci. 19, 468–472.
Jordan, W. R. and Ritchie, J. T. 1971 Influence of soil water stress on evaporation, root absorption, and internal water status of cotton. Plant Physiol. 48, 783–788.
Kanemasu, E. T., Stone, L. R. and Powers, W. L. 1976 Evapotranspiration model tested for soybean and sorghum. Agron. J. 68, 569–572.
Kanemasu, E. T., Thurtell, G. W. and Tanner, C. B. 1969 The design, calibration and field use of a standard diffusion parameter. Plant Phys. 44, 881–885.
Kriedemann, P. E. and Smart, R. E. 1971 Effects of irradiance, temperature, and leaf water potential on photosynthesis of vine leaves. Photosynthetica. 5, 6–15.
Ludlow, M. M. and Ng, T. T. 1976 Effect of water deficit on carbon dioxide exchange and leaf elongation rate of the CO_4 tropical grass *Panicum maximum* var. *trichoglume*. Aust. J. Plant Physiol. 3, 401–413.
McCree, K. J. and Davis, S. D. 1974 Effect of water stress and temperature on leaf size and on size and number of epidermal cells in grain sorghum. Crop Sci. 14, 751–755.
Meyer, W. S. and Green, G. C. 1980 Water use by wheat and plant indicators of available soil water. Agron. J. (In press).
Nkemdirim, L. C. and Haley, P. F. 1973 An evaluation of grassland evapotranspiration. Agr. Meteorol. 11, 373–383.
O'Brien, L. 1979 Genetic variability of root growth in wheat. (*Triticum aestivum* L.) Aust. J. Agric. Res. 30, 587–595.
Penman, H. L. and Schofield, K. K. 1951 Some physical aspects of assimilation and transpiration. Symp. Soc. Expt. Biol. 5, 115–129.

96

Priestley, C. H. B. and Taylor, R. J. 1972 On the assessment of surface heat flux and evaporation using large-scale parameters. Monthly Weather Rev. 100, 81–92.

Priestley, C. H. B. 1959 Turbulent transfer in the lower atmosphere. Univ. Chicago Press, Chicago. 130 p.

Ritchie, J. T. 1972 Model for predicting evaporation from a row crop with complete cover. Water Resources Res. 8, 1204–1213.

Ritchie, J. T., Burnett, E. and Henderson, R. C. 1972 Dryland evaporative flux in a subhumid climate. III. Soil water influence. Agron. J. 64, 168–173.

Scholander, P. F., Hammel, H. T., Bradstreet, E. D. and Hemmingsen, E. A. 1965 Sap pressure in plants. Science, New York. 149, 920–922.

Smith, R. E. and Parlange, J. Y. 1977 Optimal prediction of ponding. Transaction of the ASAE. 20, 19–22.

Sumayao, C. R., Kanemasu, E. T. and Baker, T. W. 1980 Using leaf temperature to assess evaporotranspiration and advection. Agric. Meteor. (In press).

Tanner, C. B. and Jury, W. A. 1976 Estimating evaporation and transpiration from a row crop during incomplete cover. Agron. J. 68(2), 239–243.

Taylor, H. M., Burnett, E. and Booth, G. D. 1978 Taproot elongation rates of soybeans. Z. Acker- und Pflanzenbau. 146, 33–39.

van Bavel, C. H. M. 1967 Changes in canopy resistance to water loss from alfalfa induced by soil water depletion. Agr. Meteor. 4, 165–176.

Watts, W. R. 1974 Leaf extension in *Zea mays*. III. Field measurements of leaf extension in response to temperature and leaf water potential. J. Exp. Bot. 25, 1085–1096.

Wiegand, C. L., Richardson, A. J. and Kanemasu, E. T. 1979 Leaf area index estimates for wheat from LANSAT and their implications for evapotranspiration and crop modeling. Agron. J. 71, 336–342.

5. Plant-water relations and adaptation to stress

NEIL C. TURNER* and JOHN E. BEGG

*Division of Plant Industry, CSIRO, Canberra City, Australia

Many of the effects of water deficits on the growth and yield of plants are clearly evident throughout arid and semi-arid regions of the world. The dramatic increases in yields of cereals during the past 40 years in regions where a regular supply of water is assured through rainfall or irrigation, and the smaller increases where water supply is limited, have only heightened awareness of the constraint that a limited water supply places on improvement of yield.

In the United Kingdom where rainfall provides a plentiful supply of water for crop growth, yields of wheat have climbed steadily over the past 35 to 40 years at an average annual rate of 67 kg ha^{-1} year^{-1} (Fig. 1a), while yield increases in Sweden and Mexico under non-limited water conditions appear to be even more spectacular than those in the United Kingdom (Evans, 1978).

By contrast, in Australia and Syria where wheat is principally a water-limited dryland crop, yield increases have been much smaller: in Australia wheat yields have increased by 13 kg ha^{-1} year^{-1}, while in Syria there has been no significant increase over the same period (Fig. 1a). In defense of Australian breeders and agronomists, it must be pointed out that there has been a rapid expansion of wheat production since 1956 into newer and more marginal areas (Fig. 1b): this contrasts with the static amount of area harvested over the same period in Syria and the United Kingdom.

Nevertheless it is clear that in water-limited environments, less advantage can be taken of improved varieties and higher fertilizer use than in more temperate environments or regions in which irrigation can be provided. A second feature of cereal production in the dryland Mediterranean regions, exemplified by Australia and Syria, is the large year-to-year variation in yield per unit area (Fig. 1a): an analysis by Russell (1973) showed that the year-to-year variation in yield of wheat in Australia was much greater than that in USA. This is undoubtedly a function of the large year-to-year variation in rainfall in Mediterranean-type environments (Harris, 1979).

In spite of smaller yield increases in dryland Mediterranean areas compared to those obtainable in more temperate regions, the Australian example in Fig. 1 does indicate that yield increases in dryland cereal production are possible in

98

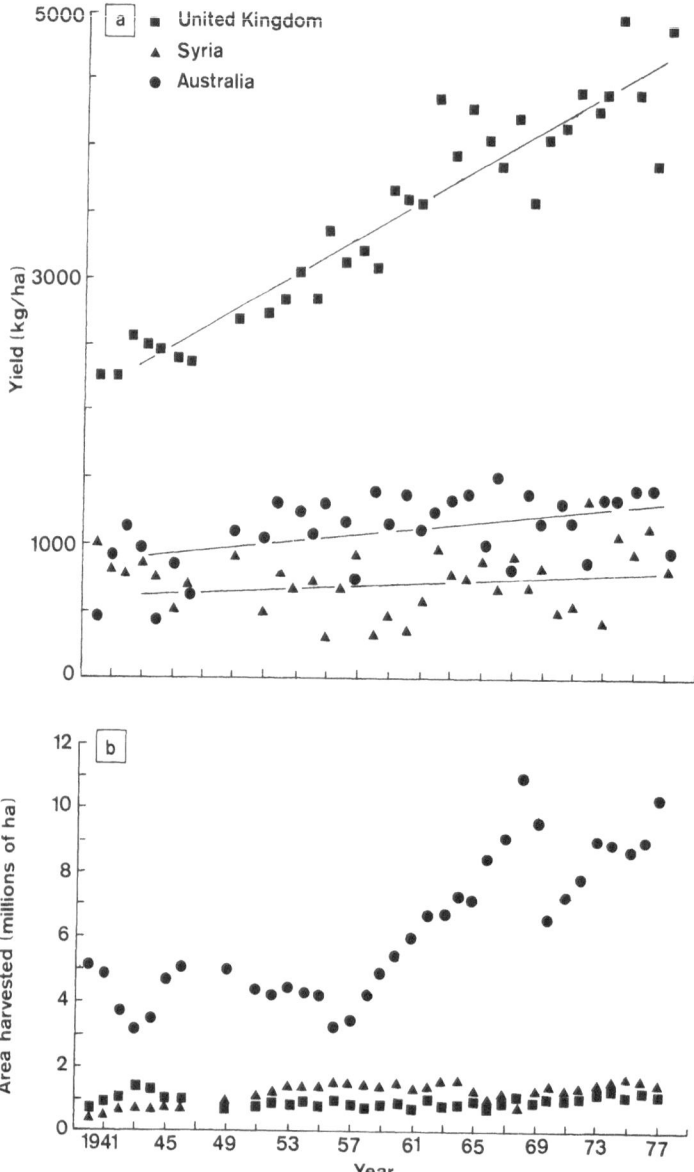

Fig. 1. National annual wheat yields (a) and annual area harvested (b) in Australia (●), Syria (▲) and the United Kingdom (■) between 1940 and 1977. Data obtained from FAO Production Yearbooks Nos 1 to 31.

semi-arid regions. However to achieve them, the crop must be designed for adversity, particularly a limited and variable water supply. One key to improved yields under such adverse conditions is the ability to adapt to prevailing environmental conditions.

In this chapter, we will describe the development of crop water deficits and the responses of the plants to those deficits. We will then explore the developmental, morphological and physiological mechanisms that plants possess to enable them to adapt to water deficits. This will lead us to formulate general design criteria and management strategies for crop plants for rainfed Mediterranean environments.

A plant, designed with certain criteria, has been termed an ideotype (Donald, 1968). In designing plant ideotypes for water-limited environments, it is imperative that the climatic regime for which the plant is being designed is understood. An ideotype designed to face stress only in the terminal stages of growth will

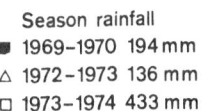

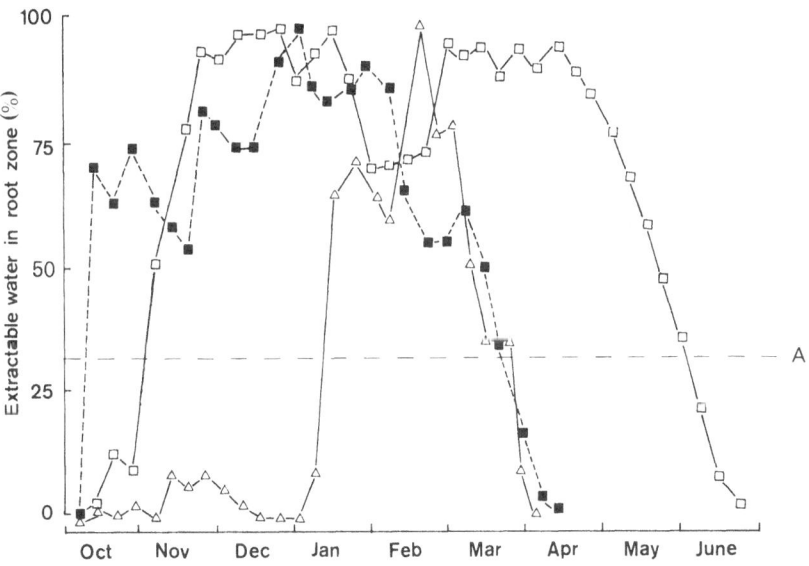

Fig. 2. The calculated extractable soil water in the root zone at Aleppo, Syria, for three seasons differing markedly in rainfall: □, 1973/74 = 433 mm (upper quartile of rainfall frequency); ■, 1969/70 = 194 mm (lower quartile); and △ 1972–1973 = 136 mm (lowest fractile). The estimated soil water percentage when actual evapotranspiration falls below potential evapotranspiration is also shown as line A.

differ from one designed to face stress that occurs unpredictably throughout its lifecycle (Turner, 1980). The soil moisture regime and climate for which we are designing our plant ideotype and management strategies is therefore briefly described.

In keeping with one of the previous chapters (Smith and Harris, 1981), we present Fig. 2 which shows the calculated soil moisture regime at Aleppo for three growing seasons from wet to dry to very dry conditions (Fig. 2). These represent years in the wettest quartile (1973–74), the driest quartile (1969–70) and the driest fractile (1972–73): the data for the median year at Aleppo were closer to the wettest quartile than the driest quartile (Smith and Harris, 1981). Assuming that actual evaporation does not fall below the potential rate until about 70 per cent of the water in the soil has been depleted (Ritchie, 1974), it is evident from Fig. 2 that once enough moisture has accumulated for germination, there is usually enough in store to maintain evaporation, and hence photosynthesis (Ritchie, 1974), throughout vegetative growth. However, there are periods in the vegetative stage and around anthesis when soil water is depleted (1969–70 and 1972–73), and this may affect leaf elongation and seed set. The incidence of rainfall decreases markedly after anthesis so that the crop has to complete its lifecycle largely on stored moisture during the period when temperature and evaporation are increasing rapidly (Harris, 1979).

DEVELOPMENT OF CROP WATER DEFICITS

The literature frequently states that water stress occurs because transpiration exceeds the rate of water absorption by the roots; sometimes referred to as the absorption lag. This statement is misleading because it implies that absorption always lags behind transpiration when plants are stressed, and that if there is no absorption lag there is no stress. It also obscures the essential nature of water movement through the plant, particularly the dynamics of the development and decay of plant water deficits on a diurnal time scale.

Water moves only from sites of high potential to those of low potential. Thus for a plant to extract water from the soil against gravity and the resistance to liquid flow through the vascular system, the potential energy of the water in the plant must be lower than that of the water in the soil. This potential difference or water potential gradient between the evaporating tissues of the plant and the soil water in the root zone increases with evaporative demand and the liquid flow resistance in the pathway. Thus water deficits occur in the tissues of all transpiring plants as an inevitable consequence of the flow of water along a pathway in which frictional resistances and gravitational potential have to be overcome.

During the early stages of a drying cycle while water is available in the root zone, the diurnal development of water deficits and their overnight recovery represent periods of both positive and negative absorption lag. As the stomata open and the evaporative demand increases, water evaporates from the meso-phyll cells of the leaves. The lowered water potential in the transpiration pathway provides the driving force for the movement of water out of adjacent tissues such as the cortex and phloem.

As a result of this loss, water deficits develop in the leaf, stem and root tissues. Thus, as a plant goes into stress during the morning, transpiration will exceed water uptake by the roots while water is drawn out of tissues surrounding the xylem, i.e. the absorption lag is positive. During the transition period between the development of deficits in the morning and the afternoon recovery period, the water potential of the vascular and surrounding tissues approaches equilibrium, as does transpiration and water uptake, i.e. there is little or no absorption lag.

However, during recovery from stress when the water deficit in the tissues is being replenished, losses from transpiration will be less than gains from water uptake and the absorption lag will be negative. Thus water deficits can occur when the transpiration rate exceeds, is equal to, or less than the rate of absorption of water by the roots.

All actively transpiring plants therefore experience some degree of short term water deficits, regardless of how well they are supplied with water. The extent of any imbalance between transpiration and water uptake is limited by the storage capacity of the plant: for crop, forage and pasture species this is usually less than one-third of the daily transpiration (Jarvis, 1975). Thus in plant water relations we are generally dealing with plants which are experiencing varying degrees-of and duration-of water deficits, and we are not simply comparing stressed and non-stressed plants.

Plants can extract water from the soil only when their water potential is lower than that in the soil. Thus water in the plant is seldom in equilibrium with water in the soil. The difference in water potential between the plant and the soil depends on the evaporative demand, the extent to which the plants can meet that demand and the water conducting properties of the soil and plant (Gardner and Nieman, 1964). Progressive changes in soil and plant water potential as the soil dries out after rain or irrigation, are presented schematically in Fig. 3. The dotted line represents the decline in soil water potential in the root zone during a drying cycle. In contrast to this steady decline in soil water potential, the plant, and particularly the leaf, exhibits marked diurnal fluctuations in water status as the evaporative demand varies day and night. There is no direct relationship between the water potential in the leaf and the water potential in the soil. The soil water

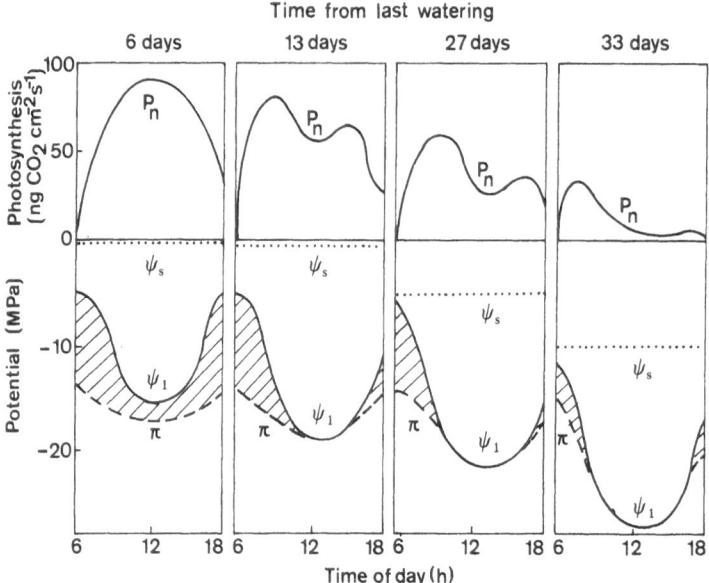

Fig. 3. Diurnal changes in photosynthesis (P_n), soil water potential (ψ_s), leaf water potential (ψ_1) and leaf osmotic potential (π) of soybean on four days between 6 and 33 days after rain or irrigation. The hatched portion gives the leaf turgor pressure. Adapted from Turner and Burch (1981).

potential merely sets the upper limit of recovery possible by the plant during the night. Thus, because of the large gradients that can exist between soil and plant water potential, it is important to relate growth to the level of stress experienced by the plant, and not to the level of stress in the soil. Indeed, in the study represented in Fig. 3, the leaf water potential did not fully recover overnight to equilibrate with the soil, possibly because partly open stomata and dry windy conditions maintained evapotranspiration (Turner et al., 1978b).

PLANT RESPONSES TO WATER DEFICITS

The development of water deficits leads to a wide range of responses by the plant, Hsiao (1973) has summarized much of the literature on these responses. It should be recognised that many of the responses reported in the literature are secondary, arising as a result of a primary process being directly affected rather than being a primary response to water deficits *per se*. For example, translocation of assimilates is reduced by water deficits, but this appears to result not from a direct effect of a water deficit on transport in the phloem, but from a reduction in the rate of photosynthesis, i.e. a reduction in the source of assimilates; or from a reduction in growth, i.e. a reduction in the sink for assimilates (Wardlaw, 1969). In this review, we concentrate on those processes that play a primary role in limiting crop productivity: in doing this, we recognise that a large number of responses to water deficits that may impinge on crop productivity are omitted.

Reduced to its simplest terms, the production of dry matter by a crop depends on the solar energy that it captures and utilizes to convert carbon dioxide and water into dry matter in the process of photosynthesis. The capture of solar energy depends on the interception of that energy by organs capable of photosynthesis. Thus, crop productivity depends on the development of leaf area (L) to intercept that radiant energy, and the rate of net photosynthesis (P) to convert it into dry matter. With the exception of forage crops, only a proportion of the above ground dry matter contributes to economic yield (Y): the distribution of assimilates (H) within the plant determines the proportion of the total that is harvested as economic yield. We therefore write

$$Y = f(L.P.H) \tag{1}$$

In this review, we briefly discuss the influence of water stress on the processes of leaf expansion and senescence, photosynthesis, and distribution of dry matter.

Leaf expansion and senescence

Both leaf expansion and senescence are known to be very sensitive to water deficits. In maize grown in controlled environments, leaf expansion declined rapidly at leaf water potentials which were lower than -0.2 MPa (0.1 MPa = 1 bar), and ceased at -0.7 to -0.9 MPa (Boyer, 1970a). Similar data are not available for wheat and barley, but results were similar for a range of tropical and temperate grasses (Turner and Begg, 1978). In the field, leaf expansion was not as sensitive to water deficits as in the glasshouse (Watts, 1974; Turner and Burch, 1981), but most of the evidence to date suggests that leaf expansion is much more sensitive to water deficits than photosynthesis, even under field conditions (Turner and Begg, 1978; Legg *et al.*, 1979).

Provided water deficits have not been too severe or prolonged, relief of stress can result in a resumption of leaf expansion at similar rates to those in well-watered plants (Boyer, 1970a; Acevedo *et al.*, 1971). Nevertheless, even a single period in which the leaf water potential fell no lower than -2.0 MPa affected the final leaf area (Acevedo *et al.*, 1971; Yegappan, 1978). Indeed, it is not simply an individual response of leaf expansion to water stress that is important in determining yield, but the integrated effects of gradually increasing short term stresses and relief of stress on the final leaf area that will determine yield.

Although leaf senescence does not appear to be as sensitive to leaf water deficits as leaf expansion (Ludlow, 1975), the senescence rate increases with water deficits in a wide range of species. Fischer and Kohn (1966) showed that the yield of wheat under dryland conditions in a Mediterranean-type climate was inversely related to the leaf area duration after anthesis, which, in turn, was related to the plant water deficit.

Photosynthesis

In addition to the effect of water deficits on leaf area development and senescence, water stress decreases the rate of net photosynthesis per unit leaf area. As mentioned above, the deficits required to influence the rate of net photosynthesis are usually greater than those required to influence leaf elongation (Turner and Begg, 1978). A recent analysis of the effects of drought on barley showed that, during the vegetative stage, water stress primarily reduced yields by a reduction in light interception arising from smaller leaves. However, for stress during the grainfilling phase, yields were reduced almost equally by reduced photosynthesis per unit area and reduced leaf area arising from leaf senescence (Legg *et al.*, 1979).

A decrease in net photosynthesis always goes hand-in-hand with a decrease in stomatal conductance (e.g. Boyer, 1970b, Turner *et al.*, 1978b; Rawson *et al.*, 1978). In most species, the immediate response of net photosynthesis rate to a water stress appears to be due to stomatal closure. More severe levels of stress were required for the biochemical and biophysical pathways of photosynthesis within the leaf to be affected (Boyer, 1970b; Slatyer, 1973). However recent evidence suggests that photosynthesis is controlling stomatal conductance (Wong *et al.*, 1979; van Keulen, 1981) and that with slowly imposed water stress, it may be the effect of water deficits in reducing photosynthesis that reduces stomatal conductance (Wong *et al.*, 1979; Osmond *et al.*, 1980).

Water deficits which are sufficient to close stomata and reduce photosynthesis also decrease dark respiration. In leaves, the decrease in the rate of dark respiration is less than the decrease in photosynthesis (Boyer, 1971), but for whole plants, the decrease is similar (Boyer, 1970a). Photorespiration is unaffected by short term water deficits (Troughton and Slatyer, 1969), but will ultimately decrease because of the effects of water deficits on the depletion of the substrates for photorespiration.

Thus far we have considered only the short term effects of water deficits on photosynthesis, respiration and stomatal behaviour. For crop production, it is the integrated effects of water stress on photosynthesis and respiration that are important. Diurnal studies of the rate of photosynthesis of soybean as water deficits developed in the field, showed initially a midday depression of photosynthesis that gradually extended into the afternoon until finally there was only a short burst of photosynthesis during the early morning (Fig. 3). The integrated daily net photosynthesis decreased linearly by 9 per cent per 0.1 MPa as the minimum daily leaf water potential decreased. However, respiration continued throughout the day and night. Thus after 33 days without rain or irrigation, there were periods of the day when net photosynthesis was positive, but plants were losing dry weight on a 24 h basis.

Distribution of assimilates

There is now considerable evidence that the assimilate distribution pathway, i.e. the phloem, is unaffected by quite severe water deficits (Wardlaw, 1967, 1969, 1971; McWilliam, 1968; Fellows and Boyer, 1978). Nevertheless, overall translocation and assimilate distribution are affected by water deficits. The effect of water deficits on photosynthesis, utilization of photosynthates, vein loading or unloading, and the differential effect of stress on organs can all influence overall translocation rates and patterns. For example, a reduction in photosynthesis and vein loading by water stress reduced the translocation of assimilates in potatoes, wheat and maize (Wardlaw, 1967, 1971; Brevedan and Hodges, 1973; Munns and Pearson, 1974; Moorby *et al.*, 1975), whereas reduced utilization of assimilates as a result of water deficits reduced translocation in darnel grass (Wardlaw, 1969).

The influence of water deficits on the distribution of assimilates depends on the stage of growth and the relative sensitivity of various plant organs to water deficits. Although the generality of high root: shoot ratios in xerophytic species has been questioned (Barbour, 1973), the root: shoot ratio of many crop and pasture species increases with stress (Begg and Turner, 1976; Turner and Begg, 1978) This could arise simply from a relatively greater decrease in shoot dry weight, but there is increasing evidence that it represents a greater allocation of the limited carbon available to roots rather than shoots as water stress develops (Bennett and Doss, 1960; Doss *et al.*, 1960; Hsiao and Acevedo, 1974; Malik *et al.*, 1979; Sharp and Davies, 1979). Such an allocation pattern would be particularly important in the vegetative phase and is the likely consequence of the greater sensitivity of leaf expansion than photosynthesis to water deficits (Wardlaw, 1969). By contrast, Wardlaw (1967) showed that grain growth of wheat was less sensitive to water deficits than photosynthesis, possibly as a result of the greater potential of wheat spikelets to adjust osmotically, compared with wheat leaves (Morgan, 1980a), and assimilates moved preferentially from the lower leaves, stems, roots and crown, to the ear when water stress occurred during grainfilling.

MECHANISMS OF ADAPTATION TO WATER DEFICITS

Many of the responses of plants to water deficits are modified by the ability of plants to adapt to water shortage. Early ecological studies (Volkens, 1884, 1887; Schrimper, 1903) identified a plethora of morphological traits in desert perennials that were considered to confer adaptive advantage in the xeric conditions of the desert over their more mesophytic cousins. More recently it has become clear that even mesophytic crop plants can adapt to water deficits and acclimate to high or low temperature (see Turner and Kramer, 1980). The adaptations of crops to water deficits have conveniently been divided into three types: developmental, morphological and physiological.

Developmental mechanisms of adaptation to water deficits

A pattern of phenological development that allows a crop to complete its lifecycle without having to face a serious water deficit would clearly be advantageous in a dry environment. The so-called desert ephemerals have a number of adaptations that enable them to do this: germination inhibitors that inhibit germination until the first soaking rain, rapid phenological development after germination and phenological plasticity that enables them to set many seeds in a year with little rainfall (Mulroy and Rundle, 1977). In agriculture, where the criterion of success is production rather than survival, the requirement for consistent seedling establishment and mechanical harvesting has led to selection against germination inhibitors and phenological plasticity for crops. However, rapid phenological development has been widely adopted for crops in Mediterranean-type environments.

In winter cereals it has been widely demonstrated that earliness leading to drought escape confers a yield advantage under terminal drought (Reitz and Salmon, 1959; Chinoy, 1960; Derera *et al.*, 1969; Bidinger, 1977; Fischer and Maurer, 1978). However in situations where late frosts occur, there is a danger that very early varieties may also suffer frost damage (Reitz and Salmon, 1959) unless frost hardiness is also incorporated into the variety. Furthermore, early varieties generally yield less under a favourable water regime. Hence there is a tradeoff between earliness to improve yields under drought and yield reductions arising from frost damage, and too rapid a phenological development in more favourable years.

Adaptation in phenological development has been demonstrated in some crops. Mild water deficits have been shown to speed the time to anthesis and maturity in wheat, possibly due to the higher plant temperatures under stress

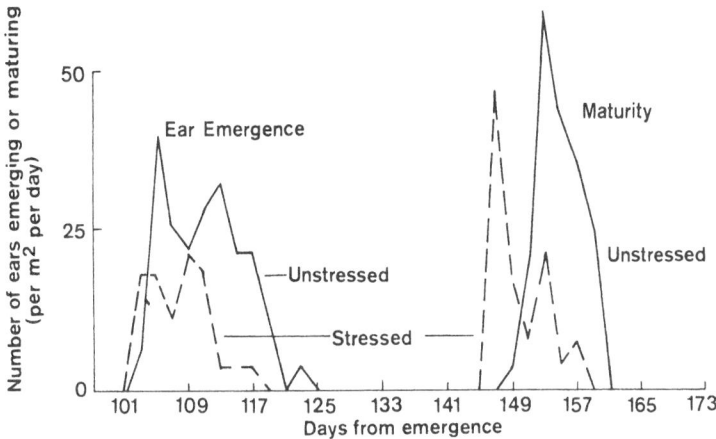

Fig. 4. Frequency distribution of the number of wheat ears that emerged and matured with time after seedling emergence when given adequate water (——) or stressed from the late vegetative stage onwards (– – –). From Turner (1966).

speeding phenological development. This is shown in Fig. 4 which presents the effects of water stress imposed in the late vegetative phase on ear emergence and maturity. Ear emergence occurred 2.6 days earlier on average and maturity occurred 5.2 days earlier on average as a result of water stress compared with the unstressed wheat. However more severe stress can delay anthesis (Angus and Moncur, 1977). It is worth investigating whether a greater degree of adaptation in phenological development to water stress is available by incorporating genetic material of wildtypes from semi-arid areas.

In addition to rapid phenological development, the year-by-year variation in available soil moisture in dryland areas (Fig. 2) suggests that developmental plasticity is highly desirable for crops in Mediterranean-type environments. Indeterminancy, branching and tillering all provide the required developmental plasticity. Indeterminancy provides the continued production of flowers and seed as long as water is available in day-neutral plants. Branching and tillering have similar effects, but the range of dates of flowering across the branches or tillers is usually more restricted than in completely indeterminate plants.

In cereals, developmental plasticity in response to water stress is provided both by the number of tillers that remain alive and produce an ear and by the number of grains set per ear. Turner (1966) showed that when water stress was applied to a wheat community in the late vegetative stage of development, i.e. when tiller number was maximal, tiller death occurred at an average rate of 11 tillers m^{-2}

day^{-1} compared to 3 tillers m^{-2} day^{-1} in well-watered wheat. Likewise the mean number of grains per ear was reduced from 33 to 22 by the same treatment.

In the indeterminate species cowpea, water stress reduced the number of pods per plant (Wien *et al.*, 1979), although it is not clear whether this was from reduced branching or a reduced number of pods per branch. This aspect of developmental plasticity is presumably a conservative trait that reduces seed number as a result of stress so as to maintain seed of a viable size for germination and emergence (Fischer and Turner, 1978).

However, it is also a wasteful trait, leading to the allocation of carbon to some tillers, flowers, and fruiting structures that do not ultimately bear seed. Although some retranslocation of soluble components from the infertile florets and tillers to the stems and florets that bear a seed undoubtedly occurs, the carbon involved in structural components can rightly be considered 'wasted'. This led Donald (1968) to suggest that uniculm wheat would be more efficient in its carbon utilization and yield. It has been argued elsewhere (Turner, 1979) that a uniculm habit may be beneficial for wheat with a regular supply of water, but that it provides little developmental plasticity for dryland wheat. Limited evidence from pot experiments indicates that two or three large tillers may provide enough seeds for high yields in dry years and enough plasticity to obtain very high yields in wet years (Jones and Kirby, 1977): this observation requires verification in field studies under drought.

Morphological mechanisms of adaptation to water deficits

The successful colonization of terrestrial environments by higher plants has occurred largely as a result of the evolution of a wide range of morphological adaptations associated with the uptake and efficient utilization of water. The fact that the basic metabolic pathways, which evolved more than two billion years ago, have persisted with relatively little change during the evolution of higher plants, points to the possibility of a dearth of genetic diversity in the biochemistry of higher plants in the context of adaptation to water deficits. However, an examination of some of the morphological adaptations may provide more fruitful ground for adapting crops to water stress environments through selection and breeding (Begg, 1980).

We need a better understanding of the various static and dynamic features of plant morphology that may enable plants to become better adapted to stress. As Passioura (1976) stated, 'it is the control of leaf area and morphology which is often the most powerful means a mesophytic plant has for influencing its fate when subject to long term water stress in the field'. Because of the enormous range of morphological diversity evident in higher plants, the discussion here is restricted primarily to aspects of leaf morphology associated with their growth and development, orientation and senescence, and to root morphology.

The expression of morphological effects can generally be traced back to earlier stages of growth when cell division and enlargement influence tissue and organ development. The sensitivity of leaf expansion to water deficits has been pointed out in the previous section. An early reduction in leaf area will reduce crop growth, particularly during establishment, when there is incomplete light interception. An important consequence of this reduction in leaf area is the associated reduction in water loss, particularly at leaf area indices less than three (Fischer and Kohn, 1966; Ritchie, 1974; Kowal and Kassam, 1978), thereby reducing the rate of water use and delaying the onset of more severe stress.

In addition to a reduction in the rate of water use, there is evidence for increased access to soil water through an increase in the root:shoot ratio (Pearson, 1966; Davidson, 1969; Hoffman et al., 1971). The effect is likely to be greatest at levels of stress which are sufficient to reduce significantly shoot growth but not photosynthesis: the continued availability of assimilates at a time of greatly reduced shoot growth permits additional root growth. Also, as mentioned above, water stress reduces leaf area by accelerating the rate of senescence of the older leaves and through its effect on leaf shedding. Hall et al. (1979) have argued that leaf senescence in cereals could confer an adaptive advantage if it were accompanied by a substantial reduction in transpiration, as the older lower leaves which senesce first are supplying relatively little carbohydrate to the developing grain.

A major disadvantage of the decrease in leaf expansion and increase in leaf senescence is that they are irreversible. Indeterminate crops, given favourable conditions, may compensate for this loss of leaf area through further vegetative development on branches or tillers. However following floral initiation, determinate crops have no scope for compensation through an increase in the number of leaves and thus the growth or yield potential that has been lost cannot be fully recovered if there is a return to more favourable conditions.

An alternative mechanism for adapting to stress without irreversibly affecting leaf area is through changes in leaf angle or orientation. Such changes can effectively reduce the radiation load on leaves and allow the plant to dissipate less energy as latent heat. An important feature of the mechanisms responsible for changes in leaf orientation is that they operate only during periods of stress; recovery is rapid when this stress is relieved, and yield may not be seriously reduced.

Changes in leaf orientation may occur as a result of a passive wilting response due to a general loss in turgor of the leaf tissue. Fig. 5 shows the effect that this has

110

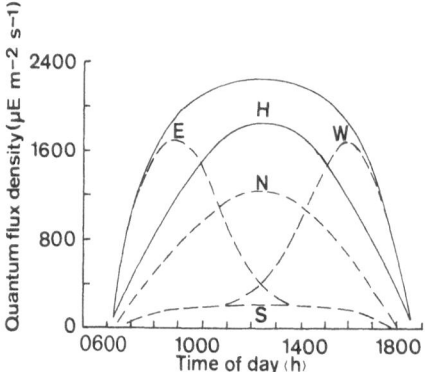

Fig. 5. Diurnal changes in quantum flux density (400-700 nm) on wilted, vertically hanging leaves orientated north (N), south (S), east (E) or west (W) and the diurnal changes in an unstressed leaf (solid line) exhibiting diaheliotropic leaf movements that orient it normal to the direct radiation. The quantum flux density on a horizontal plane (H) is shown for comparison. Data are for a cloudless day (March 19), Canberra (35 ° S). From Begg and Turner (1976).

on the level of radiation which is incident on leaves of sunflower, before and after wilting. Throughout the day wilted leaves oriented north, south, east, or west, intercepted less radiation than an unstressed diaheliotropic leaf which tracked the sun from sunrise to sunset.

Changes in leaf display can also occur as a result of a differential loss of turgor in individual cells, particularly the bulliform cells in the upper epidermis of many grasses. This results in a rolling or folding of the leaf lamina, greatly reducing the effective leaf area and resulting in a more vertical leaf orientation. Changes in turgor of the pulvini at the base of the petiole of many legumes results in active leaf movements which orient the lamina parallel to the incident radiation during water stress (Dubetz, 1969; Begg and Torssell, 1974; Shackel and Hall, 1979). In the absence of stress, the leaflets are oriented at right angles to the incident radiation and follow or track the sun in this position. However, when severe water deficits develop, the leaflets become oriented parallel to the incident irradiation. Passive leaf wilting has been shown to increase the water use efficiency of sunflower leaves (Rawson, 1979), and we conclude that leaf rolling and active leaf movement will provide similar benefits to other species.

The development of enlarged white hairs under stress is an adaptive mechanism that increases the reflection of radiation by the leaf, and decreases the conductance of water through the boundary layer of the leaf in some species (Woolley, 1964; Wuenscher, 1970; Johnson, 1975). The wax bloom noted on some sorghum leaves (Sánchez-Díaz et al., 1972) acts in a similar way, reducing net radiation, the boundary layer conductance, and transpiration (Chatterton *et al.*, 1975). The 20 per cent reduction in transpiration which the latter authors recorded suggests that the wax bloom also occluded stomata.

The effect of a water deficit on the preferential development of the root over the shoot is an adaptive mechanism that enables the crop to explore a greater soil volume for water. If soil water is available at depth, the development of a deeper root system would clearly be a useful adaptive feature (Passioura, 1974). Varietal differences in rooting depth have been demonstrated in wheat (Hurd, 1968; Derera *et al.*, 1969), soybeans (Raper and Barber, 1970), sorghum (Jordan and Miller, 1980) and tomato (Zobel, 1975; Gulmon and Turner 1978): Hurd (1968, 1974) showed that deeper rooting varieties yield better under drought stress. Alternatively, for situations in which soil water is not available at depth and plants have to survive on limited resources of stored water near the surface, the hydraulic conductance of roots could be decreased. In pot experiments, Passioura (1972) was able to show that decreasing the hydraulic conductance of wheat by reducing the seminal roots to one, substantially reduced the water loss before anthesis and increased the final weight of grain over control plants which had several seminal roots.

Physiological mechanisms of adaptation to water deficits

A range of physiological mechanisms enables plants to adapt to stress. In this chapter we concentrate on four, viz: (i) seed priming, (ii) stomatal control of water loss, (iii) osmotic adjustment and (iv) cellular tolerance of dehydration.

Henkel (1961) and Henckel (1964) proposed a method of increasing the resistance to dehydration by treating the seed before it was sown: the treatment involved allowing the seed to take up water to 30 per cent dry weight, then leaving at 10° to 25°C for 24 h followed by air drying. Subsequent work has shown that this treatment confers no increase in the physiological resistance to water stress (Jarvis and Jarvis, 1964), but seed treatment does speed up germination and root growth (May *et al.*, 1962; Keller and Bleak, 1968; Austin *et al.*, 1969; Heydecker *et al.*, 1973). This can lead to a delay in the development of water deficits which is enough to close stomata and to increased yields under drought in certain cases (Woodruff, 1969, 1973). The years in which early establishment of roots is critical, are presumably the years in which yield increases are realized.

As demonstrated by Zelitch and Waggoner (1962) and Shimshi (1963), stomatal closure can decrease transpiration and increase the efficiency of water use. This has led to several attempts to screen for higher resistance to water loss in crop plants, either by screening for low stomatal frequency or for high stomatal resistance itself. There are reports that decreased stomatal frequency decreases transpiration per unit leaf area and increases drought resistance (Dobrenz et al., 1969; Heichel, 1971; Miskin et al., 1972; Wilson, 1975), but often the successful selection for low stomatal frequency is compensated by an increase in the stomatal size or total leaf area, thereby negating any benefit of less frequent stomata on water use (Jones, 1977a, 1979).

Differences in stomatal resistance, measured directly usually by porometry, have been reported among cultivars (Kaul and Crowle, 1971; Blum, 1974; Shimshi and Ephrat, 1975; Jones, 1977b, 1979; Roark and Quisenberry, 1977), but light, humidity and water status need to be similar, and the relative ranking of cultivars may vary throughout the season (Blum, 1974; Shimshi and Ephrat, 1975; Jones, 1977b).

A large stomatal resistance, is a recognised xeromorphic character that clearly can meter out a limited water supply to a monospecific community in which competition from less conservative neighbours is unimportant. Nevertheless water is used less efficiently when stomatal resistance is large throughout the day than when stomata open in the morning when low air temperature and vapour pressure deficits are low and close at midday and during the early afternoon when air temperature and vapour pressure deficits are high. The ability of stomata to respond to humidity or leaf water potential can provide this behaviour (Cowan and Farquhar, 1977).

It has now been established unequivocally that stomata of some species respond directly to the vapour pressure deficit of the atmosphere surrounding the leaf (Schulze et al., 1972). Not all species respond directly to humidity (Rawson et al., 1977a), but all respond to leaf water potential. The responsiveness of stomata to leaf water potential varies with species (Turner, 1974). Moreover the potential at which stomata begin to close and the shape of the response curve depends on age, growth conditions, stress prehistory and rate of stress (Frank et al., 1973; Turner, 1974; Millar and Denmead, 1976; Turner et al., 1978a; Beadle et al., 1978; Jones and Rawson, 1979).

The degree of responsiveness of stomata to humidity and/or leaf water potential that is required depends on the drought conditions likely to be encountered. If only occasional days of large vapour pressure deficits or short periods of water shortage are probable, as in most temperate regions, stomata that are sensitive to small vapour pressure deficits and are highly responsive to low leaf water potentials will reduce the incidence of potentially damaging plant water deficits and prevent irreversible responses to these deficits such as tiller death, leaf senescence and a reduction in the number of seed sites. However, since stomatal closure usually results in a reduction in assimilation by the plant, stomata that are sensitive to small vapour pressure deficits and high leaf water potentials are of little benefit to crops that have to complete their lifecycle on a diminishing soil moisture supply and with increasingly large vapour pressure deficits, i.e. crops growing in Mediterranean-type climates. Under these circumstances, stomata that are responsive only to very large vapour pressure deficits and which adapt to slowly developing soil water deficits so that they remain open at low leaf water potentials will be of benefit. It has been demonstrated recently that stomata in some species do adapt to water deficits (McCree, 1974; Brown et al., 1976; Thomas et al., 1976; Turner et al., 1978a; Jones and Rawson, 1979; Ludlow, 1980) and that this behaviour is linked with the ability of leaves to adjust osmotically to low leaf water potentials (Brown et al., 1976; Turner et al., 1978a; Ludlow, 1980).

It is now well recognised that plants have the ability to lower their osmotic potential in response to slowly developing water deficits (Hsiao et al., 1976; Turner and Jones, 1980). Leaves, hypocotyls, roots and spikelets have all been shown to adjust osmotically (Greacen and Oh, 1972; Meyer and Boyer, 1972; Morgan, 1977, 1980a; Jones and Turner, 1978, 1980; Stout and Simpson, 1978; Turner et al., 1978a; Sharp and Davies, 1979) in some, but not all (Turner et al., 1978b), species. The degree of osmotic adjustment varies with species and genotype, rate of stress and degree of stress (Turner and Jones, 1980): rapid rates of stress and small degrees of stress produce a smaller osmotic adjustment than slow rates and greater degrees of stress (Jones and Turner, 1978, 1980; Jones and Rawson, 1979). Irrespective of the rate and degree of stress, the extent of osmotic adjustment does appear to be limited (Turner and Jones, 1980; Jones and Turner, 1980).

The significance of osmotic adjustment is that it helps to maintain positive turgor as water deficits develop (Hsiao et al., 1976; Turner and Jones, 1980). This enables the plant to maintain leaf expansion and photosynthetic activity at levels of stress which are not possible in its absence (Meyer and Boyer, 1972; Hsiao et al., 1976; Jones and Rawson, 1979; Steponkus et al., 1980). Moreover, it should enable plants to deplete the soil water to a lower soil water potential and allow a greater exploration of soil by roots (Sharp and Davies, 1979): the additional water made available by decreasing the soil water potential is likely to be small (Jordan and Miller, 1980), but the additional water available from the exploration of a greater soil volume could be significant. Recent evidence suggests that osmotic adjustment may have an effect on yield beyond that directly attributable to increased assimilation: Morgan (1980b) has produced evidence that suggests that abscisic acid produced in the leaves at low turgor pressures reduces seed set of wheat through its effects on pollen viability. The maintenance of turgor by osmotic adjustment of the leaves may thereby have a significant effect on yield by reducing the production of abscisic acid and maintaining pollen viability.

However, there does appear to be a cost for osmotic adjustment. Although photosynthesis is maintained in osmotically-adjusted plants, photosynthetic capacity is somewhat less in plants with low osmotic potential (Jones and Rawson, 1979). It is unclear whether this reduction arises from feedback inhibition of photosynthesis as a result of the accumulation of the soluble photosynthetic products (Neales and Incoll, 1968; Jones *et al.*, 1980), or the interference of the solutes with the activity of enzymes in the photosynthetic pathway. However, it is consistent with the observation that species with inherently low osmotic potentials have inherently low maximum rates of photosynthesis. This, together with the observation that the rate of hypocotyl extension was reduced linearly as the osmotic potential decreased in soybean, in spite of full turgor maintenance (Meyer and Boyer, 1972; Turner and Jones, 1980), suggests that inhibition of enzyme activity may be the price paid for osmotic adjustment.

Osmotic adjustment allows turgor to be maintained as the leaf water potential and water content decrease. To be effective, therefore, plants that adjust osmotically must be able to withstand low water potentials and water contents without damage to tissue. There is wide variation in the degree of desiccation that plants can tolerate without cellular injury (Sullivan and Eastin, 1974; O'Toole *et al.*, 1978). Studies with 'resurrection plants' have shown that the conditions under which dehydration occurs affect the degree of desiccation tolerance: slow rates of dehydration at high water potentials and rapid rates of drying at low water potentials appear to increase the tolerance of the tissue to desiccation (Gaff, 1980). Whether the rate of drying affects the desiccation or dehydration tolerance of crops is unknown, but the increase in sugars leading to osmotic adjustment in some species (Turner *et al.*, 1978a; Munns *et al.*, 1979; Jones *et al.*, 1980) may increase the degree of tolerance to desiccation in these species (Lee-Stadelmann and Stadelmann, 1976). Furthermore, it is not clear whether crop species differ in their tolerance, depending on their age and growth conditions. Again, since the osmotic potential of plants often decreases as plants age, and is lower in the field than in controlled environment facilities, the solutes so accumulated, particularly if they are sugars and proline, may give protection against dehydration (Lee-Stadelmann and Stadelmann, 1978; Schobert, 1977).

ROLE OF RESERVES IN ADAPTATION TO WATER DEFICITS

It is common to distinguish two sources of assimilate supply for the grain growth of cereals and legumes: (i) that from photosynthesis after anthesis and (ii) that translocated to the grain from material assimilated before anthesis and temporarily stored in the stems, leaves and roots. This is only a division of convenience since, in reality, there is probably a pool of assimilates in the stem and elsewhere to which contributions and withdrawals are made; beginning well before anthesis and continuing throughout grainfilling (Bidinger, 1977). Under conditions of adequate water supply, photosynthesis after anthesis is the major contributor to grain yield and the contribution of preanthesis assimilates to grain yield is usually less than 20 per cent (Thorne, 1966; Wardlaw and Porter, 1967; Rawson and Evans, 1971; Bidinger et al., 1977), although rice may be an exception in that it is usually between 20 and 40 per cent (Murata and Matsushima, 1975).

However under conditions of stress, as pointed out earlier assimilates move to the ear from lower leaves, stems, root and crown (Wardlaw, 1967), and pre-anthesis assimilates contribute a greater proportion of the total assimilates in the grain. The proportion of grain weight that can be attributed to preanthesis photosynthesis, stored and later transferred to the grain, will clearly depend on the timing and severity of stress and the reduction in photosynthesis after anthesis, but it has been reported to be as great as half to two-thirds in severely stressed crops (Gallagher et al., 1975; Passioura, 1976; Kobata and Takami, 1979). Fig. 6 shows diagrammatically three possible ways in which the proportion of stored preanthesis reserves can increase in the grain. In case A, the yield under stress conditions is maintained because of an absolute increase in the stem reserves transferred to the grain from 20 to 50 per cent. In case B, the yield is reduced by one-third and stem reserves transferred to the grain are increased, but not by the same degree as in A, so that the reserves contribute 60 per cent of the final grain yield. It is also possible that, as in C, the reserves transferred to the grain are similar under stress and nonstressed conditions, but as the yield is reduced, the proportion of preanthesis reserves transferred to the grain increases

Table 1. Amount and percentage of grain weight arising from assimilates stored prior to anthesis and retranslocated to the grain in two wheat and two barley cultivars. Estimates from dry weight changes in the ear and shoot from matched shoot samples. Adapted from Bidinger (1977)

Species	Cultivar	Unstressed			Stressed		
		Preanthesis assimilates translocated (mg/shoot)	Grain weight (mg/shoot)	%	Preanthesis assimilates translocated (mg/shoot)	Grain weight (mg/shoot)	%
Wheat	Yecora 70	159	2771	6	347	1619	21
	Ciano 67	8	2336	0	272	1284	21
Barley	CM 67	0	2570	0	73	1749	4
	WI 2198	0	1449	0	87	839	10

from 20 to 30 per cent. Ideally in dryland crops, we would hope that case A would apply, but, case B or C is more realistic.

Observations based on dry weight changes have shown that there are differences among cultivars in the proportion of preanthesis assimilates transferred to the grain. Table 1 gives data for two wheat and two barley cultivars. Only one wheat cultivar 'Yecora', contributed preanthesis assimilates to the grain when soil water supply was adequate, but both wheat and both barley cultivars utilized between 4 and 21 per cent of their preanthesis assimilates when given a terminal water stress. In all cases, the amount and percentage of preanthesis assimilates increased. Although the two wheat cultivars differed in the proportion of preanthesis assimilates translocated to the grain under nonstressed conditions, they were similar when stressed. Conversely, the two barley cultivars were similar when nonstressed, but differed when stressed.

Constable and Hearn (1978) have reported differences in the proportion of stem reserves transferred to the grain in two soybean cultivars. In the cultivar 'Ruse', 25 per cent of the grain yield came from stem reserves whether water deficits were imposed on the plant or not, whereas in 'Bragg' little assimilate was transferred from the stem to the grain in the irrigated plants, but 19 per cent of the dry weight of the grain was contributed by the stem when the plants were severely stressed.

Analyses of change in shoot and grain dry weights alone do not take into account possible changes in the rates of respiration arising from stress and loss of plant material by leaf shedding and decay between dry weight harvests, and, therefore, must be treated with caution. Use of ^{14}C labelling before anthesis provides a much more reliable guide to the amount of stored assimilates utilized in grainfilling. Analysing data from the same crops as in Table 1, but using ^{14}C labelling data for their calculations, Bidinger et al. (1977) estimated that the preanthesis assimilate contributed 12 to 13 per cent of the grain dry weight in the unstressed wheat and barley, 17 per cent in the stressed barley and 27 per cent the stressed wheat. Thus, in contrast to what might be expected if plant material were lost by wind or decay (Bidinger et al., 1977), the dry matter data in Table 1 underestimated the contribution of preanthesis assimilates to the grain yield: this undoubtedly arose from fact that Bidinger (1977) used carefully matched dry matter samples at successive harvests to ensure that errors arising from losses were minimised. ^{14}C labelling also reveals that although the relative contribution of preanthesis assimilates to grain yield may increase under stress, the absolute contribution is similar in stressed and unstressed plants and the proportion increases as a result of a decrease in the contribution to postanthesis photosynthates to grain yield (Passioura, 1976; Rawson et al., 1977b). Thus Fig. 6

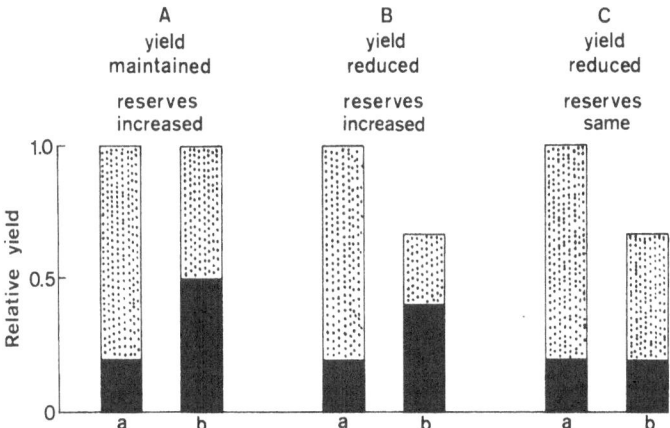

Fig. 6. Diagrammatic representation of the contribution of preanthesis assimilates (solid portion of histograms) and posthanthesis assimilates (dotted portions of histograms) to grain yield in unstressed (a) and water stressed (b) cereals. In A yield is maintained, whereas in B & C yield is reduced under stress. In A & B the amount of preanthesis assimilates in the grain is increased, whereas in C it remains the same under stress conditions compared to unstressed conditions.

shows that to date only case C and occasionally case B has been found in those studies where ^{14}C labelling has been used to verify dry weight analysis. However, the limited success in identifying wheat or barley lines which transfer more preanthesis assimilates to the grain under stress than nonstress conditions should not deter others from seeking this character in a wider range of genotypes.

IMPLICATIONS FOR BREEDING AND MANAGEMENT

An understanding of the constraint of water on yield and the adaptation of plants to water deficits, challenges both the breeder to produce a cultivar that will give a greater yield under water-limited conditions, and the agronomist to ensure that the most efficient use is made of the available water. Elsewhere it has been argued that the dry matter in the grain (Y) is function of the water passing through the crop in transpiration (T), the water use efficiency (W) and the proportion of the total dry weight that ends up in the grain (H) (Passioura, 1977; Fischer and Turner 1978), i.e.:

$$Y = f(T.W.H.) \tag{2}$$

Thus to improve the grain yield of crops in a dryland area one must increase the water passing through the crop in transpiration, increase the water use efficency and/or increase the proportion of total dry matter going to the grain. The first of these is largely in the domain of the agronomist, and the last two are largely in the domain of the breeder.

Implications for breeding

The large increases in yield of wheat and rice resulting from breeding programmes at the international research institutes in Mexico and the Philippines have been achieved by breeding and selection for high yields under conditions of adequate water supply. The increase in yield potential so achieved has been high enough for these selections to yield better under drought also (Fischer and Maurer, 1978). However, recent analyses suggest that yields of the high yielding cultivars of rice have gradually decreased as they have been more widely used in intensive cultivation, whereas locally adapted cultivars have remained stable over the same period (Evans and De Datta, 1979). Moreover, the analyses showed that in more than 25 per cent of the cases the locally-adapted cultivars of rice yielded better than the so-called high yielding ones under conditions of stress (Evans and De Datta, 1979).

The alternative approach of selecting for high yield and stability of yield under conditions of water stress likely to be encountered by the crop is expensive and time consuming. The great variability from year to year in available soil moisture (Smith and Harris, 1981) menas that selection has to be done over many seasons and selection indices and pressures vary greatly from season to season. Under subpotential yield levels, heritabilities of yield and yield components are relatively low, and selection for yield is less efficient (Roy and Murty, 1970).

Since selection for yield alone, particularly under a favourable water regime, may discard many characters useful for yield under dryland conditions, we suggest that breeding for improved dryland yields requires an approach analogous to breeding for disease resistance. As Blum (1979) has pointed out, a breeder faced with a disease problem does not attempt to solve it by selecting for yield under severe disease conditions, nor does he assume that superior yielding varieties under disease-free conditions will yield well under severe disease conditions. Rather, he recognises that disease resistance and yield are separately inherited characteristics, and disease resistance genes are transferred from sources of the resistance to a high yielding cultivar. Provided drought resistance and yield are separately inherited characters, i.e. high yielding characteristics do not necessarily confer poor drought resistance, such a procedure should be useful in conferring drought resistance to crops. The characteristics that we suggest may usefully confer drought resistance and ultimately improve yield under drought

Table 2. Desirable plant characteristics for a crop in a Mediterranean-type environment.

1. Rapid germination and early establishment of deep roots
2. Rapid phenological development
3. Developmental plasticity
4. Diaheliotropic and paraheliotropic leaf movements
5. Leaf expansion highly sensitive to water deficits
6. Stomata sensitive only to large vapour pressure deficits and insensitive to low leaf water potentials
7. Ability to adjust osmotically
8. Large transfer of assimilates from stem to grain
9. Dehydration tolerance particularly at seedling and grainfilling stages.

are presented, not within the confines of an individual crop species, but as general breeding objectives for a range of crops. The characters suggested are listed in Table 2. Ideotypes for other types of drought environments are presented elsewhere (Turner, 1980).

Because of the uncertainties of rainfall at the beginning of the season in some years (Fig. 2) and the chance of hot drying winds, rapid germination and emergence under soil moisture stress and rapid development of deep roots are desirable. Should the seedling emerge and deplete all the available soil moisture before follow-up rains occur, dehydration tolerance at this stage is warranted. Screening for emergence under soil moisture stress, rapid development of deep roots and the ability of seedlings to withstand dehydration can be easily achieved (Johnson and Asay, 1978; O'Toole and Chang, 1979; Johnson, 1980).

Rapid phenological development and developmental plasticity are required to provide some yield in very dry years and abundant seed in wet years. Where frost incidence is high, as it is in some areas of northern Africa and western Asia (de Brichambaut and Wallén, 1963, this objective will have to be modified to match phenological development with a low probability of frost injury at flowering. Because of low mean temperatures and irradiances common during the Mediterranean-type wet season (Harris, 1979), this means selection for rapid growth and phenological development at low temperatures, as achieved in sorghum (Downes, 1972). As mentioned earlier, the greatest degree of developmental plasticity occurs in indeterminate crops and there may be greater scope for these in Mediterranean-type regions. Alternatively, for cereals, the maintenance of some developmental plasticity by selection of a few large tillers appears to be appropriate.

The incorporation of diaheliotropic and paraheliotropic leaf movements enables the plant to maximize radiation interception when quantum flux densities are low in the middle of the winter and 'shed' radiation when evaporation, temperatures and radiation rise rapidly during the spring and induce severe water deficits. Additionally, if leaf expansion is very sensitive to water deficits and rapidly recovers on relief of the deficit, this will meter water use in the vegetative phase in those occasional years when water deficits occur in this phase and ensure that radiation shedding by paraheliotropic leaf movement occurs only at grain filling. This requires leaf expansion to be more sensitive to water deficits than radiation-shedding leaf movements, a characteristic observed in sunflower.

Since, in regions having a Mediterranean-type of climate, water deficits increase at the end of the growing season as evaporation and vapour pressure deficits increase, stomata need to be insensitive to large vapour pressure deficits and low leaf water potentials. Provided radiation-shedding leaf movements occur at a higher turgor than stomatal closure, a characteristic apparently present in sunflower (Rawson, 1979), water use efficiency at large vapour pressure deficits will be improved without resort to stomatal closure. The ability of leaves to adjust osmotically will ensure that stomata do not close until low leaf water potentials are reached (Turner et al., 1978a) and will enable dry matter to be produced late into the drying cycle.

Osmotic adjustment will also enable penetration of roots into deeper soil in those years when water is available at depth and will enable extraction of soil water to low soil water potentials. It is not known whether osmotic adjustment to maintain open stomata is incompatible with leaves that shed radiation by active or passive leaf movement, but our experience with sunflower indicates that it is compatible with a high sensitivity of leaf expansion to water deficits (Turner et al., 1978a and unpublished). Since osmotic adjustment allows some dehydration to occur without loss of assimilatory activity, moderate dehydration tolerance late in the life of the crop will be required.

Finally, since the degree of osmotic adjustment appears to be limited (Turner and Jones, 1980) and photosynthetic capacity is inhibited by low osmotic potentials, maximum transfer of the assimilates in the stems, roots and leaves (consistent with maintaining an erect plant) should be an objective just before maturity. Yield increases such as those for wheat shown in Fig. 1 have occurred at least partly as a result of a greater proportion of the total dry matter being partitioned to the ear (Evans, 1978). Although there is clearly an upper limit to harvest index, there is no indication at present of an asymptote and no one knows the maximum value of harvest index that is reasonable.

Implications for management

To maximize yields under dryland agriculture systems in which water is limited, management strategies should be aimed at maximizing the proportion of precipitation that passes through the crop (Equation 2). In addition to that passing through the crop, precipitation can be lost from the system through runoff, through deep percolation to soil depths below the root zone, through soil evaporation, through interception and subsequent rapid evaporation from foliage, and through transpiration by competing plants. Weed control, tillage, seed treatment, mulching and fertilization can all aid in maximizing the proportion of precipitation which passes through the crop. Other practices such as fallowing, supplementary irrigation and runoff farming can add to the water available to the crop in any one season or location.

Because transpiration is the major pathway of water loss in Mediterranean environments, weed control is of paramount importance in reducing the amount of water used. Many studies have shown that it is essential to control weeds early in the life of the crop, i.e. during establishment and early vegetative growth, particularly in the short season crops which are frequently grown in severely water-limited environments (Jordan and Shaner, 1979). Weeds not only compete directly with the crop for water in the crop's root zone, but may be deeper rooted; extracting water otherwise available later in the lifecycle of the crop. Furthermore, once established, weeds will compete with the crop for nutrients and for light as well as water. Weeds also provide a harbour for insects and diseases that attack crop plants. In addition to the use of selective herbicides, clean seed, crop rotation and tillage can all be utilized for weed control.

As well as controlling weeds, tillage can also improve the infiltration of water into the soil. This, in turn, reduces losses of water by runoff and reduces the risk of soil erosion from small storms. The improvement of infiltration into the soil from cultivation practices probably arises from an increase in the detention time as a result of an increase in soil porosity rather than an increase in surface roughness (Henderson, 1979). Tillage also has some negative effects: it increases the role of soil evaporation, it increases the risk of wind erosion, it increases the risk of erosion by water in larger storms, and increases the compaction of the subsoil and hence infiltration of water and penetration of roots into this layer. For these reasons and to effect savings in energy costs and machinery costs, there is a trend towards minimum tillage or direct drilling (or zero tillage) in which seed bed preparation is restricted to a narrow band in the zone to be planted with the crop (Baeumer and Bakermans, 1973).

The advantage of direct drilling on infiltration and soil water availability to the crop is not clear: some reports indicate an increase in infiltration, whereas others indicate a decrease (Henderson, 1979), Differences in crop residues, slope rainfall pattern and soil types may account for the reported differences in response. What is clear is that direct drilling, because it requires only one passage over a previous crop or pasture, can be used to sow a crop rapidly after rain and hence gain rapid establishment and minimize soil evaporation. It also enables effective integration of cropping and grazing systems, and by reducing the risk of soil erosion, can be used to crop land that is too steep for conventional cultivation.

Soil evaporation is a major source of water loss in semi-arid and arid ecosystems, accounting for up to 50 per cent of the total evapotranspiration in some circumstances (Fischer and Turner, 1978). Since the rate of water loss from a wet soil surface is similar to that from a free water surface or a complete crop canopy (Ritchie, 1974), the rapid establishment of a crop with full ground cover after the opening rains will minimize losses by soil evaporation.

As mentioned above, direct drilling reduces the period of exposure of bare soil to evaporative losses. Seed priming will aid this too, as seed pretreatment by wetting and air drying speeds up germination and root growth (see above). Such priming may provide considerable benefits in semi-arid agricultural systems such as those in the Mediterranean-type environments in reducing soil evaporation and improving crop competition over weeds. Clearly, also, the use of higher seeding densities and seed with a high germination percentage will aid in the establishment of uniform crops.

It is well known that mulching with crop residues reduces soil evaporation, but it is questionable whether enough crop residues are available in Mediterranean regions to have a significant effect on soil evaporation (Henderson, 1979).

Application of fertilizer to promote root growth is important in making adequate use of available soil water. However, in Mediterranean-type climates, root growth is frequently restricted by the depth of wetting (Smith and Harris, 1980). Under these circumstances, heavy use of fertilizers may actually reduce yields, as reported by several investigators (Barley and Naidu, 1964; Fischer and Kohn, 1966; Bond et al., 1971; Bolton, 1981; Lahiri, 1980), because luxurious vegetative growth induces excessive use of water early in the life of the plant, leaving inadequate soil water for grainfilling. Clearly the optimum fertilizer rate will depend on the rainfall and the state of the soil water store at the beginning of the season. As fertilizers are applied at or near seeding when the amount of expected rainfall is unknown, farmers in Mediterranean regions tend to be conservative in their use of fertilizers to reduce the danger of yield reductions from high rates of fertilization. Our analyses suggest that the application of fertilizer late in the vegetative stage should prove beneficial in wet years.

The implications for management discussed thus far have been aimed at practices to improve the proportion of precipitation passing through the crop. What of the management practices to increase the total water available to the crop? Under dryland farming systems, fallowing is the most widespread means of increasing the water available to the crop (French, 1978). As the importance of fallowing is discussed elsewhere in this volume (Bolton, 1981), it is sufficient to point out that the efficiency of fallowing depends on the soil type, soil depth, and rainfall pattern (French, 1978).

It has been pointed out that evaporation from a wet soil surface is similar to that from a crop with full ground cover. However, as the soil surface dries, the rate of evaporation decreases markedly: most of the reduction in soil water content occurs in the upper 20 to 30 cm and subsoil water is only slowly subject to evaporative loss (Henderson, 1979). Thus on land which is kept free of vegetation, i.e. weed-free fallow, rainwater from storms which are sufficent to allow penetration of rainfall to depths below 20 to 30 cm, can be stored in suitable soils for use by a subsequent crop. Because frequent small storms that penetrate only the top 20 cm of the soil will be evaporated away, particularly during the summer, fallowing is particularly beneficial in areas of winter rainfall such as the Mediterranean zones or in summer rainfall areas with infrequent heavy rainstorms. Because of losses by soil evaporation and deep percolation, fallowing increases the moisture available to crops only by about 20 per cent (Fischer and Turner, 1978), but this can be important in water-limited environments, and particularly in years when the rainfall during the growing season is less than 200 mm.

When rainfall is considerably below this figure, i.e. an average of 100 to 150 mm per year, fallowing will not provide enough water for an economic crop, and rainfall will have to be supplemented either by irrigation or runoff from a surrounding catchment. Although discussion of irrigation is outside the scope of this chapter, it must be emphasised that low cost, supplemental irrigation can benefit yields considerably if applied at stategic stages in the life history of the crop. The considerable evidence amassed by Salter and Goode (1967) indicated that irrigation to alleviate water stress in the late vegetative stage to early seed set was the most beneficial in many crops. However recent evidence by Day *et al.* (1978), suggests that additional water at any stage of development had an equal effect on the yield of barley. Where a regular but limited supply of water is available, even the use of capital-intensive drip irrigation systems may prove economic for cash crops such as cotton.

The use of runoff farming practices to supplement rainfall has long been practised in the low-rainfall zones of the Middle East (Evenari *et al.*, 1971). The system of directing the runoff from surrounding barren hills into terraced valley bottoms with deep loessal soil allowed the growth of a crop from a single rainfall event of less than 20 mm. Such a system may still be feasible when terrain and soils permit, but a modification of this system to that in which simple microcatchments collect runoff and concentrate it in an area where a crop is grown, may prove to be more easily established and maintained today.

An experimental microcatchment for tree crops (Evenari et al., 1971; Shanan and Tadmor, 1976) and an implement to establish microcatchments for grasses and cereals (Dixon, 1977) are being tested. The area of catchment per plant needs to be carefully calculated to account for rainfall incidences and patterns, soil and topographical characteristics and the crop to be grown. However, the technique is of value as it allows crops to be grown in areas where rainfall would normally be insufficient for conventional cropping.

Finally, one management practice, specifically aimed at reducing transpiration, was widely canvassed in the 1960's as a means of improving yield under dryland situations by improving water use efficiency: the use of antitranspirants (Zelitch and Waggoner, 1962; Shimshi, 1963). Although antitranspirants appear to have a future in increasing the size and fresh weight of fruits (Davenport et al., 1972, 1974), and in the improvement of the water yield of catchments (Waggoner and Turner, 1971), they appear to provide little benefit for dryland cereal crops (Mizrahi et al., 1974), presumably because any reduction in gas exchange not only reduces transpiration, but also reduces productivity (Turner, 1979).

ACKNOWLEDGEMENTS

We thank Drs. J. B. Passioura, R. A. Fischer and S. Takami for comments on the manuscript.

REFERENCES

Acevedo, E., Hsiao, T. C. and Henderson, D. W. 1971 Immediate and subsequent growth responses of maize leaves to changes in water status. Plant Physiol. 48, 631–6.

Angus, J. F., and Moncur, M. W. 1977 Water stress and phenology in wheat. Aust. J. Agric. Res. 28, 177–81.

Austin, R. B., Longden, P. C. and Hutchinson, J. 1969 Some effects of 'hardening' carrot seed. Ann. Bot. 33, 883–95.

Baeumer, K., and Bakermans, W. A. P. 1973 Zero-tillage. Adv. Agron. 25, 77–123.

Barbour, M. G. 1973 Desert dogma reexamined: root/shoot productivity and plant spacing. Am. Midl. Nat. 89, 41–57.

Barley, K. P. and Naidu, N. A. 1964 The performance of three Australian wheat varieties at high levels of nitrogen supply. Aust. J. Exp. Agric. Anim. Husb. 4, 39–48.

Beadle, C. L., Turner, N. C. and Jarvis, P. G. 1978 Critical water potential for stomatal closure in Sitka spruce. Physiol. Plant. 43, 160–5.

Begg, J. E. 1980 Morphological adaptations of leaves to water stress. In: Turner, N. C. and Kramer P. J. (eds.). Adaptation of plants to water and high temperature stress, pp. 33–42. Wiley Interscience, New York.

Begg, J. E. and Torsell, B. W. R. 1974 Diaphotonastic and parahelionastic leaf movements in *Stylosanthes humilis* H. B. K. (Townsville stylo). In: Bieleski, R. L. Ferguson, A. R., and Gresswell, M. M. (eds.), Mechanisms of regulation of plant growth, pp. 277–83. Bull. 12, Royal Society of New Zealand, Wellington.

Begg, J. E., and Turner, N. C. 1976 Crop water deficits. Adv. Agron. 28, 161–217.

Bennett, O. L., and Doss, B. D. 1960 Effect of soil moisture level on root distribution of cool-season forage species. Agron. J. 52, 204–7.

Bidinger, F. R. 1977 Yield physiology under drought stress: comparative responses of wheat and barley. Ph. D. Thesis, Cornell University, Ithaca, New York, U.S.A.

Bidinger, F., Musgrave, R. B. and Fischer, R. A. 1977 Contribution of stored pre-anthesis assimilate to grain yield in wheat and barley. Nature 270, 431–3.

Blum, A. 1974 Genotypic responses in sorghum to drought stress. I. Response to soil moisture stress. Crop Sci. 14, 361–4.

Blum, A. 1979 Genetic improvement of drought resistance in crop plants: a case for sorghum. In: Mussell, H. and Staples R. C., (eds.), Stress physiology in crop plants, pp. 429–45. Wiley Interscience, New York.

Bolton, F. E. 1981 Optimizing the use of water and nitrogen through soil and crop management. This volume, p. 231–241.

Bond, J. J., Power, J. F. and Willis, W. O. 1971 Soil water extraction by N-fertilized spring wheat. Agron. J. 63, 280–3.

Boyer, J. S. 1970a Leaf enlargement and metabolic rates in corn, soybean, and sunflower at various leaf water potentials. Plant Physiol. 46, 233–5.

Boyer, J. S. 1970b Differing sensitivity of photosynthesis to low leaf water potentials in corn and soybean. Plant Physiol. 46, 236–9.

Boyer, J. S. 1971 Nonstomatal inhibition of photosynthesis in sunflower at low leaf water potentials and high light intensities. Plant Physiol. 48, 532–6.

Brevedan, E. R. and Hodges, H. F. 1973 Effects of moisture deficits on ^{14}C translocation in corn (*Zea mays* L.). Plant Physiol. 52, 436–9.

Brichambaut, G. P. de and Wallén, C. C. 1963 A study of agroclimatology in semi-arid and arid zones of the near east. WMO Tech. Note No. 56.

Brown, K. W., Jordan, W. R. and Thomas, J. C. 1976 Water stress induced alterations of the stomatal response to decreases in leaf water potential. Physiol. Plant. 37, 1–5.

Chatterton, N. J., Hanna, W. W., Powell, J. B. and Lee, D. R. 1975 Photosynthesis and transpiration of bloom and bloomless sorghum. Can. J. Plant Sci. 55, 641–3.

Chinoy, J. J. 1960 Physiology of drought resistance in wheat. I. Effect of wilting at different stages of growth on survival values of eight varieties of wheat belonging to seven species. Phyton, Buenos Aires 14, 147–57.

Constable, G. A. and Hearn, A. B. 1978 Agronomic and physiological responses of soybean and sorghum crops to water deficits. I. Growth, development and yield. Aust. J. Plant Physiol. 5, 159–67.

Cowan, I. R. and Farquhar, G. D. 1977 Stomatal function in relation to leaf metabolism and environment. In: Jennings, D. H. (ed.), Integration of activity in the higher plant. Society for Experimental Biology, Symposia No. 31, pp. 471–505. (Cambridge University Press, Cambridge).

Davenport, D. C., Uriu, K., Martin, P. E. and Hagan, R. M. 1972 Antitranspirants increase size, reduce shrivel of olive fruits, Calif. Agric. 26(7), 6–8.

Davenport, D. C., Uriu, K. and Hagan, R. M. 1974 Antitranspirants to size peaches and replace preharvest irrigation. HortScience 9, 188–9.

Davidson, R. L. 1969 Effects of soil nutrients and moisture on root/shoot ratios in *Lolium perenne* L. and *Trifolium repens* L. Ann. Bot. 33, 571–7.

Day, W., Legg, B. J, French, B. K., Johnston, A. E., Lawlor, D. W. and Jeffers, W. de C. 1978 A drought experiment using mobile shelters: the effect of drought on barley yield, water use and nutrient uptake. J. Agric. Sci. 91, 599–623.

126

Derera, N. F., Marshall, D. R. and Balaam, L. N. 1969 Genetic variability in root development in relation to drought tolerance in spring wheats. Expl. Agric. 5, 327–37.

Dixon, R. M. 1977 New no-till device makes a good impression. Crops Soils 29(8): 16–8.

Dobrenz, A, K., Wright, L. N., Humphrey, A. B., Massengale, M. A. and Kneebone, W. R. 1969 Stomate density and its relationship to water-use efficiency of blue panicgrass (*Panicum antidotale* Retz.). Crop Sci. 9, 354–7.

Donald, C. M. 1968 The breeding of crop ideotypes. Euphytica 17, 385–403.

Doss, B. D., Ashley, D. A. and Bennett, O. L. 1960 Effect of soil moisture regime on root distribution of warm season forage species. Agron. J. 52, 569–72.

Downes, R. W. 1972 Physiological aspects of sorghum adaptation. In: Rao, N. G. P and House, L. R. (eds.), Sorghum in seventies, pp. 265–74. Oxford & I. B. H., New Delhi.

Dubetz, S. 1969 An unusual photonastism induced by drought in *Phaseolus vulgaris*. Can. J. Bot. 47, 1640–1.

Evans, L. T. 1978 The yield of crops – trends and limits. In Agriculture resources and potential. Proc. Lincoln College Centennial Seminar, May 1978. pp. 18–30. Lincoln College, Canterbury.

Evans, L. T. and De Datta, S. K. 1979 The relation between irradiance and grain yield of irrigated rice in the tropics, as influenced by cultivar, nitrogen fertilizer application and month of planting. Field Crops Res. 2, 1–17.

Evenari, M., Shannon, L. and Iadmor, N. 1971 The Negev: The challenge of a desert. Harvard University Press, Cambridge.

Fellows, R. J. and Boyer, J. S. 1978 Altered ultrastructure of cells of sunflower leaves having low water potentials. Protoplasma 93, 381–95.

Fischer, R. A. and Kohn, G. D. 1966 The relationship of grain yield to vegetative growth and post-flowering leaf area in the wheat crop under conditions of limited soil moisture. Aust. J. Agric. Res. 17, 281–95.

Fischer, R. A. and Maurer, R. 1978 Drought resistance in spring wheat cultivars. I: Grain yield responses. Aust. J. Agric. Res. 29, 897–912.

Fischer, R. A. and Turner, N. C. 1978 Plant productivity in the arid and semiarid zones. Annu. Rev. Plant. Physiol. 29, 277–317.

Frank, A. B., Power, J. F. and Willis, W. O. 1973 Effect of temperature and plant water stress on photosynthesis, diffusion resistance, and leaf water potential in spring wheat. Agron. J. 65, 777–80.

French, R. J. 1978 The effect of fallowing on the yield of wheat. I. The effect on soil water storage and nitrate supply. Aust. J. Agric. Res. 29, 653–68.

Gaff, D. F. 1980 Proplasmic tolerance of extreme water stress. In: Turner, N. C. and Kramer, P. J. (eds.), Adaption of plants to water and high temperature stress, (pp. 207–30). Wiley Interscience, New York.

Gallagher, J. N., Biscoe, P. V. and Hunter, B. 1976 Effects of drought on grain growth. Nature 264, 541–2.

Gardner, W. R. and Nieman, R. H. 1964 Lower limit of water availability to plants. Science 143, 1460–2.

Greacen, E. L. and Oh, J. S. 1972 Physics of root growth. Nature New Biol. 235, 24–5.

Gulmon, S. L. and Turner, N. C. 1978 Differences in root and shoot development of tomato (*Lycopersicon esculentum* L.) varieties across contrasting soil environments. Plant Soil 49, 127–36.

Hall, A. E., Foster, K. W. and Waines, J. G. 1979 Crop adaption in semi-arid environments. In: Hall, A. E., Canell, G. H. and Lawton, H. W. (eds.), Agriculture in semi-arid environments, pp. 148–79. Springer-Verlag, Berlin.

Harris, H. C. 1979 Some aspects of the agroclimatology of West Africa and North Africa. In: Hawtin, G. C. and Chancellor, G. J. (eds.), Food legume improvement and development, pp. 7–14. International Center for Agricultural Research in the Dry Areas, Syria, and International Development Centre, Canada.

Heichel, G. H. 1971 Stomatal movements, frequencies, and resistances in two maize varieties differing in photosynthetic capacity. J. Exp. Bot. 22, 644–9.

Henckel, P. A. 1964 Physiology of plants under drought. Annu. Rev. Plant Physiol. 15, 363–86.

Henderson, D. W. 1979 Soil management in semi-arid environments. In: Hall, A. E., Cannell, G. H. and Lawton, H. W. (eds.), Agriculture in Semi-arid Environments, pp. 224–37. Springer-Verlag, Berlin.

Henkel, P. A. 1961 Drought resistance in plants: methods of recognition and of intensification. In: Plant-water relationships in arid and semi-arid conditions. Proc. Madrid. Symp., 1959, pp. 167–74. UNESCO, Paris.

Heydecker, W., Higgins, J. and Gulliver, R. L. 1973 Accelerated germination by osmotic seed treatment. Nature 246, 42–4.

Hoffman, G. J., Rawlins, S. L., Garber, M. J. and Cullen, E. M. 1971 Water relations and growth of cotton as influenced by salinity and relative humidity. Agron. J. 63, 822–6.

Hsiao, T. C. 1973 Plant responses to water stress. Annu. Rev. Plant Physiol. 24, 519–70.

Hsiao, T. C. and Acevedo, E. 1974 Plant responses to water deficits, water-use efficiency, and drought resistance. Agric. Meteorol. 14, 59–84.

Hsiao, T. C., Acevedo, E., Fereres, E. and Henderson, D. W. 1976 Water stress, growth, and osmotic adjustment. Philos. Trans. Roy. Soc. London, Ser. B. 273, 471–500.

Hurd, E. A. 1968 Growth of roots of seven varieties of spring wheat at high and low moisture levels. Agron. J. 60, 201–5.

Hurd, E. A. 1974 Phenotype and drought tolerance in wheat. Agric Meteorol. 14, 39–55.

Jarvis, P. G. 1975 Water transfer in plants. In: De Vries, D. A. and Afgan, N. H. (eds.), Heat and mass transfer in the biosphere. I: Transfer processes in plant environment, pp. 369–94. Scripta, Washington.

Jarvis, P. G. and Jarvis, M. S. 1964 Pre-sowing hardening of plants to drought. Phyton, Buenos Aires 21, 113–7.

Johnson, D. A. 1980 Improvement of perennial herbaceous plants of drought-stressed western rangelands. In: Turner, N. C. and Kramer, P. J. (eds.), Adaption of plants to water and high temperature stress, pp. 419–33. Wiley Interscience, New York.

Johnson, D. A. and Asay, K. H. 1978 A technique for assessing seedling emergence under drought stress. Crop Sci. 18, 520–2.

Johnson, H. B. 1975 Plant pubescence: an ecological perspective. Bot. Rev. 41, 233–58.

Jones, H. G. 1977a. Transpiration in barley lines with differing stomatal frequencies. J. Exp. Bot. 28, 162–8.

Jones, H. G. 1977b Aspects of the water relations of spring wheat (*Triticum aestivum* L.) in response to induced drought. J. Agric. Sci. 88, 267–82.

Jones, H. G. 1979 Stomatal behavior and breeding for drought resistance. In: Mussell, H. and Staples, R. C. (eds.), Stress physiology in crop plants, pp. 407–28. Wiley Interscience, New York.

Jones, H. G. and Kirby, E. J. M. 1977 Effects of manipulation of number of tillers and water supply on grain yield in barley. J. Agric. Sci. 88, 391–7.

Jones, M. M. and Rawson, H. M. 1979 Influence of rate of development of leaf water deficits upon photosynthesis, leaf conductance, water use efficiency, and osmotic potential in sorghum. Physiol. Plant. 45, 103–11.

Jones, M. M. and Turner, N. C. 1978 Osmotic adjustment in leaves of sorghum in response to water deficits. Plant Physiol. 61, 122–6.

Jones, M. M. and Turner, N. C. 1980 Osmotic adjustment in expanding and fully expanded leaves of sunflower in response to water deficits. Aust. J. Plant. Physiol. 7, p. 181–92.

Jones, M. M., Osmond, C. B. and Turner, N. C. 1980 Accumulation of solutes in leaves of sorghum and sunflower in response to water deficits. Aust. J. Plant Physiol. 7, p. 193–205.

Jordan, L. S. and Shaner, D. L. 1979 Weed control. In: Hall, A. E., Cannell, G. H. and Lawton, H. W. (eds.), Agriculture in semi-arid environments, pp. 266–96. Springer-Verlag, Berlin.

Jordan, W. R. and Miller, F. R. 1980 Genetic variability in sorghum root systems: implications for drought tolerance. In: Turner, N. C. and Kramer, P. J. (eds.), Adaption of plants to water and high temperature stress, pp. 383–99. Wiley Interscience, New York.

128

Kaul, R. and Crowle, W. L. 1971. Relations between water status, leaf temperature, stomatal aperture, and productivity in some wheat varieties. Z. Pflanzenzuecht. 65, 233–43.

Keller, W. and Bleak, A. T. 1968 Preplanting treatment to hasten germination and emergence of grass seed. J. Range Manage. 21, 213–6.

Keulen, H. van 1981 Some aspects of the interaction of water and nitrogen. This volume, pp. 205–229.

Kobata, T. and Takami, S. 1979. The effects of water stress on the grain-filling in rice. Jpn J. Crop Sci. 48, 75–81.

Kowal, J. M. and Kassam, A. H. 1978 Agricultural ecology of savanna: A study of West Africa, Clarendon Press, Oxford.

Lahiri, A. N. 1980 Interaction of water stress and mineral nutrition on growth and yield. In: Turner, N. C. and Kramer, P. J. (eds.), Adaption of plants to water and high temperatures stress, pp. 341–52. Wiley Interscience, New York.

Lee-Stadelmann, O. Y. and Stadelmann, E. J. 1976 Cell permeability and water stress. In: Lange, O. L., Kappen, L. and Schulze, E. D. (eds.), Water and Plant life: problems and modern approaches, pp. 268–80. Springer-Verlag, Berlin.

Legg, B. J., Day, W., Lawlor, D. W. and Parkinson, K. J. 1979 The effects of drought on barley growth: models and measurements showing the relative importance of leaf area and photosynthetic rate. J. Agric. Sci. 92, 703–16.

Ludlow, M. M. 1975 Effect of water stress on the decline of leaf net photosynthesis with age. In: Marcelle, R. (ed.), Environmental and biological control of photosynthesis, pp. 123–34. Junk, The Hague.

Ludlow, M. M. 1980 Adaptive significance of stomatal responses to water stress. In: Turner, N. C. and Kramer, P. J. (eds.), Adaptation of plants to water and high temperature stress, pp. 123–38. Wiley Interscience, New York.

Malik, R. S., Dhankar, J. S. and Turner, N. C. 1979 Influence of soil water deficits on root growth of cotton seedlings. Plant Soil 53, 109–15.

May, L. H., Milthorpe, E. J. and Milthorpe, F. L. 1962 Pre-sowing hardening of plants to drought: an appraisal of the contributions by P. A. Genkel. Field Crop Abstr. 15, 93–8.

McCree, K. J. 1974 Changes in the stomatal response characteristics of grain sorghum produced by water stress during growth. Crop Sci. 14, 273–8.

McWilliam, J. R. 1968 The nature of the perennial response in Mediterranean grasses. II: Senescence, summer dormancy and survival in *Phalaris*. Aust. J. Agric. Res. 19, 397–409.

Meyer, R. F. and Boyer, J. S. 1972 Sensitivity of cell division and cell elongation to low water potentials in soybean hypocotyls. Planta 108, 77–87.

Millar, B. D. and Denmead, O. T. 1976 Water relations of wheat leaves in the field. Agron. J. 68, 303–7.

Miskin, K. E., Rasmusson, D. C. and Moss, D. N. 1972 Inheritance and physiological effects of stomatal frequence in barley. Crop Sci. 12, 780–3.

Mizrahi, Y., Scherings, S. G., Malis Arad, S. and Richmond, A. E. 1974 Aspects of the effect of ABA on the water status of barley and wheat seedlings. Physiol. Plant. 31, 44–50.

Moorby, J., Munns, R. and Walcott, J. 1975 Effect of water deficit on photosynthesis and tuber metabolism in potatoes. Aust. J. Plant Physiol. 2, 323–33.

Morgan, J. M. 1977 Differences in osmoregulation between wheat genotypes. Nature 270, 234–5.

Morgan, J. M. 1980a Differences in adaption to water stress within crop species. In: Turner, N. C. and Kramer, P. J. (eds.), Adaption of plants to water and high temperature stress, pp. 369–382. Wiley Interscience, New York.

Morgan, J. M. 1980b Possible role of abscisic acid in reducing seed set in water stressed wheat plants. Nature 285, 655–7.

Mulroy, T. W. and Rundel, P. W. 1977 Annual plants; adaption to desert environments. Bio Science 27, 109–14.

Munns, R. and Pearson, C. J. 1974 Effect of water deficit on translocation of carbohydrate in *Solanum tuberosum*. Aust. J. Plant Physiol. 1, 529–37.

Munns, R., Brady, C. J. and Barlow, E. W. R. 1979 Solute accumulation in the apex and leaves of wheat during water stress. Aust. J. Plant Physiol. 6, 379–89.

Murata, Y. and Matsushima, S. 1975 Rice. In: Evans, L. T. (ed.), Crop physiology: some case histories. pp. 73–99. Cambridge University Press, Cambridge.

Neales, T. F. and Incoll, L. D. 1968 The control of leaf photosynthesis rate by the level of assimilate concentration in the leaf: a review of the hypothesis. Bot. Rev. 34, 107–25.

Osmond, C. B., Winter, K. and Powles, S. B. 1980 Adaptive significance of carbon dioxide cycling during photosynthesis in water-stressed plants. In: Turner, N. C. and Kramer, P. J. (eds.), Adaptation of plants to water and high temperature stress, pp. 139–54. Wiley Interscience, New York.

O'Toole, J. C., Aquino, R. S. and Alluri, K. 1978 Seedling stage drought response in rice. Agron. J. 70, 1101–3.

O'Toole, J. C. and Chang, T. T. 1979 Drought resistance in cereals – rice: a case study. In: Mussell, H. and Staples, R. C. (eds.), Stress physiology in crop plants, pp. 373–405. Wiley Interscience, New York.

Passioura, J. B. 1972 The effect of root geometry on the yield of wheat growing on stored water. Aust. J. Agric. Res. 23, 745–52.

Passioura, J. B. 1974 The effect of root geometry on the water relations of temperature cereals (wheat, barley, oats). In: Kolek, J. (ed.), Structure and function of primary root tissues, pp. 357–63. Veda, Bratislava.

Passioura, J. B. 1976 Physiology of grain yield in wheat growing on stored water. Aust. J. Plant. Physiol. 3, 559–65.

Passioura, J. B. 1977 Grain yield, harvest index, and water use of wheat. J. Aust. Inst. Agric. Sci. 43, 117–20.

Pearson, R. W. 1966 Soil environment and root development. In: Pierre, W. H, Kirkham, D, Pesek, J. and Shaw, R. (eds.), Plant environment and efficient water use, pp. 95–126. American Society of Agronomy and Soil Science Society of America, Madison.

Raper, C. D. Jr. and Barber, S. A. 1970 Rooting systems of soybeans. I. Differences in root morphology among varieties. Agron. J. 62, 581–4.

Rawson, H. M. 1979 Vertical wilting and photosynthesis, transpiration, and water use efficiency of sunflower leaves. Aust. J. Plant Physiol. 6, 109–20.

Rawson, H. M. and Evans, L. T. 1971 The contribution of stem reserves to grain development in a range of wheat cultivars of different height. Aust. J. Agric. Res. 21, 851–63.

Rawson, H. M., Begg, J. E. and Woodward, R. G. 1977a The effect of atmospheric humidity on photosynthesis, transpiration and water use efficiency of leaves of several plants species. Planta 134, 5–10.

Rawson, H. M., Bagga, A. K. and Bremner, P. M. 1977b Aspects of adaption by wheat and barley to soil moisture deficits. Aust. J. Plant. Physiol. 4, 389–401.

Rawson, H. M., Turner, N. C. and Begg, J. E. 1978 Agronomic and physiological responses of soybean and sorghum crops to water deficits. IV: Photosynthesis, transpiration and water use efficiency of leaves. Aust. J. Plant Physiol. 5, 195–209.

Reitz, L. P. and Salmon, S. C. 1959 Hard red winter wheat improvement in the plains: a 20-year summary. U.S. Dep. Agric. Tech. Bull. 1192.

Ritchie, J. T. 1974 Atmospheric and soil water influences on the plant water balance. Agric. Meteorol. 14, 183–98.

Roark, B. and Quisenberry, J. E. 1977 Environmental and genetic components of stomatal behavior in two genotypes of upland cotton. Plant Physiol. 59, 354–6.

Roy, N. N. and Murty, B. R. 1970 A selection procedure in wheat for stress environment. Euphytica 19, 509–21.

Russell, J. S. 1973 Yield trends of different crops in different areas and reflections on the sources of crop yield improvement in the Australian environment. J. Aust. Inst. Agric. Sci. 39, 156–66.

Salter, P. J. and Goode, J. E. 1967 Crop responses to water at different stages of growth. Commonw. Agric. Bur., Farnham Royal.

Sánchez-Díaz, M. F., Hesketh, J. D. and Kramer, P. J. 1972 Wax filaments on sorghum leaves as seen with a scanning electron microscope. J. Ariz. Acad. Sci. 7, 6–7.

Schobert, B. 1977 Is there an osmotic regulatory mechanism in algae and higher plants? J. Theor. Biol. 68, 17–26.

Schulze, E.-D., Lange, O. L., Buschbom, U., Kappen, L. and Evenari, M. 1972 Stomatal responses to changes in humidity in plants growing in the desert. Planta 108, 259–70.

Shackel, K. A. and Hall, A. E. 1979 Reversible leaflet movements in relation to drought adaptation of cowpeas, *Vigna unguiculata* (L.) Walp. Aust. J. Plant. Physiol. 6, 265–76.

Shanan, L. and Tadmor, N. H. 1976 Microcatchment systems for arid zone development: a handbook for design and construction. Hebrew University of Jerusalem and Center for International Agricultural Cooperation. Rehovot.

Sharp, R. E. and Davies, W. J. 1979 Solute regulation and growth by roots and shoots of water-stressed maize plants. Planta 147, 43–9.

Shimshi, D. 1963 Effect of soil moisture and phenylmercuric acetate upon stomatal aperture, transpiration, and photosynthesis. Plant. Physiol. 38, 713–21.

Shimshi, D. and Ephrat, J. 1975 Stomatal behavior of wheat cultivars in relation to their transpiration, photosynthesis, and yield. Agron. J. 67, 326–31.

Shimper, A. F. W. 1903 Plant-geography upon a physiological basis. Clarendon Press, Oxford.

Slatyer, R. O. 1973 Effects of short periods of water stress on leaf photosynthesis. In: Slatyer, R. O. (ed.), Plant response to climatic factors. Proc. Uppsala Symp., 1970, pp. 271–6. UNESCO, Paris.

Smith, R. C. G. and Harris, H. 1981 Environmental resources and restraints to agricultural production in a Mediterranean-type environment. This volume, pp. 30–57.

Steponkus, P. L., Cutler, J. M. and O'Toole, J. C. 1980 Adaptation to water stress in rice. In: Turner, N. C. and Kramer, P. J. (eds.), Adaptation of plants to water and high temperature stress, pp. 401–18. Wiley Interscience, New York.

Stout, D. G. and Simpson, G. M. 1978 Drought resistance of *Sorghum bicolor*. I: Drought avoidance mechanisms related to leaf water status. Can. J. Plant Sci. 58, 213–24.

Sullivan, C. Y. and Eastin, J. D. 1974 Plant physiological responses to water stress. Agric. Meteorol. 14, 113–27.

Thomas, J. C., Brown, K. W. and Jordan, W. R. 1976 Stomatal response to leaf water potential as affected by preconditioning water stress in the field. Agron. J. 68, 706–8.

Thorne, G. N. 1966 Physiological aspects of grain yield in cereals. In: Milthorpe, F. L. and Ivins, J. D. (eds.), The growth of cereals and grasses, pp. 88–105. Butterworth, London.

Troughton, J. H. and Slatyer, R. O. 1969 Plant water status, leaf temperature, and calculated mesophyll resistance to carbon dioxide of cotton leaves. Aust. J. Biol. Sci. 22, 815–27.

Turner, N. C. 1966 Grain production and water use of wheat as affected by plant density, defoliation and water status. Ph. D. Thesis, University of Adelaide, South Australia, Australia.

Turner, N. C. 1974 Stomatal response to light and water under field conditions. In: Bieleski, R. L., Ferguson, A. R. and Cresswell, M. M. (eds.), Mechanisms of regulation of plant growth, pp. 423–32. Bull. 12, Royal Society of New Zealand, Wellington.

Turner, N. C. 1979 Drought resistance and adaptation to water deficits in crop plants. In: Mussell, H. and Staples, R. C. (eds.), Stress physiology in crop plants, pp. 343–72. Wiley Interscience, New York.

Turner, N. C. 1980 Designing better crops for dryland Australia: can the deserts help us? J. Aust. Inst. Agric. Sci. 46 (in press).

Turner, N. C. and Begg, J. E. 1978 Responses of pasture plants to water deficits. In: Wilson, J. R. (ed.), Plant relations in pastures, pp. 50–66. CSIRO, Melbourne.

Turner, N. C. and Burch, G. J. 1981 The role of water in plants. In: Teare, I. D. and Peet, M. M. (eds.), Crop-water relations, (in press). Wiley Interscience, New York.

Turner, N. C. and Jones, M. M. 1980 Turgor maintenance by osmotic adjustment: a review and evaluation. In: Turner, N. C. and Kramer, P. J. (eds.), Adaptation of plants to water and high temperature stress, pp. 87–103. Wiley Interscience, New York.

Turner, N. C. and Kramer, P. J. 1980 Adaptation of plants to water and high temperature stress. Wiley Interscience, New York.

Turner, N. C., Begg, J. E. and Tonnet, M. L. 1978a Osmotic adjustment of sorghum and sunflower crops in response to water deficits and its influence on the water potential at which stomata close. Aust. J. Plant Physiol. 5, 597-608.

Turner, N. C., Begg, J. E., Rawson, H. M., English, S. D. and Hearn, A. B. 1978b Agronomic and physiological responses of soybean and sorghum crops to water deficits. III: Components of leaf water potential, leaf conductance, $^{14}CO_2$ photosynthesis, and adaptation to water deficits. Aust. J. Plant Physiol. 5, 179–94.

Volkens, G. 1884 Zur Kenntnis der Beziehungen zwischen Standorten und anatomischem Bau der Vegetationsorgane. Jahrb. Bot. Gart. 3, 1–46.

Volkens, G. 1887 Die Flora der ägyptisch-arabischen Wüste, auf Gund anatomisch-ephysiolo-gischer Forschungen dartestellt. Borntrager, Berlin.

Waggoner, P. E. and Turner, N. C. 1971 Transpiration and its control by stomata in a pine forest. Conn. Agric. Exp. Stn, New Haven, Bull. 726.

Wardlaw, I. F. 1967 The effect of water stress on translocation in relation to photosynthesis and growth. I. Effect during grain development in wheat. Aust. J. Biol. Sci. 20, 25–39.

Wardlaw, I. F. 1969 The effect of water stress on translocation in relation to photosynthesis and growth. II: Effect during leaf development in Lolium temulentum. Aust. J. Biol. Sci. 22, 1–16.

Wardlaw, I. F. 1971 The early stages of grain development in wheat: response to water stress in a single variety. Aust. J. Biol. Sci. 24, 1047–55.

Wardlaw, I. F. and Porter, H. K. 1967 The redistribution of stem sugars in wheat during grain development. Aust. J. Biol. Sci. 20, 309–18.

Watts, W. R. 1974 Leaf extension in Zea mays. III. Field measurements of leaf extension in response to temperature and leaf water potential. J. Exp. Bot. 25, 1085–96.

Wien, H. C., Littleton, E. J. and Ayanaba, A. 1979 Drought stress of cowpea and soybean under tropical conditions. In: Mussell, H. and Staples, R. C. (eds.), Stress physiology in crop plants, pp. 283–301. Wiley Interscience, New York.

Wilson, D. 1975 Leaf growth, stomatal diffusion resistances and photosynthesis during droughting of Lolium perenne populations selected for contrasting stomatal length and frequency. Ann. Appl. Biol. 79, 67–82.

Wong, S. C., Cowan, I. R. and Farquhar, G. D. 1979 Stomatal conductance correlates with photosynthetic capacity. Nature 282, 424–6.

Woodruff, D. R. 1969 Studies on presowing drought hardening of wheat. Aust. J. Agric. Res. 20, 13–24.

Woodruff, D. R. 1973 Evaluation of the presowing drought hardening of wheat. Queensl. J. Agric. Anim. Sci. 30, 119–24.

Woolley, J. T. 1964 Water relations of soybean leaf hairs. Agron. J. 56, 569–71.

Wuenscher, J. E. 1970 The effect of leaf hairs of Verbascum thapsus on leaf energy exchange. New Phytol. 69, 65–73.

Yegappan, T. M. 1978 Growth physiology of Helianthus annuus (Linn.) during water stress and recovery. Ph. D. Thesis, Australian National University, Canberra, A. C. T., Australia.

Zelitch, I. and Waggoner, P. E. 1962 Effect of chemical control of stomata on transpiration of intact plants. Proc. Nat. Acad. Sci. U.S.A. 48, 1297–9.

Zobel, R. W. 1975 The genetics of root deveopment. In: Torrey, J. G. and Clarkson, D. T. (eds.), The development and function of roots, pp. 261–75. Academic Press, New York.

6. Accession, transformation, and loss of nitrogen in soils of the arid region

P. L. G. VLEK*, I. R. P. FILLERY and J. R. BURFORD

*I.F.D.C., Muscle Shoals, Alabama, U.S.A.

Nitrogen is a key element in improving crop productivity throughout the world. It has attracted considerable research attention both as a plant nutrient and as an environmental pollutant and several reviews have appeared (Bartholomew and Clark, 1965; Nielsen and MacDonald, 1978; West and Skujins, 1978; NRC, 1978). Nearly all of the research on nitrogen as a plant nutrient has been conducted in irrigated or humid ecosystems where spectacular improvements in crop yield have been achieved through the use of large amounts of N.

In arid and semi-arid regions, limited water resources and low crop productivity have discouraged the widespread use of N and consequently limited the research interest in this element. However, with an ever-increasing demand for food, it is now realized that arid and semi-arid regions will need to be exploited to the fullest extent. This may be achieved only through a better understanding of the various components of these ecosystems and their interactions (Russell, 1977). With the introduction of improved crop varieties which are responsive to fertilizer, applied nitrogen has become an important input in these ecosystems.

Nitrogen is one of the most complex and mobile plant nutrients. Fig. 1 is a general diagram representing the various cyclic N-processes operating in a soil-plant system. In the internal nitrogen cycle, immobilization and mineralization are continuously causing changes in the mineral nitrogen reserves of the soil. Simultaneously, mineral N may be converted to gaseous forms due to denitrification or release of ammonia from the soil solution.

An external cycle between the soil and the atmosphere depletes the soil's gaseous nitrogen content by emission of ammonia, nitrous oxide, or nitrogen gas, while enrichment of the nitrogen in the soil takes place through biological nitrogen fixation, nitrogen deposition, or nitrogen fertilization. Mineral nitrogen in the soil can also be depleted through the uptake of nitrogen by the crop, whereas the return to the soil of the non-harvested crop will add to the organic nitrogen pool. Finally, alternate wetting and drying conditions in the soil will tend to move mineral nitrogen up and down the profile. Where excessive wetting prevails, mineral nitrogen may leach beyond the reach of the crop roots. The

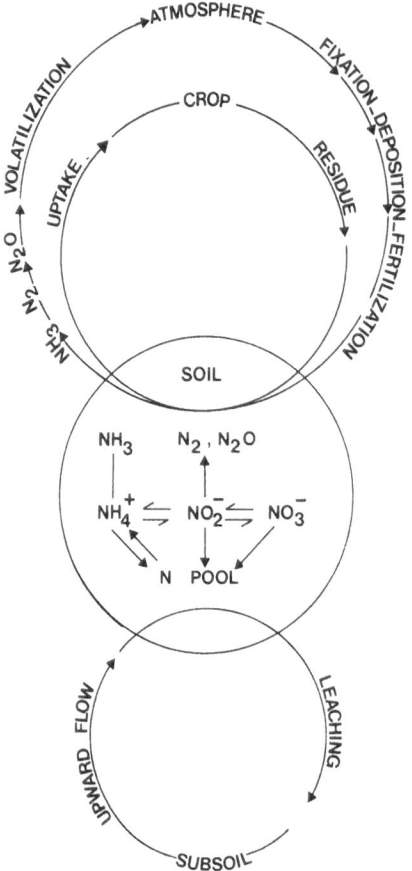

Fig. 1.   Nitrogen cycles in agricultural ecosystem.

extent to which these various processes take place in a particular soil, depends on the agroclimatic conditions under which the soil is found. Table 1 contains a typical illustration of the differences in nitrogen inputs, pools, internal fluxes, and losses as affected by the type of bioclimatic system.

Fig. 2 depicts typical seasonal patterns of climate, plant growth, and nitrogen flows in the semi-arid winter rainfall ecosystem. To a great extent, prevailing ambient temperatures and the available soil moisture determine the biological activity in the soil. It is when the availability of mineral nitrogen and of soil moisture coincide that nitrogen uptake by the crop is possible and is reflected in a concomitant increase in plant biomass.

Table 1. Estimated nitrogen balances for some bioclimatic types

	Desert[a]	Tundra[b]	Deciduous Forest[c]	Tropical Forest[d]
Inputs (kg ha^{-1} yr^{-1})	16.1			
Wetfall				1.4
NO_3-N		0.5	4.0	
NH_4-N		0.5	5.3	
NH_3		5		
Total N			13.0	27.0
Nitrogen fixation	3.6	2–90		88.0
Pools (kg ha^{-1})				
Plant tissue	41	90	492	1230
Detritus	68	32	119	126
Soil organic matter	162	3350	5080	
NH_4-N		72	75	
NO_3-N		0	3	
Internal fluxes (kg ha^{-1} yr^{-1})				
Uptake	12.7	11.0	124.0	102
Mineralization	12.7	11.0	115.6	
Losses (kg ha^{-1} yr^{-1})				
Stream flow				29.0
NO_3-N			0.4	
NH_4-N			1.1	
Total N		+ 10.0	3.1	
Denitrification	14.2	0		56.0
Ecosystem residence Time (yr)	19		1370	

[a] West (1975).
[b] Rosswall et al. (1975).
[c] Henderson and Harris (1975).
[d] Edmiston (1970).

The objective of the farmer is to maximize plant biomass production or yield through manipulation of the factors that influence the availability of soil moisture and plant nutrients in the soil. This chapter aims to review the state of the art with respect to the nitrogen cycles that act upon, or in, the soil: plant system in rainfed agriculture of Mediterranean-type environments and the extent to which they can be manipulated.

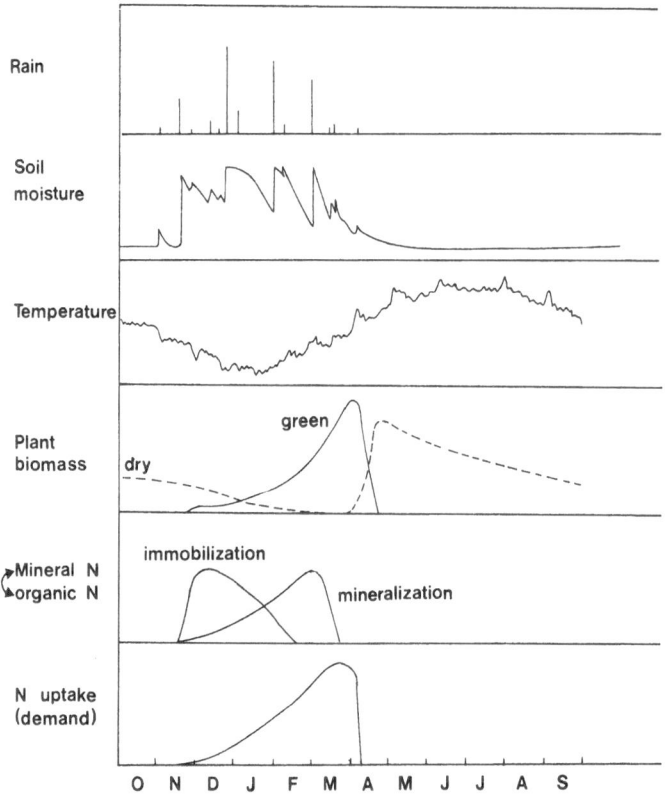

Fig. 2. Annual weather, plant, and soil-N patterns of the arid-Mediterranean climate of the Northern Negev (from: Noy-Meir and Harpaz, 1977).

NITROGEN CYCLING IN ARID SOILS

Mineralization

The various transformations that nitrogen undergoes between entering and leaving the soil system are complex. Despite a voluminous literature, our understanding of the various processes is limited. In particular, we are uncertain about rates of mineralization and immobilization of nitrogen in soils because they are closely linked to the turnover of organic matter which usually contains more than 95 per cent of the soil's nitrogen.

The timing and extent of the net release of inorganic nitrogen from organic matter, determine the availability of mineral N for uptake by the crop or for utilization by competing microorganisms. The dynamics of nitrogen supply are particularly important in the rainfed agricultural systems of Mediterranean-type climates where nitrogen fertilization to overcome soil nitrogen deficits is still uncommon and is not without economic risk.

Soil organic matter has no well-defined composition, and attempts to characterize it on the basis of some of its identifiable components have met with varied success (Bremner, 1965). As a result, most studies concerned with net mineralization of soil nitrogen have made no attempt to distinguish between the various organic compounds in soils, and have expressed the change in total soil nitrogen as a simple first order rate equation.

$$\frac{dN}{dt} = -kN + A, \tag{1}$$

where k is a decomposition constant, N is the nitrogen content of a given mass of soil at time t, and A is an accretion constant giving the amount of nitrogen added to the given mass of soil per unit time (Greenland, 1971). After several cycles of a certain cropping pattern, the nitrogen level will tend to fluctuate about a mean, and the amount of mineral nitrogen released by the soil will equal the amount of nitrogen annually returned to the soil in the form of organic matter. Greenland demonstrated the usefulness of his model in calculating what lengths of crop and pasture were required to maintain a certain equilibrium nitrogen level in the soil.

A more refined equation developed by Russell (1975) allows for year-to-year variation of the rate constant and includes a factor to account for additions of manure. The equation can be written:

$$\frac{dN}{dt} = -k_1(t)N + k_2 + k_3(t)Y(t), \tag{2}$$

where N is soil organic nitrogen, $k_1(t)$ is a time-dependent decomposition coefficient, k_2 represents a constant addition of N not associated with cropping, Y(t) is plant biomass at time t and $k_3(t)Y(t)$ is the addition of N from plant residues which depends on the nature of the sequential crop. Russell used this equation and three more restricted variants to analyze the relationship between yield and equilibrium soil nitrogen content for a series of long-term cropping system experiments with and without additions of manure.

Although useful as a tool to analyze the effect of cropping patterns and cultural practices on the long-term behavior of organic nitrogen in the soil, these rate equations lack the necessary sensitivity to predict seasonal fluctuations in the availability of inorganic N.

Attempts have been made to follow the short-term nitrogen dynamics in soil by simulation of the various transformations in mechanistic models (Tanji and Gupta, 1978). These models are based on the assumption that microbially mediated processes are kinetically first order in nature:

$$\frac{d[\text{org N}]}{dt} i = -k_1[\text{org N}]_i + k_2[\text{NH}_4^+] + k_3[\text{NO}_2^-] + k_4[\text{NO}_3^-] \qquad (3)$$

(Tanji and Gupta, 1978). Various models incorporate environmental factors to allow for seasonal variations in the rate constants (k_1, k_2, k_3, k_4) of the transformations (Beek and Frissel, 1973; Hagin and Amberger, 1974; Donegian and Crawford, 1976; Stanford and Smith, 1972; Watts, 1975). Incorporation of seasonality is particularly important for Mediterranean-type climates.

The extent and rate of organic matter decomposition depend on the quality of the material being added to the soil either in the form of manure or vegetation. The C : N ratio of the organic material added gives an impression of the degree to which net mineralization of nitrogen can take place. Fig. 3 shows the relationship between the original C : N ratio of the organic material and the expected net mineralization from this material. It is seen from this curve that, with C : N ratios of 25 for the organic material added to the soil, immobilization and mineralization are about in equilibrium. Net mineralization is expected at C : N ratios lower than 25 while net immobilization occurs at C : N ratios of more than 25 (Harmsen and Kolenbrander, 1965; Debnath and Hajra, 1972; Russell, 1973). An

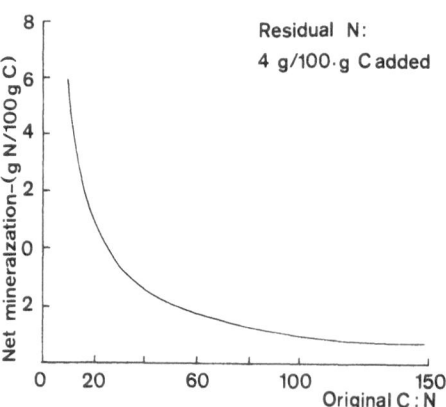

Fig. 3. Calculated net mineralization from added organic material as a function of the originalC/N-ratio, assuming residual N is 4% of original C.

elegant description of organic carbon and nitrogen compounds in soils, provided by Oades and Ladd (1977), brings out the complex nature of soil organic matter.

Models on N behavior in soils rarely distinguish between the various fractions of organic matter. However, Beek and Frissel (1973) made a distinction between proteins (the most easily decomposable component) and the remainder of soil organic matter consisting of sugars, cellulose, lignin, and biomass. Only the decomposition of proteins was considered to be influenced by soil temperature and moisture. In their model, the rate of biomass production was linked to the decomposition of organic matter. Consequently, the production of mineral N was expressed as a function of organic matter decomposition, biomass production, and the C : N ratios of the decomposing substrate and biomass. Alternative models dealing with the decomposition of the various soil organic matter fractions were developed by Hagin and Amberger (1974) and Van Veen and Frissel (1976).

Seasonal fluctuations of the soil conditions in Mediterranean-type climates are generally extreme, particularly at higher altitudes or latitudes. Temperatures may range from below $0\,^\circ$C to as high as $60\,^\circ$C while moisture content may range from field capacity to below the conventional wilting point. Such variation in the environment has a great impact on the dynamics of nitrogen transformations in soil. For instance, during dry periods carbon decomposition exceeds nitrogen mineralization (Birch, 1960), resulting in a decreased C : N ratio which will favor net mineralization during the subsequent wet season. If temperatures are favorable, the onset of the rainy season will be accompanied by a flush of mineral nitrogen in the soil (Hardy, 1946; Birch, 1960). If, in a winter rainfall climate, the early rains coincide with low soil temperatures, the mineral nitrogen flush may be delayed until early spring. A better understanding of the kinetics of mineralization in dryland agriculture of Mediterranean-type climates would help to predict the availability of mineral N and allow for timely correction of nitrogen deficiencies through applications of fertilizers.

Ammonification

Ammonification of organic N in soil is affected by a number of factors, many of which are related to biological activity. Myers (1975), studying the temperature effect on ammonification in a tropical soil, found it to fit an Arrhenius-type equation with a maximum at about $50\,^\circ$C. The lower temperature limit for ammonification is generally around freezing (Sabey *et al.*, 1956; Stanford *et al.*, 1973). The effect of temperature on ammonification appears to be rather uniform among soils (Stanford and Epstein, 1974). Mahendrappa *et al.* (1966) point out that the temperature effect on nitrogen mineralization may vary from one climatic zone to another.

The optimum soil potential for ammonification ranges from 10 to 50 kPa (0.1 to 0.5 bar) (Miller and Johnson, 1964; Stanford and Epstein, 1974), while the rate of ammonification declines linearly with decreasing water content (Miller and Johnson, 1964; Reichman *et al.*, 1966; Stanford and Epstein, 1974). Robinson (1957) found little evidence of ammonification below the permanent wilting point (1.5 MPa) while Miller and Johnson (1964) and Reichman *et al.* (1966) found ammonification to proceed at matric suctions exceeding 1.5 MPa. Cameron and Kowalenko (1976) demonstrated the importance of a temperature: water content interaction term in quantifying microbially mediated ammonification. The effect of soil factors such as pH, salinity, and texture on ammonification has been studied, but good fundamental relationships have not yet been established (Nyborg and Hoyt, 1978; Laura, 1973, 1974).

Nitrification

Ammonium N mineralized from organic matter can be either assimilated by microorganisms and plants or oxidized to NO_3^- in the process termed nitrification. It appears that the traditional view that *Nitrosomonas* alone is responsible for NH_4^+ oxidation is no longer tenable. *Nitrosolobus* and *Nitrosopira* were the dominant NH_4^+ oxidizers in a range of soils examined by Soriano and Walker (1973). It has been suggested (Verstraete, 1979) that nitrification under low CH_4 conditions, and where the competition from chemolithotrophs (nitrifiers) is nonexistent due to poor environmental conditions, could well be due to methylotrophs. As with mineralization and immobilization, a vast amount of literature has accumulated on nitrification, albeit for temperate soil conditions.

Where inorganic N is derived chiefly from the ammonification of organic N, nitrification will rarely limit the rate of production of NO_3^-. Only at low and high pH (Morrill and Dawson, 1967), at high water potentials and temperatures (see Focht and Verstraete, 1977), or at water potentials below 1.5 MPa (Justine and Smith, 1962) might NH_4^+ or NO_2^- accumulate following ammonification. Thus, it is not surprising that the level of NH_4^+ and NO_2^- found in soils is generally low (Alexander, 1965).

As a rule, the optimum temperature for nitrification in soil falls between 25 and 35 °C. The rate of nitrification drops rapidly below 15 °C to almost zero at 0 °C (Alexander, 1965). There have been occasional reports of appreciable NO_3^- accumulation in soils at temperatures above 40 °C (Focht and Verstraete, 1977). However, it is not understood whether the production of NO_3^- follows heterotrophic nitrification which involves soil organic N or NH_4^+ oxidation by chemo-lithotrophs acclimated to nitrify at high soil temperatures (Mahendrappa et al., 1966).

The optimum soil water potentials for nitrification agree closely with those recorded for ammonification. The limitation of high water potentials reflects the obligate requirement for O_2 and the need for adequate gaseous exchange between the soil and the surrounding atmosphere (Alexander, 1965; Focht and Verstraete, 1977). Nitrate is not formed in air-dried soil nor is it produced at low soil moisture levels. However, the point at which nitrification ceases has not been well established. Justine and Smith (1962) have found low rates of nitrification at -1.5 MPa but no activity at -11.5 MPa. Little is known about the intermediate range of potential.

Nitrifiers also exhibit a remarkable ability to survive desiccation, at least at laboratory temperatures (Alexander, 1965), but whether this capability extends to the field where high temperatures often accompany desiccation is yet unknown. Kowalenko and Cameron (1976) demonstrated the existence of a temperature: water content interaction on nitrification in soils subjected to a range of mesophilic temperatures and water potentials above -1.5 MPa. Research is urgently needed to characterize the nitrification response to a range of thermophilic temperatures and low water potentials, for a range of soils representative of the arid zone.

The rates of NH_4^+ and NO_2^- oxidation often follow first order kinetics. Substrate inhibition of nitrification occurs but is pH dependent. The use of ammonium or ammonium-forming fertilizer such as urea in arid zone soils may lead to high concentrations of mineral N and, in the case of urea, to a temporary increase in soil pH (Hauck and Stephenson, 1965; Pang et al., 1975 a, b; Christianson et al., 1979). Thus, NO_2^- might accumulate where high NH_4^+ and associated pH inhibit Nitrobacter activity (Alexander, 1965). There is much less likelihood that ammonium oxidizers will be subjected to end-product inhibition, although high concentrations of NO_1^- (500–2000 mg l^{-1}) have been found to inhibit Nitrosomonas at low pH (Focht and Verstraete, 1977).

Denitrification

Several recent comprehensive reviews on denitrification are available (see Broadbent and Clark, 1965; Payne, 1973; and Focht and Verstraete, 1977). In brief, denitrification refers to the biological reduction of NO_3^- and NO_2^- to gaseous forms of N, namely N_2O and N_2. The most accepted pathway for denitrification is outlined by Payne, 1973:

$$NO_3^- \rightarrow NO_2^- \rightarrow NO \rightarrow N_2O \rightarrow N_2.$$

There appears to be little doubt that significant N loss occurs via denitrification in temperate soils. Key requirements include the development of anoxic microsites, a supply of oxidizable carbon, and the presence of NO_3^- or NO_2^- (Broadbent and Clark, 1965, Focht and Verstraete, 1977).

The reduction of the nitrogen oxides occurs in the absence of O_2 with each acting as an alternate electron acceptor to O_2. Energy generation is coupled to the reduction of NO_3^-, NO, and N_2O (Focht and Verstraete, 1977). A large number of NO_3^- respiring organisms ($NO_3^- \rightarrow NO_2^-$) have been isolated, but most are incapable of reducing NO_2^- further to N_2 (Payne, 1973). Focht and Verstraete (1977) list only 14 genera that are capable of NO_2^- reduction. Several reports also exist of organisms capable of reducing NO_2^- to N_2O but not to N_2 gas.

Since denitrification involves NO_3^- (or NO_2^-) and oxidizable C as substrate, it is not surprising that denitrification kinetics are complex and difficult to model. It is now recognized that the rate of denitrification is affected by the NO_3^- level when C is not limiting and NO_3^- levels are lower than 40 mg l^{-1} (Stanford et al., 1975a), while no effect of NO_3^- is found in C-limited systems or at NO_3^- concentrations above 40 mg l^{-1} (Starr and Parlange, 1975). Reduced denitrification rates have also been found with increased NO_2^- concentration (Galsworthy and Burford, unpublished data; Fillery, unpublished data), due possibly to a suppression in soil respiration at the higher NO_3^- concentrations (Kowalenko et al., 1979).

Several attempts have been made to describe the kinetics of denitrification using Michaelis-Menten expressions (Bowman and Focht, 1974; Doner et al., 1974). The application of Michaelis-Menten kinetics to the study of denitrification is discussed by Kohl et al. (1976). Clearly one expression alone is unlikely to describe denitrification because several different enzyme-controlled steps are involved. This feature may pose a problem if attempts are made to describe denitrification kinetics within model systems.

Satisfactory correlations between denitrification activity and soluble and mineralizable carbon (Burford and Bremner, 1975) and available C as assessed by glucose equivalent (Stanford et al., 1975a) have been reported. Since these studies represent denitrification activity following rewetting of previously airdried soil, such relationships may predict adequately the potential for denitrification in arid soils following precipitation. Doner (1975) has stressed that the solubilization of organic matter following rewetting of soils is a dominating factor on the transient state denitrification kinetics. The addition of crop residues, the decomposition of root material (Woldendorp, 1962), and the presence of exudates about living roots (Stefanson, 1972; Volz et al., 1976) are additional factors that will alter the availability of C and concomitantly the rate of denitrification.

Organic C level can also indirectly affect the rate of denitrification by altering the soil respiratory activity and consequently the extent of anaerobiosis. The development of anoxic microsites will be heavily dependent on the O_2 consumption rate, the O_2 diffusion rate, which is primarily affected by soil moisture status, and the geometry of the soil. Several attempts have been made to develop models which describe the aeration status of soils in relation to moisture contents, respiratory activity, and geometry (Greenwood, 1963; Smith, 1978).

Very few denitrification studies have been made on arid lands, possibly because of the general notion that anaerobiosis is rare in the soils from this region. However, it was recognized two decades ago that, under field conditions, poor O_2 supply to soil aggregates could result in localized anaerobiosis and denitrification (Allison et al., 1960; Burford and Millington, 1968; Dowdell and Smith, 1974). Soil moisture plays such an important role in governing soil aeration that production of gaseous N_2O often responds rapidly to incidental heavy rainfall (Burford and Millington, 1968). Cawse and Sheldon (1972) reported a rapid reduction of nitrate following rewetting of soils to a wide range of moisture contents well below saturation while nitrifiers continued to be active simultaneously. Some aspects of the variability patterns of aeration parameters of a soil were discussed by Fluhler et al. (1976a, b) in relation to simultaneous nitrification and denitrification.

The effects of temperature on the rate of denitrification are well characterized. Some workers have observed optimum rates at 35 °C (Bremner and Shaw, 1958; Stanford et al., 1975b), but Nommik (1956) and Keeney et al. (1979) report an optimum of about 65 °C. An exponential increase of rate is typically observed between 15 °C and 30 °C, but, below 12 °C, the exponential increase does not generally hold. Denitrification can occur at temperatures close to freezing (Stanford et al., 1975b). The reports of rapid denitrification at thermophilic temperatures are significant in studies of denitrification in 'hot' arid soils. Keeney et al. (1979) reported complete reduction of NO_3^- (100 μN g^{-1}) to gaseous N_2 within 12 h, when a silt loam soil was incubated under He at 60 °C. The soil in question was not amended with carbon nor air-dried before incubation. Lower rates of loss were found at 40 °C, yet 60 per cent of the NO_3^- was reduced within a 48 h period. Keeney et al. (1979) also report that the surprisingly high optimum for denitrification (65 °C) (Table 2) was likely due to a combination of NO_3^- reduction to NO_2^- and chemodenitrification reactions involving NO_2^-.

Table 2. Nitrogen balance at end of the incubation*

Temperature	Total incubation time	Nitrite + Nitrate-N	Gaseous N	Sum
°C	days	% of initial NO_3-N in system**		
7	16	91 ± 2	11 ± 0	102 ± 2
15	16	93 ± 5	12 ± 1	104 ± 2
25	16	61 ± 6	44 ± 7	104 ± 1
40	4	34 ± 2	63 ± 7	97 ± 8
50	4	0	125 ± 4	125 ± 4
60	4	0	134 ± 3	134 ± 3
65	4	0	127 ± 6	127 ± 6
67	4	0	143 ± 11	143 ± 11
70	4	0	109 ± 31	109 ± 31
75	4	98 ± 7	0	98 ± 7

* Keeney et al., 1979.
** 6.09 mg NO_3^--N 50 g soil.

Soil pH is an additional factor that may affect cycling of N via denitrification in the arid zone. It is generally accepted that denitrification is reduced at low pH, but the pH range for denitrification is normally much wider than that noted earlier for nitrification (Focht and Verstraete, 1977). Temperature and pH also affect the composition of the gaseous products of denitrification. A higher $N_2O : N_2$ ratio is observed at lower temperatures (Bailey, 1976; Nommik, 1956; Keeney et al., 1979) and in acid soils (Nommik, 1956; Blackmer and Bremner, 1978), while N_2 dominates at the higher temperatures and at close to neutrality. The inhibitory effect of NO_3^- on N_2O reduction is more pronounced at lower soil pH, but acid soils can reduce N_2O to N_2 in the absence of NO_3^- (Blackmer and Bremner, 1978).

ATMOSPHERE-SOIL INTERACTIONS

Nitrogen deposition

Ecosystems of the arid regions depend to a great extent on nitrogen inputs from the atmosphere to compensate for losses of nitrogen from the soil: plant systems. The most important mechanisms of atmospheric N input are biological N fixation and N deposition through resorption, precipitation, and fallout. Re-

Table 3. Summary of some levels of gaseous and particulate forms of nitrogen in the atmosphere*

Site and Measurement Interval	NH_3	N_2O	NO	NO_2	Aerosol** NH_4^+	NO_3^-
	ppb(v/v)				$\mu g\ m^{-3}$	
Nonurban air***						
Temperate land	2–6	300–330	2	4		
Tropical land	6–15	300–330	2	4	2.0	0.1–0.4
Oceans	0.4	300–330	0.2	0.5	0.4	0.02
Annual average (U.S.)					0.19	1.10
24-hr maximum (U.S.)					1.52	3.24

* NRC, 1978.
** Recent findings indicate that values for nitrate aerosols may include substantial amounts of nitric acid vapor.
*** Background levels calculated by sources cited from available data, with measurements over various time intervals.

latively little is known about the role and magnitude of N deposition, but interest in the subject is increasing as a result of recent emphasis on anthropogenic influences on the environment.

The nitrogen compounds in the earth's atmosphere comprise gases such as NO, NO_2, N_2O and NH_3, as well as particulate forms including NO_3^- and NH_4^+ aerosols. Table 3 summarizes some of the pertinent atmospheric levels of these compounds. Nitrogen concentrations in rainfall are highly variable due to a host of factors that are fairly well understood qualitatively (Brezonik, 1976). These factors include climatic conditions, frequency of rainfall, and edaphic and other natural factors. Anthropogenic sources can contribute significantly to the nutrient load in rain. Atmospheric transport processes can move air masses hundreds of kilometers a day. Since the residence times of ammonia and nitrogen oxides are several days, local emissions of N can become dispersed throughout a region.

Global terrestrial inputs from resorption, precipitation, and fallout are presently estimated at 66 to 200 million t yr^{-1} (NRC, 1978) and are thus of the same order of magnitude as the input from biological N fixation (Burns and Hardy, 1975). Uncertainties in these estimates are large, particularly for N deposition, because data on atmospheric loading through woodburning is scarce. Brezonik (1976) estimated N deposition to range from 10 to 20 kg N $ha^{-1}yr^{-1}$ over large areas of USA, with rates outside the range of 5 to 30 kg N $ha^{-1}yr^{-1}$ being exceptional. Deposition of gaseous forms of N, primarily NH_3, may constitute a major portion of the terrestrial inputs, either through sorption by the soil (Rodgers, 1978) or through foliar absorption by the plant (Denmead et al., 1976) in NH_3-enriched atmospheres (Farquhar et al., 1979).

Regional variation in nitrogen deposition in USA shows lowest deposition rates of less than 5 kg N ha^{-1} yr^{-1} in the arid regions of the West. This low rate of deposition may be associated with the alkalinity of the soils in this region, preventing extensive NH_3 sorption by the soil. West (1975) estimates the N deposition for arid regions on a worldwide basis to be 12.5 kg ha^{-1} yr^{-1}; 3.5 times the input through biological fixation (see Table 1). N deposition by washout alone averaged 40 g ha^{-1} of NH_4-N for each millimeter of rain for seven stations in Israel, amounting to contributions from 4 to 20 kg ha^{-1} yr^{-1} (Yaalon, 1964). A review of data in the literature by Harpaz (1975) suggests that the annual input in rain for semi-arid climates averages about 5 kg N ha^{-1} and is of the same order of magnitude as reported values of annual fixation by nonsymbiotic microorganisms.

Fertilizer inputs

The chemical synthesis of nitrogenous fertilizers is an exclusively human contribution to N fixation. Few figures are available for data on the relative importance of fertilizer N inputs in rainfed agriculture in the arid zone of the Middle East and North Africa. For the 25 countries considered in this region, the total 1977 nitrogen consumption amounted to 2.298 million tonnes (Table 4) with a total area of more than 65 million ha in food crop cultivation (FAO, 1979). If all this were applied to food crops, the mean yearly fertilizer input would be around 35 kg N ha^{-1}. However, such calculations are misleading since the regional N use is dominated by the consumption in Egypt, Pakistan, Iran, and Turkey where the bulk of the nitrogen is applied on irrigated crops. For instance, Egypt with 99 per cent of the 2.5 million ha of cultivated land under intensive cultivation applies an average of about 120 kg N ha^{-1} per cropping season. With a cropping intensity of nearly two, it consumes about 0.5 million tonnes of N or about 20 per cent of the regional total. A better perspective of the present fertilizer use in arid zone agriculture is obtained from fertilizer statistics of some representative countries, collected by a joint exploration mission on regional development of fertilizer industry in Arab states (UNDP, 1977).

The Syrian Arab Republic has 6 million ha of potential arable land of which one-third is suited for intensive agricultural development with ample water resources. Wheat and barley are the major crops, and they are grown under a wide range of conditions. With decreasing rainfall, fallow is an increasingly

Table 4. Nitrogen production and consumption (1977) and projected consumption (1985) for Middle East and North Africa (metric tons)*

	N-production 1977/78	N-consumption 1977/78	N-consumption 1985 projection	Potential N-production in 1985**	1977 N Use (kg ha^{-1}) Agricultural area	Arable land and perm. crops
Algeria	41 831	66 400	142 500	462 700	1.5	8.8
Chad	0	2 940	4 000	0	.1	.4
Egypt	195 171	459 504	594 500	523 300	162.3	162.3
Libya	0	16 700	25 800	308 400	1.8	6.6
Mali	0	7 900	9 000	0	.2	.8
Mauritania	0	1 900	2 200	0	–	9.5
Morocco	14 600	75 200	139 000	918 000	3.7	9.6
Niger	0	200	500	0	–	
Sudan	0	32 000	159 000	30 600	1.0	4.3
Tunisia	5 000	17 200	36 000	154 200	2.2	3.9
Afghanistan	37 639	37 014	50 800	32 900	2.7	4.6
Cyprus	0	12 000	20 200	0	22.9	27.8
Iran	177 850	189 206	385 700	510 900	7.0	11.9
Iraq	125 460	45 000	70 000	461 500	4.8	8.5
Israel	50 555	38 395	46 500	44 100	30.8	89.3
Jordan	0	1 400	3 700	0	1.0	1.0
Lebanon	1 200	13 700	25 300	0	38.3	39.4
Oman	0	200	1 700	93 000	.2	5.6
Pakistan	307 674	550 800	837 000	907 200	21.8	27.1
Saudi Arabia	92 800	4 708	6 850	92 400	.1	4.2
Syria	24 180	57 200	70 900	177 500	4.1	10.4
Turkey	187 300	665 700	1 189 000	940 700	12.0	23.8
United Arab Emirates	0	500	1 400	306 700	2.4	45.5
Yemen AR	0	1 894	2 850	0	.2	1.2
Yemen DEM	0	100	1 200	0	–	.4
Total	1.261 310	2.297 761	3.825 600	5.964 200		

* FAO, 1979.
** Based on 70 per cent operating rate and allowing for losses and industrial usage.

important practice. The 1977 consumption of N was 42 500 t (principally as ammonium nitrate), mostly used on cash crops such as cotton and vegetables in the areas with good water supply. With cash crops already receiving near optimum fertilizer rates, part of the planned growth in N consumption (1985 projection, 220 000 t of N) is likely to take place on grain crops (UNDP, 1977).

More detailed information is available for Morocco with 7.3 million ha of land. Nitrogen consumption in 1975 was 62 000 t with an average use of 11 kg N ha^{-1}. Again, most of the fertilizer was used ȯn irrigated lands and rainfed crops in higher rainfall areas. As much as 44 per cent of the fertilizer was used on cereal areas covering three-fourths of the cultivated land, half of which is in the > 400-mm rainfall zone. Similar calculations for Tunisia, where irrigation systems are sparse, yield an average N use of 3 kg ha^{-1} for cereal crops; more than 80 per cent is used in the northern 60 per cent of the cultivated land with 400 to 600 mm rain. Thus, fertilizer use in true arid-zone agriculture is almost negligible, although good responses to N and P can be obtained in rainfed wheat and barley production (UNDP, 1977).

Approximate figures for the nitrogen fertilizer dominating in the region in 1977, were urea, 40 per cent and ammonium nitrate 35 per cent with minor contributions from ammonium sulfate ($\sim 10\%$) and ammonium phosphates ($\sim 10\%$). This situation is expected to shift more in favour of urea since most of the planned production capacity in the region is urea. How much of this future production is for local use, will depend to a large extent on national policies for agricultural development and the world fertilizer situation (Harris and Harre, 1979).

Volatilization losses

Nitrogen losses to the atmosphere from agricultural soils are comprised of gaseous emissions of NH_3, N_2O, NO_x, and N_2, although incidental losses of particulate nitrogenous compounds due to wind erosion do occur in arid climates. The relative contribution of wind erosion appears limited, and generally leads to particulate fallout in the vicinity of the source. However volatilization losses from arid-zone arable lands are considered to be significant (West, 1975).

The relative importance of ammonia volatilization and denitrification depends on a host of factors. The complexity of the processes involved and the paucity of direct measurements, make generalizations on the magnitude of these losses risky.

NH_3 volatilization. The volatilization of NH_3 can be regarded as a chain of events, the overall rate of which can be controlled by any one link in the chain

represented by

$$NH_4^+ \rightarrow NH_3(aq) \rightarrow NH_3(g) \rightarrow NH_3(atm),$$

where NH_4^+(aq) depends on soil cation-exchange reactions (Fenn and Kissel, 1976; Gasser, 1964), soil moisture content (Ernst and Massey, 1960; Fenn and Escarzaga, 1976), and net mineralization. Conversion from NH_4^+ in solution to aqueous NH_3 is an extremely rapid (first-order) process with a rate constant of 24.6 sec^{-1} (Emerson et al., 1960) and is thus rarely limiting. The concentration of NH_3(aq) changes proportionally with ammoniacal N, approximately linearly with temperature (Craswell and Vlek, 1980), and increases about 10 fold per unit increase in pH up to pH 9 (Vlek and Stumpe, 1978).

Equilibrium between aqueous NH_3 and gaseous NH_3 is governed by Henry's Law with K_H ($K_H^\circ = 0.0164$) a function of temperature (Beutier and Renon, 1978). Whether equilibrium between NH_3(aq) and NH_3(g) is maintained, depends on the rate of NH_3(g) evasion from solution and the rate of NH_3(g) transfer away from the source-sink interface. In still air, the partial pressure gradient of NH_3 determines the rate of gas dispersion, whereas in natural environments, temperature gradients and wind accelerate the transfer resulting in dispersion coefficients 10 to 600 times higher than the diffusion coefficient in still air (Inoue et al., 1975). Bouwmeester and Vlek (1980) developed a theoretical (diffusion) model in an effort to assess the effect of various chemical and environmental parameters on the rate of NH_3 loss from flooded soils. A typical application of the model is presented in Fig. 4 giving the variations in NH_3 volatilization rate (Q) with change in wind velocity, water temperature, fetch, and pH for floodwater containing 100 ppm of total ammoniacal N (TAN). Development of a similar model for dryland conditions that incorporates the role of soil moisture, would greatly increase our understanding of the transfer of NH_3 from these soils.

The actual rate of ammonia loss from a soil or water body serving as a source thus mainly depends on the NH_3(aq) concentration at the source-air interface (Du Plessis and Kroontjes, 1964), which in turn depends on the total ammoniacal N concentration (TAN), pH, temperature, and moisture content of the soil. The TAN level in the soil or soil solution varies with time due to soil N transformation (immobilization, mineralization), soil water dynamics (rainfall or evapotranspiration), and natural or anthropogenic ammoniacal input (deposition and fertilization).

Measurements of ammonia volatilization losses have concentrated on losses from ammonia-fertilized fields or from highly polluted environments such as feedlots or eutrophic lakes. A notable exception is the recent work by Denmead et

150

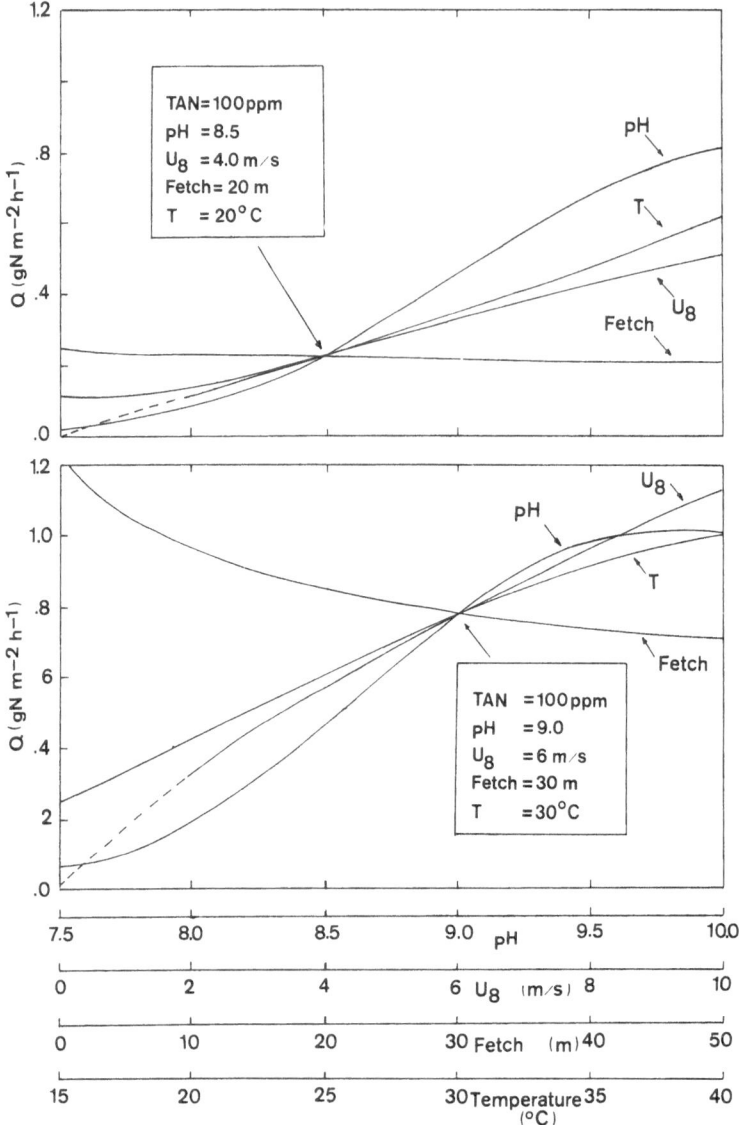

Fig. 4. Variation of NH_3 volatilization rate from shallow water with change in wind velocity (u_8), temperature (T), Fetch (F), and pH.

Table 5. Summary of N-source effect on NH_3 loss

Fertilizer	Soil pH	Soil CEC	N Rate	N Loss*	Source	Airflow
		meq 100 g^{-1}	kg ha^{-1}	%		
Urea	6.3	NA	168	19.3	Kresge and Stachell (1959)	NA
Ammonium sulphate	6.3	NA	168	1.1		
Ammonium nitrate	6.3	NA	168	0.6		
Urea	6.7	6.6	500	36.0	Martin and Chapman (1950)	NA
Ammonium sulphate	6.7	6.6	500	4.0		
Ammonium nitrate	6.7	6.6	500	1.0		
Urea	7.1	14.6	500	18.0		
Ammonium sulphate	7.1	14.6	500	2.0		
Ammonium nitrate	7.1	14.6	500	1.0		
Urea	7.5	16.2	500	16.0		
Ammonium sulphate	7.5	16.2	500	19.0		
Ammonium nitrate	7.5	16.2	500	7.0		
Urea	7.7	13.9	500	14.0		
Ammonium sulphate	7.7	13.9	500	14.0		
Ammonium nitrate	7.7	13.9	500	7.0		
Urea	7.8	21.0	110	17.0	Meyer, et al. (1960)	NA
Ammonium nitrate	7.8	21.0	110	4.0		
Urea	7.7	17.0	450	13.0	Harding et al. (1963)	27 l min^{-1} 432 liter headspace
Ammonium sulphate	7.7	17.0	450	8.0		
Ammonium nitrate	7.7	17.0	450	1.0		
Urea	8.0	21.8	110	6.8	Gasser (1964)	NA
Ammonium sulphate	8.0	21.8	110	9.6		
Urea	8.0	15.1	110	10.6		
Ammonium sulphate	8.0	15.1	110	12.1		
Ammonium sulphate	7.6	58.0	110	33.0	Fenn and Kissel (1976)	8-9 l min^{-1} 0.57 liter headspace
Diammonium phosphate	7.6	58.0	110	33.0		
Ammonium nitrate	7.6	58.0	110	18.0		

* Determined in a forced-draft system.

al. (1974, 1976) who measured NH_3 losses in pastures. These reports are relevant to arid-zone agriculture mainly from a standpoint of methodology and understanding of the principles governing the rate of ammonia volatilization. Still conditions appear to prevail in many of the NH_3 volatilization measurements, conducted in the past, causing great difficulty in interpretation. Ammonia volatilization measurements should thus be conducted in undisturbed fields (Denmead *et al.*, 1977) or in systems proven to be representative of a nonrestrictive environment (Kissel *et al.*, 1977; Vlek and Craswell, 1979).

With the eventual introduction of fertilizers, NH_3 losses from applied fertilizers will need to be given full attention. Table 5 summarizes records collected by various researchers using the forced-draught system. Although there is much variation in the data, a trend with respect to ammonium source and soil pH is easily discernible. Irrespective of soil pH, urea is lost from all soils because, upon hydrolysis, it serves as an effective alkaline buffer (Vlek and Stumpe, 1978). Ammonium sulphate becomes less efficient as a source of N with increasing soil pH and is inferior to urea in soils containing $CaCO_3$ as pointed out by Avnimelech and Laher (1977), Musa (1978), and Fenn and Kissel (1975). From the standpoint of NH_3 volatilization, fertilizers for the low rainfall areas of the Mediterranean should preferably be nitrate carriers.

Ammonia volatilization on a global scale is estimated at 170×10^6 t annually or a yearly average of 10 kg N ha^{-1} (Burns and Hardy, 1975). Measurements of ammonia volatilization from natural arid ecosystems are lacking, possibly reflecting the inadequacy of suitable techniques and the general notion that NH_3 volatilization under these conditions is not an important loss mechanism (Husz, 1977; Noy-Meir and Harpaz, 1977). Conditions favoring NH_3 volatilization from arid or semi-arid ecosystems with winter rain occur only at the end of the growing season (high temperatures and late spring rains) and in urine patches throughout the summer. Sheep urine generally contains 0.92 per cent N, of which 76 per cent is urea (Doak, 1952). One sheep excretes about 7 kg N annually, leaving urine patches that may contain N at about 75 kg N ha^{-1}. Future studies will need to assess the magnitude of NH_3 loss.

NO_x and N_2 volatilization. The loss of nitrogen from the plant: soil system in the form of nitrogen or nitrogen oxides has been studied in detail. The processes bringing about such losses which are generally labelled 'denitrification' were discussed in the previous section. More recently, nitrous oxide production has also been found to occur during nitrification (Bremner and Blackmer, 1977; Freney *et al.*, 1979), a process generally confined to oxidized environments. There appears to be little doubt that nitrogen and nitrogen oxide gases primarily form as a result of nitrate and nitrite reduction, be it chemically or biochemically (Broadbent and Clark, 1965; Focht and Verstraete, 1977).

The gaseous products of NO_3^- and NO_2^- reduction emitted from soils are highly variable. Nitrogen gas is commonly considered the principal gaseous product, while nitrous oxide can be a major product in partially oxidized soils, particularly following heavy N fertilization (Rolston, 1977), and in soils of low pH (Nommik, 1956). Many researchers have failed to identify NO as an intermediate, though recent evidence shows that NO at low concentrations in the field (~ 0.01 ppm) can pass unoxidized from the soil to the atmosphere (Galbally and Roy, 1978). At high concentrations (> 100 ppm), NO is readily oxidized to NO_2 which might explain why the latter was found to predominate in some laboratory studies. Under reduced conditions, NO is readily converted to N_2O (Nelson and Bremner, 1970).

Assessment of the magnitude of N_2O and N_2 emissions from soils poses a formidable problem. One of the major experimental problems is that N_2 is a major constituent of air, while analysis of N_2O in ambient air is far from simple. Estimates of NO, N_2O and N_2 loss from field soils have been based largely on deficit measurements in a nitrogen balance. A complete nitrogen balance takes into account all fluxes into and out of a soil that are known to be significant, while the deficit is assumed to be due to N_2O and/or N_2 volatilization. The fluxes to be considered are those presented in Fig. 1. The introduction of ^{15}N tracer techniques has greatly facilitated nitrogen balance studies during the past few decades. This sensitive technique is particularly suited to study the fate of chemical fertilizer nitrogen added to soil (Hauck and Bremner, 1976) but can also be used to follow the fate of ^{15}N spiked plant residues or manure through the plant: soil system (Westerman and Tucker, 1979).

Estimates of global denitrification losses are scanty, and mostly made by inference. Estimates vary widely between ecosystems (see Table 1; Hauck, 1971). Besides, denitrification losses are far from constant in time due to changing environmental conditions, primarily temperature and water content (Nommik, 1956; Craswell and Martin, 1974; Westerman and Tucker, 1978). These factors are of particular interest to arid and semi-arid winter rainfall ecosystems where soil moisture and temperatures go through annual cycles represented in Figure 2.

Hauck (1971, 1977) summarized N-balance studies reported in the literature and tentatively concluded that N deficits of 20 to 30 per cent are representative. Whether these deficits are actually N_2 and N_2O loss in all cases is questionable, particularly since losses through NH_3 volatilization and N leaching are generally assumed to be negligible. The reliability of estimates obtained by difference is, at best, no better than the reliability of the other measurements in the nitrogen balance.

Some studies more relevant to Mediterranean-type environments have ap-

154

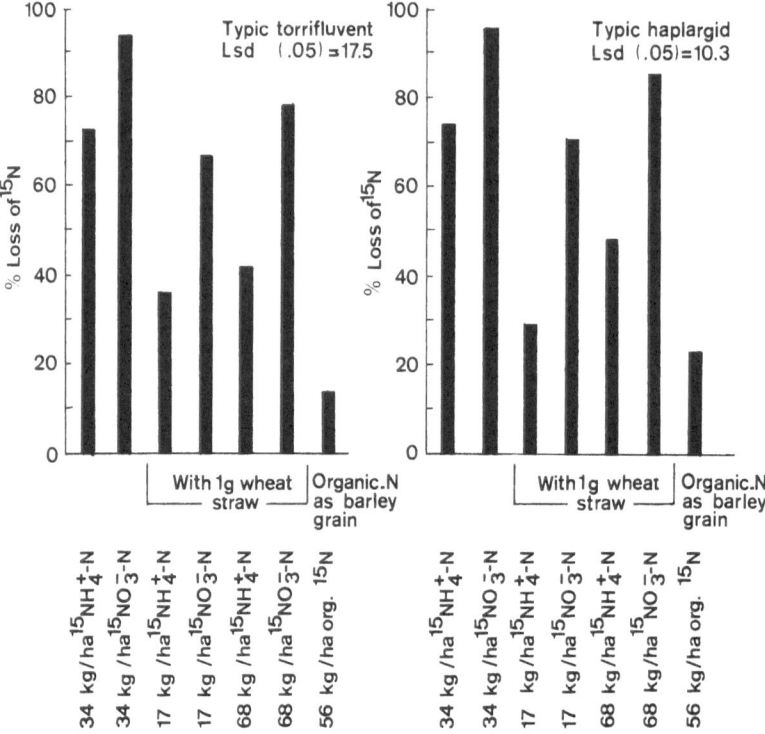

Fig. 5. Loss of ^{15}N from the top 10 cm of two U.S.-desert soils during 12 months following application to field microplots (from: Westerman and Tucker, 1979).

peared recently (Eberhardt and Skujins, 1973; Westerman and Tucker, 1979; Craswell, 1978; Ganry et al., 1978). Westerman and Tucker studied denitrification from Sonoran Desert soil using ^{15}N. They traced nitrogen applied to undisturbed microplots (diameter 4.8 cm) with N added at rates up to 83 kg/ha in the form of NO_3^-, NH_4^+, organic N, or combinations thereof. Following an initial wetting equivalent to 41 mm of rain, rainfall during the first month was 32.5 mm and was not significant during the following 11 months. Fig. 5 shows the significance of denitrification losses 12 months from the time of application of N. Most of these losses occurred during the first three months. Contrary to earlier belief, denitrification of added mineral N and organic N following mineralization was rapid and extensive, particularly if added as nitrate. In a similar study, Craswell (1978) failed to observe denitrification losses from black clay soil amended with NO_3^- and maintained at a soil water content of $0.56\,\mathrm{g\,g^{-1}}$ ($10\,\mathrm{kPa}$), but ^{15}N loss, assumed due to denitrification, was rapid at $0.72\,\mathrm{g\,H_2O\,g^{-1}}$ soil ($\cong 80$ per cent saturation), provided soil temperature was favorable. Ganry et al.

155

Table 6. Nitrogen balance studies in semi-arid regions

	N applied, kg ha^{-1}	Fate of N-15				Note
		% Soil	% Grain + straw uptake	% Loss	Grain and straw % Ndff**	
Wheat (Myers and Paul, 1971)	56 NH$_4$NO$_3$	39.4	25.2	35.4	43.0	Silt loam
	112 NH$_4$NO$_3$	45.3	21.4	33.3	55.0	straw levels
	46 oat straw	79.1	11.3	9.6	19.0	
	112 NH$_4$NO$_3$	28.2	51.1	20.7	51.5	Clay; Two straw levels
Wheat (Craswell and Martin, 1975)	112 Ca(NO$_3$)$_2$	31.6	43.4	25		Clay, wet
	112 Ca(NO$_3$)$_2$	31.3	62.7	6		Clay, dry
(Craswell and Strong, 1976)	96 Ca(NO$_3$)$_2$	30.8	59.0	10.2	51.6	Clay, moist
		54.6	37.9	7.5	44.2	Clay, dry
Wheat (Olson et al., 1979)	50 (NH$_4$)$_2$SO$_4$	36.4	44.2	19.4	22.2	Typic Argiudoll
	100 Fall	28.7	47.9	23.4	36.7	
	50 (NH$_4$)$_2$SO$_4$	33.4	46.0	20.6	26.4	
	100 Spring	23.2	57.1	19.7	44.2	
Barley (Korenkov et al., 1975)	90 NH$_4$OH	29	41	30	44	Sandy loam
	90 (NH$_4$)$_2$SO$_4$	24	45	31	50.0	
Pearl millet (Ganry et al., 1978)	90 Urea	35.4	17.9	46.7	31.7	Ferruginous sand
	150 Urea	31.6	18.6	49.8	45.3	levels of these straw

* Nitrogen derived from fertilizer at harvest.

(1978) report ^{15}N losses from a urea-fertilized sandy soil in Senegal to vary from 39 per cent to 57 per cent of the applied N during a growing season, receiving 392 mm of rain. Most of these losses were attributed to denitrification.

With soils containing fairly large quantities of mineral N following the winter rainfall of the Mediterranean climate, denitrification losses may be substantial. Whenever such soils are cultivated the standing crop serves as an effective competitor to the nitrate reducers in the soil (Ganry *et al.*, 1978). Table 6 provides a summary of complete ^{15}N-balance studies conducted in arid and semi-arid agricultural systems. Nitrogen losses vary greatly, but the limited data collected to date do not permit identification the processes which are responsible for this variation. Properly conducted field studies are needed to assess the magnitude of gaseous loss from arid and semi-arid agricultural systems as a function of climate and soil type.

SOIL: SUBSOIL INTERACTIONS

Movement of soluble nitrogen compounds through soil has been studied extensively over the past few decades. Excellent reviews on the subject have appeared (Harmsen and Kolenbrander, 1965; Gardner, 1965; Thomas, 1970). Relatively little attention has been given to N movement in rainfed agricultural systems of arid and semi-arid regions, possibly because leaching of N beyond the crop rooting zone is often considered negligible (Noy-Meir and Harpaz, 1977; Van Keulen, 1974; Westerman and Tucker, 1979). However, even in the absence of excessive leaching of N, information concerning the upward and downward movement of N in the root zone is a prerequisite for the management of arid and semi-arid soils and crops (Nielsen, 1976).

Nitrogen movement up and down the soil profile may induce a surface accumulation of nitrate or loss from the root zone, depending on the distribution, timing, and intensity of rain events as well as soil characteristics such as soil depth, texture, and structure (Wetselaar, 1961, 1962). Relatively dry soils have extremely low conductivities in the topsoil, thus preventing water (and solute) movement to the soil surface. If the capillary conductivity is large, as is the case in wet soils, the upward water movement due to evaporation will be rapid; leading to accumulations of nitrate in the topsoil. Conversely, the limited capacity of an initially wet profile to store additional water will generally lead to rapid leaching of nitrate.

Although water infiltration into dry soil is relatively fast, solutes are moved more efficiently (per unit of water flow) when water moves slowly and the water content is low. Thus, even though more water enters the soil than evaporates, soluble salts tend to accumulate near the soil surface (Nielsen, 1976). This is because infiltration occurs primarily through macropores that are relatively devoid of nitrate, while nitrate-rich micropores are responsible for capillary rise.

The total loss of nitrogen due to leaching is equivalent to the sum of the fluxes of the individual nitrogen compounds at the root zone: substratum interface, i.e.

$$J_{NT} = \sum J_W C_N + \sum J_D \tag{4}$$

with J_{NT}, the total nitrogen flux; J_W, the flux of water into the subsoil; and C_N, the concentration of the individual nitrogen solutes in the percolating water. Diffusion is represented by J_D. Native soil nitrogen in well-drained soils is generally leached as nitrate, with only a small fraction in the form of ammonium (Low, 1973; Myers, 1975).

Ammoniacal N not only has a greater affinity for the soil-exchange complex but is susceptible to nitrification as well. Fertilizer N leached from freely drained soils is lost largely as NO_3^-, while appreciable quantities of NH_4^+-N and urea-N may be leached from urea-fertilized soils (Terman and Allen, 1970). Small quantities of N may be drained from soils in the form of dissolved gas, N_2O, and N_2 (Dowdell et al., 1979).

Gardner (1965) reviewed the knowledge about nitrogen movement in soils. He pointed out that the mathematics of the dispersion of ions simultaneously involved in various reactions are complex. Chemical transformations, biological immobilization, mineralization, and denitrification can alter the concentration of solutes on their way out of the rootzone (Ardakani et al., 1973; Starr et al., 1974; Misra et al., 1974a, b, c). Carrying the existing mathematical models beyond the laboratory column studies that they were tested on, is usually difficult because of the anisotropic properties of field soils.

The inapplicability of mathematical approaches to predict the leaching of nitrogen from root zones has stimulated the development of empirical models. Burns (1975, 1976) developed a simple equation to predict leaching of topdressed or incorporated nitrate, assuming layer-to-layer piston flow when the water content exceeds field capacity. The model neglects plant uptake and denitrification but seems accurate if such conditions prevail. Terry and McCants (1970) utilized a multiple regression analysis to predict leaching of ions. Observing that leached ions are generally normally distributed with depth, they developed a regression model with the quantity of percolated water, soil porosity. CEC, and various waterholding indices as independent variables. The model was tested for various soils in column studies and worked well.

158

Generalizations concerning the movement of N in the soil for an entire agroclimatic zone such as the Mediterranean dry region is of little value. Actual losses depend largely on the level of mineral N in the profile at the latter part of the rainy season. Rainfall of high intensity during this period, when capillary conductivities are relatively large, may cause leaching of mineral N beyond the root zone, particularly if the soil is shallow. Average precipitation values for wet days (> 2 mm rainfall) for Mediterranean-type climates do not exceed 20 mm and are generally lower than 12.5 mm (Harris, 1978; Bell, 1979). Yet, in these higher latitude desert areas with winter cyclonic precipitation, maximum daily falls between 25 per cent and 60 per cent of the mean annual total have been recorded at most stations (Bell, 1979; Rainey, 1977).

Hingston (1974) estimated water drainage during the first month of wheat growth on five coarse-textured sandplain soils in Western Australia with mean annual rainfall ranging from 300 to 550 mm. He concluded that only in one out of the five soils was leaching below 90 cm likely. Tourte *et al.* (1964) showed that under semi-arid summer rainfall conditions in Senegal, leaching of indigenous soil nitrogen occurred only early in the drainage (rainy) period, when mineralization exceeded immobilization, and was greatly affected by the degree of competition from root uptake. For instance, the amount of N leached beyond 40 cm

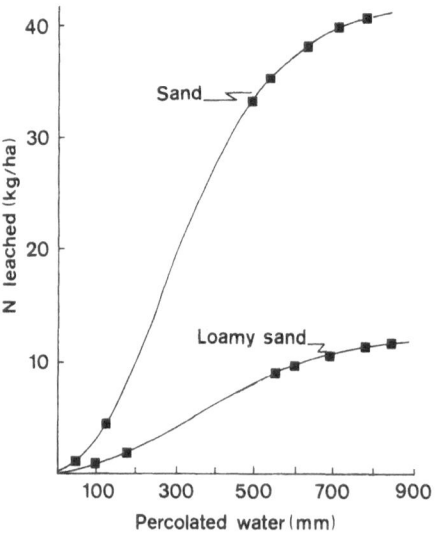

Fig. 6. Accumulated N loss due to leaching as a result of water percolation through field lysimeters filled with two Galician soils planted in *Lolium perenne* (from: Diaz-Fierros *et al.*, 1973).

soil depth amounted to 1.6 kg ha^{-1} under a mixed crop of millet and green manure and 6.4 kg ha^{-1} under natural fallow. Loss under bare fallow was 45.6 kg N ha^{-1}. The addition of chemical fertilizer [KNO_3 or $(NH_4)_2SO_4$] greatly increased the amount of nitrogen lost (Blondel, 1971).

Although not entirely representative, some data from fertilized sandy soils in Spain serve as an illustration of N leaching in a winter rainfall area (Fig. 6). There is a close relationship between the amounts of water and nitrogen leaving the root zone. The actual amount lost varies with soil type, in this case primarily C : N ratio of the organic matter fraction (Diaz-Fierros *et al.*, 1973). Similar results were reported by Kissel *et al.* (1974) for a swelling clay soil. However, of the ^{15}N-labelled NO_3^- added to an Australian cracking clay soil at 10 cm depth, and receiving about 188 mm of rain over a 16-week fallow period, more than 95 per cent was recovered within the top 40 cm of the profile (Craswell and Martin, 1975). Essentially similar results were found in column experiments, except that N displacement in the presence of the crop was substantially reduced (Craswell and Strong, 1976). Model experiments such as these, conducted on representative soils under a variety of agroclimatic conditions of the Mediterranean region, will help to select soil, crop, and fertilizer management systems that optimize utilization of indigenous soil N and fertilizer N.

SOIL: CROP INTERACTION

The critical factor in the management of the nitrogen economy of soils in arid and semi-arid regions is the assessment of whether there will be enough net accumulation of mineral N for the requirements of crops. The N mineralized will depend not only on the abundance of soil organic matter but also on the incorporation of residues from a previous crop. Information about shortfalls in nitrogen is needed, by empirical experimentation if not otherwise, so that the balance can be provided by the application of nitrogen as fertilizer.

The demands for nitrogen by crops in arid Mediterranean-type climates vary widely with location, cropping system, and year-to-year weather conditions. For instance, large fluctuations in dry matter production of the natural vegetations of the northern Negev resulted in total N uptake varying from nil to 100 kg N ha^{-1} (Van Keulen, 1977) with an estimated average of 50 kg N ha^{-1} (Noy-Meir and Harpaz, 1977). These estimates are high compared with N uptake by native vegetation of the Northern Plains of the U.S.A. where Black (1968) recorded an average N uptake of 15 kg N ha^{-1}, reflecting a lower productivity level in the latter region. Wallace *et al.* (1978) estimated a yearly uptake of 10 to 30 kg N ha^{-1} for the native vegetation in the northern Mohave desert.

A typical wheat crop of the dryland farming systems in the northern Negev is estimated to absorb ~ 25 kg N ha^{-1} (Noy-Meir and Harpaz, 1977) for a dry matter production of about 2.5 t ha^{-1}, which compares with $\cong 20$ kg N ha^{-1} for a similar dry matter yield for the American Great Plains (Ramig and Rhoades, 1962). Dry matter yields of 5 t ha^{-1} reported for the Sudan, required 50 kg N ha^{-1} (Khalifa et al., 1977). Nitrogen demands by barley seem to be slightly higher (30–40 kg N ha^{-1}) with less dry matter produced (Luebs and Laag, 1967; Power et al., 1973).

Supplemental irrigation and/or application of fertilizer N may significantly increase N uptake by these crops. For instance, in dryland wheat experiments (coordinated by IAEA, 1974) in five countries during the 1970–71 growing season, the average N uptake in the control plots was 34.6 kg N ha^{-1} while N fertilization at 120 kg N ha^{-1} more than doubled the N content of the above-ground plant parts (72.5 kg N ha^{-1}). In similar studies, in seven other countries with wheat irrigated at 50 per cent or 75 per cent available water depletion, nitrogen uptake averaged 64.1 kg/ha with a slightly higher uptake (68.5 kg N ha^{-1}) for the higher moisture treatments. Varma (1976) reported N uptake of 60.1 kg N ha^{-1} for wheat fertilized at a level of 75 kg N ha^{-1} if water stressed as compared to 70.7 kg N ha^{-1} uptake if properly irrigated. Comparable figures for barley were 44.2 and 46.9 kg N ha^{-1}, respectively. Nitrogen uptake by leguminous crops varies between species. Some typical levels were given by Varma (1976) for mung (124.5 kg N ha^{-1}), pea (56.6), and cowpea (90.7). The contribution of the soil N to the N-status of these crops is difficult to assess and will, in general, depend on the success of inoculation.

The actual N uptake by a crop may be limited by the nitrogen-supplying power of the soil. Probably the most conclusive study establishing soil-N as a major constraint to dry-matter production in a Mediterranean winter rainfall climate was reported by Van Keulen (1977). At a northern Negev location (average rainfall, 250 mm) computer simulation indicated that in six out of 12 years, nitrogen availability limited plant dry-matter production, while water was the primary yield-limiting factor in the remaining years. The principles of these findings have been shown to apply to agricultural cropping situations as well. The dependence of crop response to N-fertilization on the availability of moisture in arid-zone agriculture is well documented (Bauer et al., 1965; Singh and Ramakrishna, 1977; Ramig and Rhoades, 1962; Russell, 1967). Fig. 7 from Ramig and Rhoades serves as an illustration of this relationship.

Many attemps have been made to identify a 'nitrogen availability index' that adequately describes the dynamic nitrogen status and yield potential of soils and predicts N-response to fertilizer (Scarsbrook, 1965). Incubation methods are

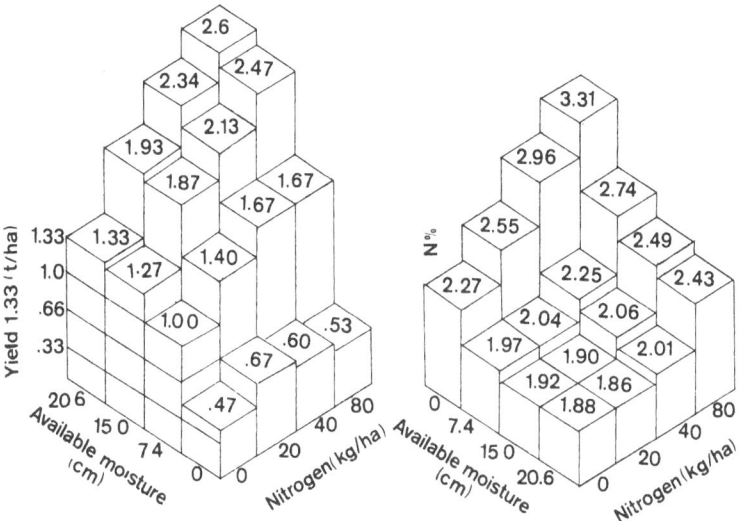

Fig. 7. Effect of preplanting soil moisture and rate of N fertilization on the yield and %N in the grain of wheat following wheat (from: Ramig and Rhoades, 1962).

most widely used despite their limitations (Bremner, 1965). Various chemical methods have been proposed for assessing available N in soils (Bremner, 1965; Keeney and Bremner, 1966; Stanford and Smith, 1976) while others (Storrier *et al.*, 1971) used a combination of readily and potentially available N as an index of yield potential. Taylor *et al.* (1974) compared several biological and chemical soil nitrogen indices on as many as 778 locations under a range of arid and semi-arid conditions. They found indices of readily available nitrogen (nitrate or total mineral nitrogen) to separate the low and high-yielding sites under all rainfall conditions in the semi-arid region, but the tests for potentially available N did not. Similar results were found by Tejeda and Gogan (1970) for the Mediterranean environment of Chile.

Recently, attention has been drawn to the benefits of residual N in the subsoil, particularly for late-season uptake by deep-rooting crops in dry areas (Cooke, 1979). Often therefore, the amount of nitrate at depth is an even better index of how well the N-demands by the following crop can be met (Soper *et al.*, 1971; Taylor *et al.*, 1974). Cultural practices such as crop rotation, tillage, manuring, fallowing, and time of planting all help to determine the extent to which the availability of N in the soil profile meets the needs of the crop (Dewey and Nielson, 1969; Haas and Evans, 1957; Russell, 1968).

162

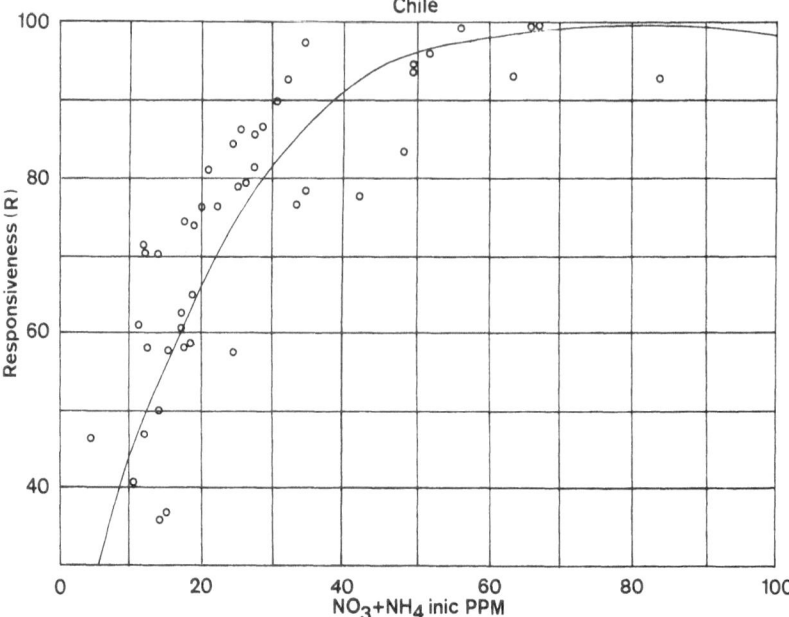

Fig. 8. Relation between extractable N at preplanting and yield responsiveness (R) to N application for irrigated wheat in Chile (from: FAO, 1973).

$$R = \frac{\text{check yield}}{\text{max yield with fertilizers}} \times 100\%$$

Extensive research conducted by FAO (1973) in the Mediterranean climate of central Chile confirmed the usefulness of the mineral N soil test (0 to 15 cm) as a predictive tool for fertilizer response in dryland agriculture. Fig. 8, taken from this report, shows the check yields (no N applied) as a percentage of the maximum yield obtained from an N response curve for irrigated wheat. Whenever the soil contained more than 40 ppm mineral N before planting, the response to fertilizer N was less than 10 per cent. In the absence of supplemental irrigation the reliability of such predictions is less. However, Taylor et al. (1974) used soil nitrate levels in the top 30 cm of soil as an index of available soil N for studying possible N response in dryland wheat production in Southern New South Wales. They concluded that nitrate levels would restrict yields on most farms whenever nitrate concentrations fell below 20 ppm. In a later communication, they reported that profitable response to N (34 kg N ha^{-1}) was generally obtained if soil nitrate was less than 8 ppm, provided seasonal conditions were better than average (Taylor et al., 1978).

The ability of soils in arid environments to provide the necessary nitrogen to meet the potential demand may depend to a great extent on the quantities of nitrogen and water stored in the soil profile before planting. Under Mediterranean climates, there is usually a cool moist winter suitable for the establishment and development of cereal crops, but the crop matures in an environment of diminishing soil water supplies and increasing transpiration demand due to increasing temperatures and radiation. Increased vegetative growth results in greater evaporative demand, and a more rapid diminution of soil water reserves so that the crop may be left with insufficient water to complete its growth (Russell, 1967). This type of situation indicates the need for caution in the use of N fertilizer. For instance, Haas and Evans (1957) presented data for the Great Plains where, over a 3-year period at 13 locations, a consistent response to nitrogen was found at only two locations, while at other locations a response was obtained in some years.

A summary of IAEA coordinated research results (Table 7) demonstrates the variability in N response for various countries with arid, winter rainfall climates. Of the 13 experiments, all conducted under low moisture conditions, 11 experiments gave a substantial response in grain yields when fertilized with 120 kg N

Table 7. Grain yield and nitrogen-15 recovery from various nitrogen sources

Country	Grain yield, t ha^{-1}				N Recovery, % of applied		
	Control	Urea	AN	AS	Urea	AN	AS
	1970/71						
Lebanon**	.98	3.6	3.6		33	37	
Egypt**	3.8	5.6	5.9		36	38	
Morocco	2.0	2.3	1.8		23	21	
Pakistan**	1.8	4.4	4.7		48	49	
Turkey	1.1	3.7	4.0		38	42	
Greece	1.4	1.9	2.1		31	42	
Uruguay	.86	1.9	1.8		30	28	
	1971/72						
Lebanon**	3.0	6.5	7.3	6.6	52	47	49
Egypt**	3.8	4.9	4.5	4.2	53	47	39
Pakistan**	2.0	5.4	5.3	5.2	58	65	60
Iran**	2.4	3.8	4.7	3.6	25	24	26
Turkey	3.0	5.1	4.3	4.3	48	40	36
Greece	2.7	2.7	2.7	3.3			

* Source: IAEA, 1974. N applied at rate of 120 kg ha^{-1} (one-half broadcast at planting and one-half at tillering).
** Irrigated when 75 per cent available moisture lost.

ha^{-1} (IAEA, 1974), indicating that the availability of moisture was not the primary yield constraint in most cases.

Thus, nitrogen uptake by the crop depends on a host of environmental and cultural factors (Russell, 1968). Moreover, nitrogen removal in the harvested crop depends on the farming system and management practices of the region or individual farmer. For extensive wheat farming in the northern Negev, Noy-Meir and Harpaz (1977) estimated that 15 of the 25 kg N taken up by the crop is present in the straw and returned to the land. If straw is consistently removed, only 7 out of 17 kg N remains in the field. On land grazed by sheep, they estimated that 22 out of 39 kg N in the crop either remains on the field (16 kg) or returns in the form of animal droppings (6 kg). In dryland wheat experiments coordinated by IAEA, the harvested straw of the check plots contained an average of 12 kg N ha^{-1} (range 6 to 24) which would be lost if not returned to the field.

Although straw residues may contain appreciable amounts of N, the straw may be of a sufficiently high C : N ratio that this N may not be mineralized for some time. Further, decomposition of the bulk of carbon in the straw may result in large immobilization of the inorganic N mineralized from soil organic matter that would otherwise be available for crop uptake. Grazing of stubble by livestock, a common practice in the Mediterranean region, will alleviate this condition at the risk of accelerated volatilization of N. If burning of stubble were practised, the subsequent crop might benefit from an improved N status due to lowered immobilization; however, potential organic matter will be lost. In the long run, the advantages of stubble burning are more than offset by a progressive reduction in yield as shown for Canadian wheat-growing conditions by Dormaar et al., 1979. If the natural soil fertility is to be sustained, proper management of crop residues should be incorporated in the package of cultural practices and livestock management recommended to the farmer.

Cultivation of arid land has a pronounced effect on the fertility status of soils. The nitrogen level generally attains a dynamic equilibrium under a given set of climatic conditions and agricultural practices. A change in these conditions will upset this equilibrium, sometimes with catastrophic consequences (Owen, 1977). The long term nitrogen status of soils in the arid and semi-arid Great Plains of USA was followed at 14 different locations for periods between 30 and 43 years (Haas and Evans, 1957). Total soil nitrogen levels in the top 30 cm of soil declined as much as 24 to 60 per cent with an average loss of N of 39 per cent over a 36-year period. Organic carbon losses were slightly greater. The major loss of N appeared to have taken place during the first 10 to 20 years of cultivation.

The cropping system imposed on the land greatly affects the nitrogen dynamics in the soil. Haas and Evans (1957) found that land which was alternately

Table 8. Removal of nitrogen in grain and straw and measured loss of soil nitrogen, under narrow cereal rotations, in some long-term experiments*

Site	Rotation**	Period (years)	N Removed in Produce (kg ha^{-1})	Measured loss of soil N (kg ha^{-1})
Werribee, Victoria	F,W	35	375	290
	W,W		265	24
Rutherglen, Victoria	F,W	36	435	435
	W,W		260	50
Waite Inst., South Australia	F,W	36	685	3140
	W,W		585	2265
Chapman, W., West Australia	F,W	23	115	415
	W,W		150	140
Wongan Hills, West Australia	F,W	23	165	80
	W,W		125	25

  * From: Clarke and Russell, 1977.
 ** F = fallow; W = wheat.

cropped with small grain and fallowed, lost more nitrogen (29 per cent) than continuously cropped land (24 per cent). Similar findings have been reported from Australia as shown in Table 8. The incorporation of legume-based pasture in the rotation has been quite successful in maintaining the nitrogen status in the arid regions of Australia (Clarke and Russell, 1977), while legumes used as green manure were ineffective in reducing the loss of N in the Great Plains of USA (Haas and Evans, 1957). If one is proposing a management strategy for low-input farming, one should optimize possibly 'free' benefits such as biological N fixation wherever climatic factors permit such practices. These benefits have been estimated to range from 20 to 180 kg ha^{-1} yr^{-1} in semi-arid environments of Australia (Clarke and Russell, 1977).

I he effect of total soil nitrogen loss on wheat yields was analyzed by Haas and Evans (1957) for five locations with four cropping systems using multiple regression analysis including moisture availability and cropping period (Table 9). The decline in yield was greatest under continuous cropping or rotation without manure. Manuring in the rotation prevented some of the yield loss while alternate fallowing improved the yield with time. No clear relationship between total N decline and yield decline was apparent. Apparently, the long term management of the soils has an influence on soil productivity factors other than the soil nitrogen status. An example of the effect of rotation on the yield trends of wheat under Australian conditions (620 mm rainfall annually) is given in Fig. 9.

Table 9. Comparison of the percentage change in soil nitrogen with the percentage change in yield* of wheat under four cropping systems at six locations**

Location	Time period	Continuous		Alternate fallow		Rotation***			
		Soil-nitrogen change	Yield change	Soil-nitrogen change	Yield change	Without manure		With manure	
						Soil-nitrogen change	Yield change	Soil-nitrogen change	Yield change
	(years)			%					
Mandan, North Dakota	30	−19	37	−26	34	−34	−14	0	49
Dickinson, North Dakota	40	−44	−26	−50	−19	−49	−33	−39	−6
Havre, Montana	31	−44	−4	−42	66	−48	−13	−26	−36
Sheridan, Wyoming	30	−21	−40	−27	−5	−30	−13	−2	10
Archer, Wyoming	34	−32	−47	−33	5	−37	−43	−19	−50
Hays, Kansas	30	−10	−24	−7	−18	−35	−9	−22	−26
Average	33	−28	−17	−31	11	−39	−21	−18	−10

* The percentage yield change was computed by dividing the calculated yield change for the period by the calculated yield in the first year and multiplying by 100.
** Haas and Evans, 1957.
*** Rotation at Hays: Kafir, fallow, wheat; other locations: corn, wheat, oats.

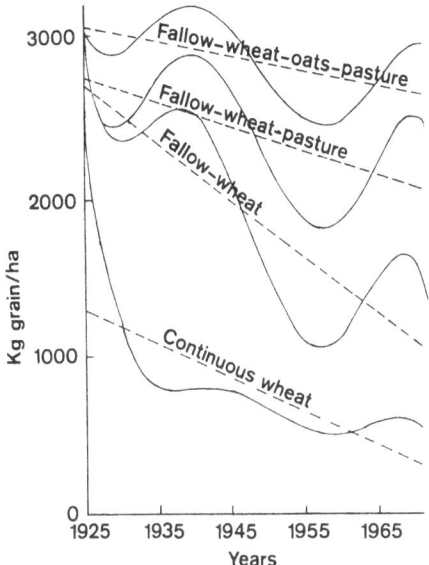

Fig. 9. Yield trends of wheat from several rotations in the permanent rotation experiment at the Waite Institute (from: Clarke and Russell, 1977).

While much is understood about nitrogen transformations, the major problem is predicting the relative contribution of each component to the nitrogen economy of crops and pastures; predictability is influenced by the year-to-year variability in moisture, with very marked consequences for plant demand for nitrogen, the net mineralization (or soil supply), and the losses of N that will occur by leaching or as gases. Because rainfall variability affects other yield-determining parameters, e.g., diseases, pests, erosion, the recommended management system for a Mediterranean-type climate will always have some uncertainty.

REFERENCES

Alexander, M. 1965 Nitrification. In: Bartholomew and Clark, F. E. (eds.), Soil Nitrogen, pp. 307–343. American Society of Agronomy. Madison, Wisconsin.

Allison, F. E., Carter, J. N., and Sterling, L. D. 1960 The effect of partial pressure of oxygen on denitrification in soil. Soil Sci. Soc. Am. Proc. 24, 283–285.

Ardakani, M. S., Rehbock, J. T. and McLaren, A. D. 1973 Oxidation of nitrite to nitrate in a soil column. Soil Sci. Soc. Am. Proc. 37, 53–56.

Avnimelech, Y. and Laher, M. 1977 Ammonia volatilization from soils: equilibrium considerations. Soil Sci. Soc. Am. J. 41, 1080–1084.

Bailey, L. D. 1976 Effects of temperature and root on denitrification in a soil. Can. J. Soil Sci. 56, 79–87.

Bartholomew, W. V., and Clark, F. E. (eds.) 1965 Soil Nitrogen. American Society of Agronomy, Madison, Wisconsin.

Bauer, A., Young, R. A. and Ozbun, J. L. 1965 Effects of moisture and fertilizer on yields of spring wheat and barley. Agronomy J. 57, 354–356.

Beek, J., and Frissel, J. J. 1973 Simulation of nitrogen behavior in soils. Pudoc, Wageningen, The Netherlands, 67 pp.

Bell, F. C. 1979 Precipitation. In: Goodall, D. W. and Perry, R. A. (eds.), Arid-land ecosystems I. Cambridge University Press, Cambridge, Great Britain.

Beutier, D., and Renon, H. 1978 Representation of ammonia-hydrogen sulfidewater, ammonia-carbon dioxide-water, and ammonia-sulfur dioxide-water, vaporliquid equilibriums. Ind. Eng. Chem. Proc. Des. Dev. 17, 220–230.

Birch, H. S. 1960 Soil drying and soil fertility. J. Trop. Agr. (Trinidad) 37, 3–10.

Black, A. L. 1968 Nitrogen and phosphorus fertilization for production of crested wheat grass and native grass in northeastern Montana. Agronomy J. 60, 213–216.

Blackmer, A. M., and Bremner, J. M. 1978 Inhibitory effect of nitrate on reduction of N_2O to N_2 by soil microorganisms. Soil Biol. Biochem. 10, 187–191.

Blondel, D. 1971 Contribution a l'etude du lessivage de l'azote en sol sableux (Dior) au Senegal. L'Agron Trop. 26, 688–696.

Bouwmeester, R. J. B., and Vlek, P. L. G. 1980 Rate control of ammonia volatilization from shallow water. Atmospheric Environment (in press).

Bowman, R. A., and Focht, D. D. 1974 The influence of glucose and nitrate concentrations upon denitrification roles in sandy soils. Soil Biol. Biochem. 6, 297–301.

Bremner, J. M. 1965 Nitrogen availability indexes. In: Black, C. A. et al. (eds.), Methods of soil analysis. Agronomy No. 9. ASA, Madison, Wisconsin.

Bremner, J. M. 1965 Organic nitrogen in soils. p. 93–150. In: Bartholomew, W. V., and Clark, F. E. (eds.), Soil Nitrogen. American Society of Agronomy, Madison, Wisconsin.

Bremner, J. M., and Blackmer, A. M. 1977 Nitrous oxide: Emission from soils during nitrification of fertilizer nitrogen. Science 199, 295–296.

Bremner, J. M., and Shaw, K. 1958 Denitrification in soil. II. Factors affecting denitrification. J. Agr. Sci. 51, 40–52.

Brezonik, P. L. 1976 Nutrients and other biologically active substances in atmospheric precipitation. J. Great Lakes Res. 2, Suppl. 1, 166–186.

Broadbent, F. E., and Clark, F. 1965 Denitrification. In: Bartholomew, W. V. and Clark, F. E. (eds.), Soil Nitrogen. American Society of Agronomy, Madison, Wisconsin, p. 343–359.

Burford, J. R., and Bremner, J. M. 1975 Relationships between the denitrification capacities of soils and total water-soluble and readily-decomposable soil organic matter. Soil Biol. Biochem. 7, 389–394.

Burford, J. R., and Millington, R. J. 1968 Nitrous oxide in the atmosphere of a red brown earth. 9th International Congress of Soil Science, Vol. II, 505–511.

Burns, I. G. 1975 An equation to predict surface-applied nitrate. J. Agric. Sci. 85, 443–454.

Burns, I. G. 1976 An equation to predict the leaching of nitrate uniformly incorporated to a known depth or uniformly distributed throughout a soil profile. J. Agric. Sci. 86, 305–313.

Burns, R. C., and Hardy, W. F. 1975 Nitrogen fixation in bacteria and higher plants. Springer-Verlag. New York.

Cameron, D. R., and Kowalenko, C. G. 1976 Modeling nitrogen processes in soil: mathematical development and relationships. Can. J. Soil Sci. 56, 71–78.

Cawse, P. A., and Sheldon, D. 1972 Rapid reduction of nitrate in soil remoistened after air drying. J. Agric. Sci. Camb. 78, 405–412.

Christianson, D. B., Hedlin, R. A. and Cho, C. M. 1979 Loss of nitrogen from soil during nitrification of urea. Can. J. Soil Sci. 59, 147–154.

Clarke, A. L., and Russell, J. S. 1977 Crop sequential practices. In: Russell, J. S. and Greacen, E. K. (eds.), Practice and crop production in a semi-arid environment. University of Queensland Press, St. Lucia, Queensland, Australia.

Cooke, G. W. 1979 Some priorities for British soil science. J. of Soil Science, 30, 187–213.

Craswell, E. T. 1978 Some factors influencing denitrification and nitrogen immobilization in a clay soil. Soil Biol. Biochem. 10, 241–245.

Craswell, E. T., and Martin, A. E. 1974 Effect of moisture content on denitrification in a clay soil. Soil Biol. Biochem. 6, 127–129.

Craswell, E. T., and Martin, A. E. 1975 Isotopic studies of the nitrogen balance in a cracking clay, II. Recovery of nitrate ^{15}N added to columns of packed soil and microplots growing wheat in the field. Aust. J. Soil Res. 13, 53–61.

Craswell, E. T., and Strong, W. M. 1976 Isotopic studies of the nitrogen balance in a cracking clay, III. Nitrogen recovery in plant and soil in relation to the depth of fertilizer addition and rainfall. Aust. J. Soil Res. 14, 75–83.

Craswell, E. T., and Vlek, P. L. G. 1980 Research to reduce losses of fertilizer nitrogen from wetland rice soils. Fertiliser Association of India (FAI) Annual seminar, DEc 4–6, 1980, New Delhi, India.

Debnath, N. C., and Hajra, J. N. 1972 Transformations of organic matter in relation to mineralization of carbon and nutrient availability. J. Indian Soc. Soil Sci. 20, 95–102.

Denmead, O. T., Simpson, J. R. and Freney, J. R. 1974 Ammonia flux into the atmosphere from a grazed pasture. Science 185, 609–610.

Denmead, O. T., Freney, J. R. and Simpson, J. R. 1976 A closed ammonia cycle within a plant canopy. Soil Biol. Biochem. 8, 161–164.

Denmead, O. T., Simpson, J. R. and Freney, J. R. 1977 The direct field measurement of ammonia emission after injection of anhydrous ammonia. Soil Sci. Soc. Am. J. 41, 1001–1004.

Dewey, W. G., and Nielson, R. F. 1969 Effect of early summer seeding of winter wheat on yield, soil moisture, and soil nitrate. Agronomy J. 61, 51–55.

Diaz-Fierros, F., Macias Vasquez, F. and Guitan Ojea, F. 1973 Influencia del tipo de fertilizacion sobre el balance de nitrigeno en suelos de Galicia. Anales Edaf. Agrobiol.: 939–954.

Doak, D. W. 1962 Some chemical changes in the nitrogenous constituents of urine when voided on pastures. J. Agric. Sci. 32, 162–171.

Donegian, Jr., A. S., and Crawford, W. H. 1976 Report to U.S. Environmental Protection Agency, Environmental Protection Technology Series, EPA 600–2–76–043 On modeling pesticides and nutrients in agricultural lands, 318 pp.

Doner, H. E. 1975 Disappearance of nitrate under transient conditions in columns of soil. Soil Biol. Biochem. 7, 257–259.

Doner, H. E., Volz, M. G. and McLaren, A. D. 1974 Column studies of denitrification in soil. Soil Biol. Biochem. 6, 341–346.

Dormaar, J. F., Pittman, V. T. and Spratt, E. D. 1979 Burning crop residues: Effect on selected soil characteristics and long-term wheat yields. Can. J. Soil Sci. 59, 79–86.

Dowdell, R. J., Burford, J. R. and Crees, R. 1979 Losses of nitrous oxide dissolved in drainage water from agricultural land. Nature 278, 342–343.

Dowdell, R. J., and Smith, K. A. 1974 Field studies of the soil atmosphere II. Occurrence of nitrous oxide. J. Soil Sci. 25, 231–238.

Du Plessis, M. C. F., and Kroontje, W. 1964 The relationship between pH and ammonia equilibria in soils. Soil Sci. Soc. Am. Proc. 28, 751–754.

Eberhardt, P. J., and Skujins, J. 1973 ^{15}N studies of the nitrogen cycle in the Curlew Valley Desert. Annual Rep. Desert Biome., US/IBP. Logan, Utah.

Edmiston, J. 1970 Preliminary studies of the nitrogen budget of a tropical rain forest. p. 811–815. A tropical rainforest, a study of irradiation and ecology at Elverde, Puerto Rico. Odun, H. T., Tigeon, R. S. (eds.) Clearinghouse for Federal, Scientific, and Technical Information, National Bureau of Standards, PID–24270, Winfield, Virginia.

Emerson, M. T., Grunwald, F. and Kromhout, R. A. 1960 Diffusion control in the reaction of ammonium ion in aqueous acid. J. Chem. Phys. 33, 547–555.

Ernst, J. W., and Massey, H. F. 1960 The effects of several factors on volatilization of ammonia formed from urea in the soil. Soil Sci. Soc. Am. Proc. 24, 87–90.

170

FAO, 1973 Reconocimiento e investigacion de los suelos, Chile Produtividad y manejo de los suelos Chilenos. AGL:/CHI 18, Informe technico 2, Food and Agriculture Organization of the United Nations, Rome, Italy.

FAO, 1979 FAO Fertilizer Yearbook, 1978. FAO, Rome, Italy.

Farquhar, G. D., Wetselaar, R., and Firth, P. M. 1979 Ammonia volatilization from senescing leaves of maize. Science, Vol. 203, 1257–1258.

Fenn, L. B., and Escarzaga, R. 1976 Ammonia volatilization from surface applications of ammonium compounds on calcareous soil: V. Influence of soil water content and method of nitrogen application. Soil Sci. Soc. Am. J. 40, 537–541.

Fenn, L. B., and Kissel, D. E. 1975 Ammonia volatilization from surface application of ammonium compounds on calcareous oils: 4. Effect of calcium carbonate content. Soil Sci. Soc. Am. Proc. 39, 631–633.

Fenn, L. E., and Kissel, D. E. 1976 The influence of cation exchange capacity and depths of incorporation on ammonia volatilization from ammonium compounds applied to calcareous soils. Soil Sci. Soc. Am. J. 40, 394–398.

Feth, J. H. 1964 Nitrogen compounds in natural waters – a review. Water Resour. Res. 2, 41–58.

Fluhler, H., Ardakani, M. S., Szuszkiewicz, R. E. and Stolzy, L. H. 1975 Field measured nitrous oxide concentrations, redox potentials, oxygen diffusion rates, and oxygen partial pressures in relation to denitrification. Soil Sci. 118, 173–179.

Fluhler, H., Stolzy, L. H. and Ardakani, M. S. 1976 A statistical approach to define soil aeration in respect to denitrification. Soil Science 122, 115–123.

Focht, D. D., and Verstraete, W. 1977 Biochemical ecology of nitrification and denitrification. Adv. in Microb. Ecology I, Plenum Press.

Freney, J. R., Denmead, O. T. and Simpson, J. R. 1978 Soil as a source or sink for atmospheric nitrous oxide. Nature 273, 530–532.

Freney, J. R., Denmead, O. T. and Simpson, J. R. 1979 Nitrous oxide emission from soils at low moisture contents. Soil Biol. Biochem. 11, 167–173.

Galbally, I. E., and Roy, C. R. 1978 Loss of fixed nitrogen from soils by nitric oxide exhalation. Nature 275, 734–735.

Ganry, S., Guiraud, G., and Dommergues, Y. 1978 Effect of straw incorporation on the yield and nitrogen balance in the sandy soil pearl-millet cropping system of Senegal. Plant and Soil 50, 647–662.

Gardner, W. R. 1965 Movement of nitrogen in soil. In: Bartholomew, W. V., Clark, F. E. (eds.), Soil nitrogen. American Soc. of Agron., Madison, Wisconsin.

Gasser, J. K. R. 1964 Some factors affecting losses of ammonia from urea and ammonium sulfate applied to soil. J. Soil Sci. 15, 258–272.

Greenland, D. J. 1971 Changes in the nitrogen status and physical condition of soils under pastures, with special reference to the maintenance of the fertility of Australian soils used for growing wheat. Soils Fertilizers 34(4), 237–251.

Greenwood, D. J. 1963 Nitrogen transformations and the distribution of oxygen in soil. Chemistry and Industry 799–803.

Haas, H. J., and Evans, C. E. 1957 Nitrogen and carbon changes in Great Plains soils as influenced by cropping and soil treatments. USDA Technical Bulletin No. 1164. 111 pp.

Hagin, J., and Amberger, A. 1974 Contribution of fertilizers and manures to the N- and P-load of waters, a computer simulation. Final report to Deutsche Forsch. Gemeinschaft from Technion, Israel. 123 pp.

Harding, R. B., Endleton, P. W., Jones, W. W. and Ryan, T. M. 1963 Leaching and gaseous losses of nitrogen from some nontilled California soils. Agron. 55, 515–518.

Hardy, F. 1946 Seasonal fluctuations of soil moisture and nitrate in humid tropical climates. J. Trop. Agr. (Trinidad) 23, 40–49.

Harmsen, G. W. and Kolenbrander, G. J. 1965 Soil inorganic nitrogen. In: Bartholomew, W. V., Clark, F. C. (eds.), Soil nitrogen. American Soc. of Agron., Madison, Wisconsin.

Harpaz, Y. 1975 Simulation of the nitrogen balance in semiarid regions. Ph. D. Thesis, Hebrew University, Jerusalem.

Harris, G. T., and Harre, E. A. 1979 World fertilizer situation and outlook. 1978–85. Technical Bulletin IFDC–T–13. Muscle Shoals, Alabama, U.S.A.

Harris, H. C. 1978 Some aspects of the agroclimatology of the Middle East and North Africa. Internal Report ICARDA. Aleppo, Syria.

Hauck, R. D. 1971 Quantitative estimates of nitrogen cycle processes: concept and review. In: Nitrogen-15 Soil-Plant Studies. IAEA–PL–341/6, Vienna, Austria.

Hauck, R. D. 1977 Nitrogen deficits in ^{15}N balance studies. Denitrification seminar, The Fertilizer Institute, Washington, D.C.

Hauck, R. D., and Bremner, J. M. 1976 Use of tracers for soil and fertilizer nitrogen research. Adv. Agron. 28, 219–266.

Hauck, R. D., and Stephenson, H. F. 1964 Nitrification of nitrogen fertilizers. Effect of nitrogen source, size, and pH of the granule and concentration. J. Agr. Food Chem. 13, 486–492.

Henderson, G. S., and Harris, W. S. 1975 An ecosystem approach to characterization of the nitrogen cycle in a deciduous forest watershed, p. 179–193. Forest Soils and Forest Land Management, Proceedings of the Fourth North American Forest Soil Conference, B. Bernier and P. H. Winget, Les Presses de l'universite Laval, Quebec, Canada.

Hingston, F. J. 1974 Seasonal distribution of mineral nitrogen with particular reference to leaching. Aust. J. Exp. Agr. Animal Husb. 14, 815–821.

Husz, G. St. 1977 Agro-ecosystems in South America. In: M. J. Frissel (ed.). Agro-ecosystems 4 (special issue) 244–276.

IAEA. 1974 Isotope studies on wheat fertilization. Technical Report Series No. 157. International Atomic Energy Agency, Vienna.

Inoue, K., Uehijima, Horie, T. and Iwakere, S. 1975 Studies of energy and gas exchange within crop canopies. X. Structure of turbulence in rice crop. J. Agric. Meteor. 31, 71–82.

Justine, J. K., and Smith, R. L. 1962 Nitrification of ammonium sulfate in calcareous soil as influenced by combinations of moisture, temperature, and levels of added nitrogen. Soil Sci. Soc. Amer. Proc. 26, 246–250.

Keeney, D, R., and Bremner, J. M. 1966 Characterization of mineralizable nitrogen in soils. Soil Sci. Soc. Am. Proc. 30, 714–719.

Keeney, D. R., Fillery, I. R. and Marx, G. P. 1979 Effect of temperature on the rate and gaseous N products of denitrification in a silt loam soil. J. Soil Sci. Soc. Am. 43, 1124–1128.

Keulen, H. van. 1975 Simulation of water use and herbage growth in arid regions. Pudoc, Wageningen, The Netherlands.

Keulen, H. van. 1977 On the role of nitrogen in semiarid regions. Stikstof No. 20, 23–28.

Khalifa, M. A., Akasha, M. H. and Said, M. B. 1977 Growth and N-uptake by wheat as affected by sowing date and nitrogen in irrigated semi-arid conditions. J. Agric. Sci. 89, 35–42.

Kissel, D. E., Brewer, H. L. and Arkin, D. S. 1977 Design and test of a field sampler for ammonia volatilization. Soil Sci. Soc. Am. J. 41, 1133–1138.

Kissel, D. E., Ritchie, J. T. and Burnett, E. 1974 Nitrate and chloride leaching in swelling clay soil. J. Environ. Qual. 3, 401–404.

Kohl, D. H., Vitayathil, F. Whitlow, P., Shearer, G. and Chien, S. H. 1976 Denitrification kinetics in soil systems: the significance of good fits to mathematical forms. Soil Sci. Soc. Am. Proc. 40, 249.

Korenkov, D. A., Romanyuk, L. F., Varyushkina, N. M. and Kirpaneva, L. I. 1975 Use of Stable N^{15} isotope to study the balance of fertilizer nitrogen in field lysimeters on sod-podzolic sandy loam soil. Soviet Soil Science 7, 244–249.

Kowalenko, C. G., and Cameron, D. R. 1976 Nitrogen transformations in an incubated soil as affected by combinations of moisture content and temperature and adsorption-fixation of ammonium. Can. J. Soil Sci. 56, 63–70.

Kowalenko, C. G., Ivarson, K. C. and Cameron, D. R. 1979 Effect of moisture content, temperature, and nitrogen fertilization on carbon dioxide evolution from field soils. Soil. Biol. Biochem. 10, 417–423.

172

Kresge, C. B., and Satchell, D. P. 1960 Gaseous loss of ammonia from nitrogen fertilizers applied to soils. Agron. J. 52, 104–107.

Laura, R. D. 1973 Effects of sodium carbonate on carbon and nitrogen mineralization of organic matter added to soil. Geoderma 9, 15–26.

Laura, R. D. 1974 Effects of neutral salts on carbon and nitrogen mineralization of organic matter in soils. Plant and Soil 41, 113–127.

Low, A. J. 1973 Nitrate and ammonium nitrogen concentration in water draining through soil monoliths in lysimeters cropped with clover or uncropped. J. Sci. Fd. Agric. 24, 1489–1495.

Luebs, R. E., and Laag, A. F. 1967 Nitrogen effect on leaf area, yield, and nitrogen uptake of barley under moisture stress. Agron. J. 59, 219–222.

Mahendrappa, M. K., Smith, R. L. and Christianson, A. T. 1966 Nitrifying organisms affected by climatic regions in western United States. Soil Sci. Soc. Am. Proc. 30, 60–62.

Martin, J. P., and Chapman, H. D. 1951 Volatilization of ammonia from surface fertilized soils. Soil Sci. 71, 25–34.

Meyer, R. D., Olson, R. A. and Rhoades, H. F. 1961 Ammonia losses from fertilized Nebraska soils. Agron. J. 53, 241–244.

Miller, R. D., and Johnson, D. D. 1964 Effect of soil moisture retention on carbon dioxide evolution, nitrification, and nitrogen mineralization. Soil Sci. Soc. Am. Proc. 28, 644-647.

Misra, C., Nielsen, D. R. and Biggar, J. W. 1974a Nitrogen transformations in soils during leaching. 1. Theoretical considerations. Soil Sci. Soc. Am. Proc. 38, 289–293.

Misra, C., Nielsen, D. R. and Biggar, J. W. 1974b Nitrogen transformations in soil during leaching. 2. Steady-state nitrification and nitrate reduction. Soil Sci. Soc. Am. Proc. 38, 294–299.

Misra, C., Nielsen, D. R. and Biggar, J. W. 1974c Nitrogen transformations in soil during leaching. 3. Nitrate reduction in soil columns. Soil Sci. Soc. Am. Proc. 38, 300–304.

Morrill, L. G., and Dawson, J. E. 1967 Patterns observed for the oxidation of ammonium to nitrate by soil organisms. Soil Sci. Soc. Am. Proc. 31, 757.

Musa, M. M. 1968 Nitrogenous fertilizer transformations in the Sudan Gezira soils. 1. Ammonia volatilization losses following surface application of urea and ammonium sulfate. Plant and Soil 28, 413–421.

Myers, R. J. K. 1975 Temperature effects on ammonification and nitrification in a tropical soil. Soil Biol. Biochem. 7, 83–86.

Myers, R. J. K., and Paul, E. A. 1971 Plant uptake and immobilization of ^{15}N-labelled ammonium nitrate in a field experiment with wheat. In: Nitrogen-15 in soil-plant studies. IAEA, Vienna, Austria.

National Research Council. 1978 Nitrates: An Environmental Assessment Environmental Studies Board, Commission on National Resources, National Academy of Sciences, Washington, D.C.

Nelson, D. W., and Bremner, J. M. 1970 Gaseous products of nitrate decomposition in soils. Soil Biol. Biochem. 2, 203–215.

Nielsen, D. R. 1976 Solute and water movement under arid and semiarid conditions. In: Efficiency of water and fertilizer use in semiarid regions. IAEA Technical Document No. 192. Vienna, Austria.

Nielsen, D. R., and MacDonald, J. G. (eds.) 1978 Nitrogen in the environment. Academic Press, New York.

Nommik, H. 1956 Investigations on denitrification in soil. Acta Agri. Scand. VI, 195–228.

Noy-Meir, I., and Harpaz, Y. 1977 Agro-Ecosystems in Israel. In: M. J. Frissel (ed.), Agro.Ecosystems 4, (special issue), 143–167.

Nyborg, M., and Hoyt, P. B. 1978 Effects of soil acidity and liming on mineralization of soil nitrogen. Can. J. Soil Sci. 58, 331–338.

Oades, M. M., and Ladd, J. N. 1977 Biochemical properties: carbon and nitrogen metabolism. In: Russell, J. S., Greacen, E. L. (eds.), Soil Factors in Crop Production in a Semiarid environment, University of Queensland Press, St. Lucia, Australia.

Olson, R. V., Murphy, L. S., Moser, H. C. and Swallow, C. W. 1979 Fate of tagged fertilizer nitrogen applied to winter wheat. Soil Sci. Soc. Am. J. 43, 973–975.

Owen, W. F. 1977 Living within the limits of aridity. Ekistics 258, 291–296.

Pang, P. C., Cho. C. M. and Hedlin, R. A. 1975a Effects of nitrogen concentration on the transformation of band-applied nitrogen fertilizers. Can. J. Soil Sci. 55, 23–27.

Pang, P. C., Cho, C. M. and Hedlin, R. A. 1975b Effects of pH and nitrifier population on nitrification of band-applied and homogeneously mixed urea nitrogen in soils. Can. J. Soil Sci. 55, 15–21.

Payne, W. J. 1973 Reduction of nitrogenous oxides by microorganisms. Bacteriol. Rev. 37, 409.

Power, J. F., Alessi, J. Reichman, G. A. and Grunes, D. L. 1973 Recovery, residual effect, and fate of nitrogen fertilizer sources in a semiarid region. Agron. J. 65, 765–768.

Rainey, R. C. 1977 Rainfall: scarce resource in 'opportunity country.' Phil. Trans. R. Soc. Lond. 278, 439–453.

Ramig, R. E., and Rhoades, H. F. 1962 Interrelationships of soil moisture level at planting time and nitrogen fertilization on winter wheat production. Agron. J. 54, 123–127.

Reichman, G. A., Grunes, D. L. and Viets, F. G. Jr. 1966 Effect of soil moisture on ammonification and nitrification in two northern plains soils. Soil Sci. Am. Proc. 30, 363–366.

Robinson, J. D. 1957 The critical relationship between soil moisture content in the region of the wilting point and mineralization of native soil nitrogen. J. Agr. Sci. 49, 100–105.

Rodgers, G. A. 1978 Dry deposition of atmospheric ammonia in Rothamsted in 1976 and 1977. J. Agric. Sci., Camb. 90, 537–542.

Rolston, D. E. 1977 Measuring nitrogen loss from denitrification. Calif. Agric. 31, 12–13.

Roswall, T. et al. 1975 Stohdalen, Abisco, Sweden. In: Rosswall, T., Heal, O. W. (eds.), Structure and function of tundra ecosystems, p. 265–294. Ecol. Bull. 20, Stockholm.

Russell, E. W. 1973 Soil condition and plant growth, 10th ed., Longmans, London.

Russell, J. S. 1967 Nitrogen fertilizer and wheat in a semi-arid environment 1. Effect on yield. Austr. J. Exp. Agr. Anim. Husb. 7, 453–462.

Russell, J. S. 1968 Nitrogen fertilizer and wheat in a semi-arid environment 3. Soil and cultural factors affecting response. Austr. J. Exp. Agr. Anim. Husb. 8, 340–348.

Russell, J. S. 1975 A mathematical treatment of the effect of cropping systems on soil organic nitrogen in two long-term sequential experiments. J. Soil Sci. 120, 37–44.

Russell, J. S. 1977 Introduction. In: Russell, J. S., Greacen, E. L. (eds.), Soil factors in crop production in a semi-arid environment. University of Queensland Press, St. Lucia, Queensland, Australia.

Sabey, B. R., Bartholomew, W. V., Shaw, R. and Pesek, J. 1956 Influence of temperature on nitrification in soils. Soil Sci. Soc. Am. Proc. 20, 357–360.

Scarsbrook, C. E. 1965 Nitrogen availability. In: Bartholomew, W. V., Clark, F. E. (eds.), Soil Nitrogen. Agronomy No. 10. ASA, Madison, Wisconsin.

Singh, R. P. and Ramakrishna, Y. S. 1975 Moisture use efficiency of dryland crops as influenced by fertilizer use. II Rabi cereals. Annals. of Arid Zone 14, 263–267.

Smith, K. A. 1978 A model of the extent of anaerobic zones in aggregated soils, and its application to estimates of denitrification. Abstracts 11th Congress Int. Soc. Soil Sci., Edmonton, Canada, p. 304.

Soper, R. J., Racz, G. J. and Fehr. R. E. 1971 Nitrate nitrogen in the soil as a means of predicting the fertilizer requirements of barley. Can. J. of Soil Sci. 51, 45 000.

Soriano, S., and Walker, N. 1973 The nitrifying bacteria in soils of Rothamsted classical fields and elsewhere. J. Appl. Bact. 36, 523–529.

Stanford, G., Dzienia, S. and Van der Pol, R. A. 1975b) Effect of temperature on denitrification rate in soils. Soil. Sci. Soc. Am. Proc. 39, 867–870.

Stanford, G., and Epstein, E. 1974 Nitrogen mineralization-water relations in soils. Soil Sci. Soc. Am. J. 38, 103–107.

Stanford, G., Frere, M. H. and Schwaninger, D. H. 1973 Temperature coefficient of soil nitrogen mineralization. Soil Sci. 115, 321–323.

Stanford, G., and Smith, S. J. 1972 Nitrogen mineralization potentials of soils. Soil Sci. Soc. Am. Proc. 36, 465–472.

Stanford, G., and Smith, S. J. 1976 Estimating potentially mineralizable soil nitrogen from a chemical index of soil nitrogen availability. Soil Science 122, 71–76.

Stanford, G., van der Pol, R. A. and Dzienia, S. 1975a Denitrification rates in relation to total and extractable soil carbon. Soil Sci. Soc. Amer. Proc. 39, 284–289.

Starr, J. L., Broadbent, F. E. and Nielsen, D. R. 1974 Nitrogen transformations during continuous leaching. Soil Sci. Soc. Am. Proc. 38, 283–289.

Starr, J. L., and Parlange, J. Y. 1975 Nonlinear denitrification kinetics with continuous flow in soil columns. Soil Sci. Soc. Am. Proc. 39, 875–880.

Stefanson, R. C. 1972 Soil denitrification in sealed soil-plant systems. I. Effect of plants, soil water content, and soil organic matter content. Plant Soil 37, 113–117.

Storrier, R. R., Manly, A. T., Spence, T. B. and Smith, A. N. 1971 Wheat yields and indices of available soil nitrogen in southern New South Wales. A preliminary evaluation. Austr. J. Exp. Agric. Animal Husb. 11, 295–000.

Tanji, K. K., and Gupta, S. K. 1978 Computer simulation modeling for nitrogen in irrigated cropland. In: Nielsen, D. R. and MacDonald, J. G. (eds.), Nitrogen in the Environment, Vol. 1 Nitrogen Behavior in Field Soil. Academic Press, New York.

Taylor, A. C., Storrier, R. R. and Gilmour, A. R. 1974 Nitrogen needs of wheat, I. Grain yield in relation to soil nitrogen and other factors. Austr. J. Exp. Agric. and Animal Husb. 14, 241–248.

Taylor, A. C., Storrier, R. R. and Gilmour, A. R. 1978 Nitrogen needs of wheat II. Grain yield response to nitrogenous fertilizer. Austr. J. Exp. Agric. and Animal Husb. 18, 118–128.

Tejeda, H. S., and Gogan, G. L. 1970 Metodos para determinar nitrogino disponible en suelos con diferente origen y contenido de materia organica. Santiago de Chile, Agricultura Technica, 30, 57–64.

Terman, G. L., and Allen, S. E. 1970 Leaching of soluble and slow-release N and K. fertilizers from Lakeland sand under grass and fallow. Soil and Crop Sci. Soc. of Florida Proc. 30, 130–140.

Terry, D. L., and McCants, G. D. 1970 Quantitative prediction of leaching in field soils. Soil Sci. Soc. Am. Proc. 34, 271–276.

Thomas, G. W. 1970 Soil and climatic factors which affect nutrient mobility. In: Nutrient mobility in soils: Accumulation and losses. Engelstad, O. P. (ed.) SSSA special publication No. 4. Soil Sci. Soc. of Amer., Madison, Wisconsin.

Tourte, R., Vidal, P., Jacquinot, L., Faucher, J. and Nicou, R. 1964 Bilan d'une rotation quadriennale sur sole de regeneration au Senegal. L'Agr. Trop. 12.

UNDP, 1977 Report of joint UNDP/AFCFP/IDCAS/UNIDO exploration mission on regional development of fertilizer industry in Arab states. REM/75/018. Mission co-ordinated by International Fertilizer Development Center. Muscle Shoals, Alabama.

Varma, S. K. 1976 Nitrogen content, uptake and grain quality as affected by soil moisture stress and applied nitrogen and phosphorus in some cereal and leguminous crops under semiarid conditions. In: Sen, S. P., Abrol, Y. P. and Sinha, S. K. (eds.), Nitrogen assimilation and crop productivity. Ass. Publ. Co., New Delhi.

Veen, J. A. Van, and Frissel, M. J. 1976 Computer simulation model for the behavior of nitrogen in soil and leaching to groundwater. Pudoc, Wageningen, 26 pp.

Verstraete, W. 1979 Nitrification. A paper prepared for the SCOPE-UNEP meeting, Osterfarnebo, Sweden. September 16–22, 1979. F. Clark (ed.).

Vlek, P. L. G., and Craswell, E. T. 1979 Effect of nitrogen source and management on ammonia volatilization losses from flooded rice soil systems. Soil Sci. Soc. Am. J. 43, 352–358.

Vlek, P. L. G., and Stumpe, J. M. 1978 Effect of solution chemistry and environmental conditions on ammonia volatilization losses from aqueous systems. Soil Sci. Soc. Am. J. 42, 416–421.

Volz, M. G., Ardakani, M. S., Schulz, R. K., Stolzy, L. H. and McLaren, A. D. 1976 Soil nitrate loss during irrigation: enhancement by plant roots. Agron. J. 68: 621–627.

Wallace, A., Romney, E. M., Kleinkopf, G. E. and Soufi, S. M. 1978 Uptake of mineral forms of nitrogen by desert plants. In: West, N. E. and Skujins, J. J. (eds.), Nitrogen in desert ecosystems. US/IBP Synthesis series No. 9. Dowden, Hutchinson, and Ross, Inc. Stroudsburg, Pennsylvania.

Watkins, S. H., Strand, R. F., DeBell, D. S. and Esch, J. Jr. 1972 Factors influencing ammonia losses from urea applied to Northwestern forest soils. Soil Sci. Soc. Am. Proc. 36, 354–357.

Watts, D. G. 1975 Ph. D. Thesis. Utah State University, Logan. 187 pp.

West, N. E. 1975 Nutrient cycling in desert ecosystems. In: Short-term ecosystem Dynamics in Arid Lands. Cambridge University Press, London.

West, N. E., and Skujins, J. 1978 Nitrogen in desert ecosystems. US/IBP Synthesis Series No. 9, Dowden, Hutchinson, and Ross, Inc. Stroudsburg, Pennsylvania.

Westerman, R. L., and Tucker, T. C. 1978 Factors affecting denitrification in a Sonoran desert soil. Soil Sci. Soc. Am. J. 42, 596–599.

Westerman, R. L., and Tucker, T. C. 1979 In-active transformations of nitrogen-15 labeled materials in Sonoran Desert soils. Soil Sci. Soc. Am. J. 43, 95–100.

Wetselaar, R. 1961 Nitrate distribution in tropical soils I. Possible causes of nitrate accumulation near the soil surface after a long dry period. Plant and Soil 15, 110–120.

Wetselaar, R. 1962 Nitrate distribution in tropical soils III. Downward movement and accumulation of nitrate in the subsoil. Plant and Soil 16, 19–31.

Woldendorp, J. W. 1962 The quantitative influence of the rhizosphere on denitrification. Plant and Soil 17, 267–270.

Yaalon, D. H. 1964 The concentration of ammonia and nitrate in rain water over Israel in relation to environmental factors. Tellus 16, 200–204.

7. Nitrogen and plant production

R. NOVOA* and R. S. LOOMIS

*INIA, Castilla, Santiago, Chile

Green plants play a unique role among living organisms through their ability to reduce carbon in photosynthesis. While reduced carbon provides the energy source for all life, nitrogen must be viewed as the central element because of its role in substances such as proteins and nucleic acids which form the living material. Proteins serve as enzyme catalysts in metabolic pathways, as structural elements of cytoplasm and membranes and as carriers in transport functions. Nucleic acids provide the means for codification, storage and translation of genetic information. In those and other organic materials, nitrogen appears in a chemically reduced state. Such organic nitrogen constitutes 1.5 to 5 per cent of the dry weight of higher plants, and 80 to 90 per cent of that is in protein.

Despite the great abundance of dinitrogen gas in the atmosphere, the element is commonly deficient in agricultural soils in the oxidized (nitrate) and reduced (ammonium) forms which can be absorbed by plants. In aerated soils, bacteria quickly transform ammonium ions to nitrate, and the oxidized form is the principal in one plant nutrition. As a result of nitrogen's critical roles and low supply, the management of nitrogen resources is an extremely important aspect of crop production. It has also been an area for intensive research at all levels from yield behavior to cell physiology.

As agronomists, we are concerned particularly with designing strategies to increase the efficiency of nitrogen use in crop systems. This chapter focuses on aspects which are related to the uptake and assimilation of nitrogen into amino acids and proteins, and their subsequent interaction in growth and development. These events are highly dynamic and complex, and are not yet fully understood. Our intent here is to provide a broad review of those subjects for crop physiologists, geneticists and agronomists. We begin with the biochemical and cellular level and integrate towards field performance.

ESSENTIAL FIRST STEPS: NITRATE AND AMMONIUM TO PROTEIN

At the cellular level, interest will be focused on the transport of nitrate and ammonium ions across membranes, reduction of nitrate and nitrite, amino acid synthesis and its regulation, synthesis and degradation of protein, the importance of compartmentation, and the costs of assimilation. The serious student of nitrogen use efficiency should study the early classical work in this field (e.g. McKee, 1962) and more recent reviews (Hewitt *et al.*, 1976; Huffaker and Rains, 1978; Miflin and Lea, 1977).

The concentration of nitrate nitrogen in soil solution is generally low, 1 mM or less, with 0.1 mM (100 μM) not uncommon. Ammonium ions are generally present in much lower amounts, and are tightly bound to the cation exchange complex. In aerated soils, nitrate is the dominant form in plant nutrition. It is absorbed first into the free-space of roots (cell wall spaces) and then across membranes into the plant cells themselves.

Membrane transport is by two mechanisms: 'passive' and 'active'. Passive uptake depends on the permeability of the membrane. It occurs by diffusion along electrochemical gradients without involvement of metabolic energy. The electrical potential of the cytosol of root cells is generally negative (ca $-100\,\text{mV}$) relative to the external solution, and the gradient usually favours the passive loss of nitrate from the cell back to the external solution. Passive uptake can occur only when a fairly large concentration of nitrate is present in the soil solution, and cell levels are very low.

Active uptake processes overcome the unfavorable electrochemical gradient through the expenditure of metabolic energy (ATP). Active uptake is probably accomplished by 'carriers' as outlined by Epstein (1972), visualized as protein units in the membrane which associate with the ion from the free space and then discharge it into the interior space. Uptake responses show a diminishing-returns relationship to increasing ion concentration in the external solution and in analogy to enzyme behavior, the reponses can be fitted by the Michaelis-Menten equation (Barley, 1970). The parameters of that equation are Km (the external concentration sufficient for a half maximal rate of uptake) and Vmax (the maximal rate). Observed values of Km and Vmax are highly variable among species and environments: Km for grass species including cereals, generally falls in the range 10 to 100 μM for nitrate while 2.5 to 8 moles g^{-1} (fresh wt of plant) hr^{-1} is a representative range for Vmax (Novoa, 1979).

Ammonium can enter into amino acid synthesis directly, but nitrate must first be reduced to the ammonium form. That is accomplished in at least two stages: nitrate to nitrite is mediated by the nitrate reductase system; the second stage, nitrite to ammonium is carried out by nitrite reductase. In roots, the reduction occurs in the cytosol, employing NADH as the reducing agent. A total of eight electrons and eight H^+ ions is required per nitrate. Since plants obtain 24 electrons and $24\,H^+$ from the complete oxidation of a glucose molecule, 0.33

mole or 60 g glucose (photosynthate) must be expended as each mole of nitrate (14 g N) is reduced. This is a very expensive process.

In leaves, nitrate reductase also operates in cytosol with NADH, but nitrite reductase is in the chloroplasts and uses, as an electron carrier, ferrodoxin which is coupled to the light reaction of photosynthesis. That means that the glucose cost is less in leaves. At times when carbon dioxide fixation and reduction are limited by carbon dioxide supply (i.e. at 'light saturation'), surplus reductant could be available and that might allow nitrite reduction to occur without affecting carbon reduction; at low light, the processes could be competitive.

Some of the nitrogen absorbed by roots is assimilated into amino acids in those organs, but most seems to be translocated to the leaves in the transpiration stream. Thus, most amino acid synthesis occurs in leaves. The basic scheme in Fig. 1 illustrates the enzyme pathways for the formation of the five, first-formed amino acids and amides. All other amino acids are formed by transamination from these products.

In leaves, most of the steps except for asparagine synthesis are found in chloroplasts. In roots, the activities seem to be located in the cytoplasm. In both

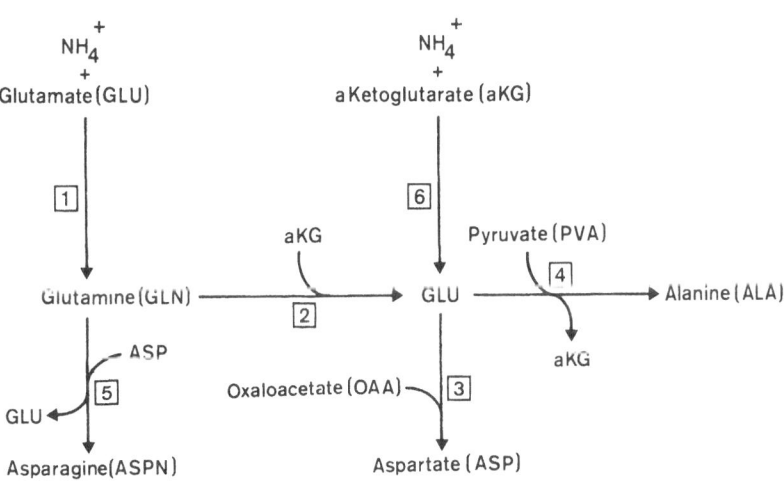

Fig. 1. Biochemical pathways for nitrogen assimilation. Glutamine is synthetized by glutamine synthetase (1) and is converted to glutamate by glutamate synthetase (2). Aspartate is formed by transanimation from glutamate by aspartate aminotransferase (3); alanine arises in a similar way through the action of alanine aminotransferase (4). Asparate can be transaminated into asparagine by asparagine synthetase (5). Glutamate can also be formed by glutamic dehydrogenase (6) but that reaction seems relatively unimportant in plants.

tissues, the amino acids then enter into to protein synthesis. Roots export very few amino acids, whereas leaves export via phloem to other tissues which are active in protein synthesis.

Those cellular processes are regulated in a number of ways. Enzyme levels and activities, and supplies of ATP, reductant and carbon skeletons all are variables. Nitrate reductase, for example, is subject to 'substrate induction' by nitrate and to 'feedback repression' by glutamine. Nitrate supply is particularly important. Nitrate reductase activity is affected only by the concentration of nitrate in the cytosol. An additional and sometimes large pool of nitrate exists in the vacuole.

Uptake and release by vacuoles is an important factor in regulation. In leaves, release is stimulated by light thus assuring that most of the reduction occurs during the photoperiod. For whole plants, the supplies of nitrate and sugar and the relative roles of leaves and roots are paramount. Enzyme kinetics seem to be less important.

We have mentioned the cost of nitrate reduction and expressed it in terms of glucose equivalents. Other costs are encountered in amino acid synthesis, for skeletons, reductant and energy, and for polymerization into proteins. Penning de Vries et al. (1974) also included membrane transfers, including nitrate uptake, at one ATP per transfer (36 ATP are generated in the respiration of one glucose) and the maintenance of the enzyme machinery. Starting with nitrate and glucose, they found that 1 g glucose will yield only 0.45 g of an average protein. Nearly 90 per cent of the difference in weight represents carbon lost in respiration during the generation of the necessary reductant and ATP for amino acid synthesis. From this, we distinguish a component of the dark respiration of plants which is coupled with biosynthesis. The rest of the dark respiration goes mostly for maintenance of the accumulated biomass, with a major part of that due to protein turnover (Penning de Vries, 1975).

NITROGEN IN THE WHOLE PLANT

The cell physiology which we have described, operates inside of whole plants with varying proportions of roots, stems and leaves, growing in a fluctuating environment. It turns out that the whole plant controls its biochemistry as much or more than the reverse. This is fortunate because it allows us to manage nitrogen metabolism through crop manipulations.

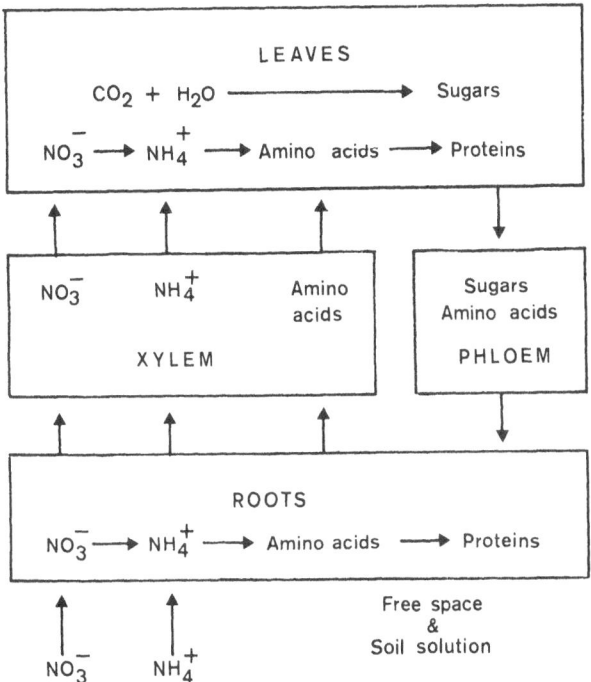

Fig. 2. Basic scheme of nitrogen uptake and reduction and protein formation in a higher plant.

Nitrogen uptake and movement within the plant

As we have seen, some nitrate is tranformed into amino acids in the roots but most of the reduction occurs in leaves. Amino acids formed there can move back to roots or to other organs via the phloem (Pate *et al.*, 1979). A scheme of these movements is shown in Fig. 2. This movement of nitrogen can be an important aspect of the root:shoot functional balances which are observed in plants (Radin, 1978).

The uptake of nitrate is influenced by many factors including temperature, pH and nitrate concentration of the external solution (Bassioni, 1971). The rate of transpiration and the presence of ammonium are also important. Vmax of the active mechanism increases exponentially as temperature increases (van den Honert and Hooymans, 1955) with a Q_{10} near 2 (Hallmark and Huffaker, 1978). In corn, the maximum rate is observed between 30 °C and 40 °C (Ezeta and Jackson, 1975). Increasing pH tends to decrease nitrate uptake. However, ammonium uptake is increased. When both ions are present, ammonium acts as an inhibitor of nitrate uptake (Minnoti *et al.*, 1969; Rao and Rains, 1976a). The effect appears to be non-competitive (D. W. Rains, pers. comm.).

The role of transpiration has been studied intensively, but simple generalizations are not possible. Within the root, it might act to enhance transpor across the cortex or to reduce the internal concentration by increasing transport to the shoot. The experiments of Hylmo (1953) and Brouwer (1954) show enhancement of nitrate uptake by transpiration. Van den Honert and Hooymans (1955) found no such influence on ammonium uptake. Shaner and Boyer (1976a) found that nitrate supply to the xylem was fairly constant, regardless of transpiration rate With ryegrass, Luxmoore and Millington (1971) observed that the uptake rate was essentially the same in the light and the dark, while transpiration was quite different.

Many researchers (Brouwer, 1965; Kramer, 1969; Pitman, 1977) have analyzed transpiration-ion uptake relations in general terms. Pitman's conclusion that transpiration can have a strong influence whenever passive uptake is dominant seems to be sound. Transpiration can have an indirect effect on ion uptake through the mass flow of ions to the root. For potassium, Tinker (1969) found that mass flow contributed not more than only 13 per cent of the total flux.

It now seems clear that the active transport system for nitrate is induced by nitrate (Jackson *et al.*, 1973; Ashley *et al.*, 1975). However, it is not clear whether the effect is direct or indirect via nitrate reductase. Rao and Rains (1976b) pointed out that absorption can be limited by a high concentration of nitrate in the cytosol. In that case, uptake will increase as nitrate reduction increases. Nitrate reductase is induced by nitrate and one could interpret that as an induction of the uptake system. Other authors (Shaner and Boyer, 1976a; Chantarotwong *et al.*, 1976; Aslam *et al.*, 1979) have concluded that uptake has more influence in regulating reduction than the reverse. Uptake and reduction are independent processes but both depend on the same variables which define the current state of the system e.g., nitrate concentrations in external solution and cytosol. As a result, the rates of the two processes will always be correlated in some manner

The light environment of the shoot may influence ion uptake (Rains, 1968 Higinbotham, 1973). This effect is also indirect and may be due to increased demand for nitrogen as the photosynthesis supply increases, to greater energy supply in the roots, to effects of light on membrane permeability in the leaves, or to increased transpiration. Certainly growth seems to strongly influence ion uptake, presumably by creating a 'sink demand.' This leads us to important questions about source-sink relations within the whole plant.

Shoot: Root Interactions

A central problem in integration, at the whole plant level, is to establish what consequences the metabolic activities in one organ have on the balance of the system. For example, growth of roots into new volumes of soil may increase nutrient supply but it requires photosynthate that might have been used in shoot growth. Brouwer (1962a, 1962b) developed the idea of a functional balance between the activities of shoots and roots in which each organ is dependent on certain unique functions of the other. Leaves supply carbohydrates and lose water while roots supply the water and nutrients.

Several lines of evidence indicate that leaves have first priority in use of the carbon supply, largely because of their proximity to the photosynthetic source (Cook and Evans, 1976). Roots, on the other hand, are considered to have a similar priority in nutrient and water use. The balance operates to restrict root or shoot growth depending upon whether shoot- or root-supplied factors are more limiting at the moment. Edwards and Barber (1976), Barta (1976) and Raper *et al.* (1977) reported data which supported a role of nitrogen in such relationships. They observed that the relative growth rate of the roots and the relative accumulation rate of nitrogen, were highly correlated in tobacco, cotton and soybean. Also, Radin's (1977) data on cotton plants showed that the relative growth rate of roots was very close to the relative rates of increase in the content of nitrate and reduced nitrogen. The experiments of Troughton (1977) with *Lolium perenne* show that the relative growth rate of roots increased if the nitrogen concentration in the external solution was diminished, while a small decrease was observed when nitrogen supply was increased. In both cases, however the relative growth of shoots was decreased as the nitrogen supply was reduced.

Schuurman and Knot (1974) observed that nitrogen increased the shoot: root ratio in *Lolium multiflorum*; Lemaire (1975) attributed this to the occurrence of maximum root growth at lower nitrogen levels than that for shoot growth. At nitrogen concentrations above 3.5 mM, shoot growth continued while root growth was depressed. In cotton, an increase in nitrogen and water supply suppressed root growth but stimulated leaf and boll growth (Aung, 1974).

Neales *et al.* (1963), from experiments with wheat, suggested that leaves influence the nitrogen content of grains indirectly by promoting the contribution of nitrogen from roots to grain and culm before any direct role of leaves as a source of reduced nitrogen. Lower leaves normally serve as the source of photosynthates for roots as shown by Tanaka (1958) in rice, and Pate (1966) in peas. In that case, the shading of lower leaves might reduce the carbohydrate supply to roots and impair nitrate uptake unless the supply is then met by upper leaves. A low supply of carbohydrate and the strong nitrogen demands of growing organs appear to be partially responsible for the initiation of senescence in older leaves.

Nitrogen demand of growth

The demand for nitrogen is determined by the growth rate and the nitrogen composition of the new tissues. In the field, both growth rate and tissue composition will vary with nitrogen and water supply, plant competition and other environmental factors. The maximum demand for nitrogen will be achieved under non-limiting conditions for photosynthesis when growth rate approaches its genetic potential.

We can start with the idea that a minimum concentration of protein, or ratio of reduced nitrogen to carbon, is required in biomass for unimpaired plant growth. Stanford and Hunter (1973) estimated that an average nitrogen content of 1.4 per cent in the biomass of mature wheat is required for maximum grain yield. They point out the remarkable constancy of that value for many wheat varieties. Van Keulen's (1977) conclusion, after analysis of many experiments, was that the minimum level to which nitrogen can be diluted in mature small grains was 1.6 per cent. From the value 1.6 per cent, and assuming a wheat biomass yield of $15\,000\,kg\,ha^{-1}$ (around $6000\,kg$ grain) it is possible to estimate a nitrogen requirement of $240\,kg$ nitrogen ha^{-1}. If uptake efficiency equals 80 per cent of the available nitrogen, then that supply must be around $300\,kg\,ha^{-1}$. That is a reasonable estimate of efficiency. For example, Spiertz and van der Haar, (1978) obtained a biomass yield of $17\,000\,kg\,ha^{-1}$ with only $100\,kg\,ha^{-1}$ of added nitrogen. Uptake efficiency was about 90 per cent in the fertilized plots.

Reports of efficiencies of 50 per cent or less are not uncommon, but in our experience, such figures involve either miscalculation or poor technology. The most common miscalculation is to base the recovery estimate on grain nitrogen alone; ignoring the nitrogen accumulated into roots and other residues which recycles to soil humus and thus is available to subsequent crops. Poor technology includes such practices as fertilization beyond the needs of the crop, fertilization well in advance of the crop season, and excessive leaching through over-irrigation.

There is no simple relationship between the nitrogen content of grain and grain yield (Evans *et al.*, 1975). Much depends on the extent of dilution by starch during grainfilling. Part of the carbon and nitrogen in grain comes from current assimilation activities of leaves, and part is remobilized from senescing leaf and stem tissues. That movement is thus associated with decline in photosynthesis and root activity. The nitrogen concentration (1.4 to 1.6 per cent) in mature biomass reflects the final balance which is achieved between grain and straw fractions. In the experiments of Thompson *et al.* (1975), maximum grain yields with semi-dwarf wheats were obtained with a protein content of 13 per cent in grain. This corresponds to about 2.3 per cent nitrogen, using 5.75 as an appropriate crude protein factor for wheat grain.

Our hypothetical mature wheat crop therefore consists of:

	Aboveground biomass	Residue	Grain
Dry matter (kg ha^{-1})	15 000	9000	6000
Organic nitrogen (%)	1.6	1.1	2.3
Organic nitrogen (kg ha^{-1})	240	100	140

In this example, each ton of grain contains 23 kg nitrogen. Bhatia and Rabson (1976) took smaller factors for the standard chemical composition of grain and found nitrogen requirements ranging from 10 kg nitrogen per ton of rice to 16 kg for wheat and rye.

It is possible to use this type of analysis in the development of a 'stress factor' for the prediction of grain yields with varying nitrogen supply. For example, Greenwood's (1976) factor was used by Rojas (1979) to estimate reductions in rice yield due to nitrogen shortage. Van Keulen (pers. comm.) uses a limit function for that purpose in his dynamic models.

The critical role of leaves

Leaves play an important role in nitrogen metabolism. They are generally the most important organs in amino acid synthesis and nitrogen storage, both because large amounts of nitrogen are required in leaf growth and because nitrogen deficiencies strongly limit leaf growth and photosynthesis.

The partitioning of nitrate reduction between roots and leaves has been shown to vary with species. Pate (1973) and Raven and Smith (1976) indicated that cereals reduce nitrate in both roots and shoots, while roots were shown to be more important in *Vicia, Pisum, Phaseolus* and *Lupinus*. Coic (1971) concluded that corn reduces more nitrate in roots than in leaves, while the reverse was true for tomato. In wheat, up to 40 per cent of the absorbed nitrate may be reduced in roots (Ashley *et al.*, 1975).

Nitrate reduction in leaves depends on transport of nitrate from the roots.

Cotransport of cations of potassium, calcium and magnesium is required for electrochemical balance. Radin (1978) postulated that division of nitrate assimilation between roots and leaves depends on the supply of carbohydrate and the ability of roots to transport nitrate into the xylem. His experiments with soybean showed that nitrate transport is restricted, so nitrate reduction in soybean leaves is limited by substrate rather than by enzyme. Transport was less restricted in cotton. With nitrogen deficiency, the supply of carbohydrate in roots and the fraction of nitrate remaining there, increased in both species.

A practical consequence of nitrate reduction in roots is that the test of nitrate content of leaves and petioles is less useful as a tool for defining the current nitrogen status of the plant. Most of the available evidence concerning the fractions of nitrate which are reduced in leaves and roots is rather indirect. Due to the potential benefits of photo-assimilation, more effort needs to be given to assessing the relative importance of leaves in reducing nitrate.

Nitrate reductase activity of leaves shows diurnal and seasonal changes. Nicholas *et al.* (1976) showed that the diurnal pattern in soybean leaves involved an increase in activity during the day and a decrease during the night. Those changes can be associated with greater nitrate fluxes in the xylem during the day as shown by Shaner and Boyer (1976a, b) for corn and/or to the movement of stored nitrate out of the leaf vacuoles under the stimulus of light (Aslam *et al.*, 1976). Harper and Hageman (1972) found with soybeans that nitrate reductase activity was maximal in mid-season just after full bloom and was associated with high nitrate levels and a high nitrogen demand for grainfilling.

In wheat, the seasonal profile of leaf nitrate reductase activity showed a value of 5.9 units of nitrate reduced g^{-1} (f.w.) h^{-1} before booting and a value of 8.2 from booting to full grain (Dalling *et al.*, 1975). All cultivars showed a peak activity at the booting to ear emergence stage, followed by a decline during grainfilling. Those changes paralleled the total nitrogen content of the plant.

It is difficult to evaluate the importance of vacuolar storage for plant production. Some plants can accumulate very large concentrations of nitrate e.g. 1 per cent nitrogen in dry matter of sugar beet leaves. The release rate seems to vary with species and is subject to requirements for electrochemical balance i.e., it must be replaced in the vacuole by another anion if neutrality and osmotic potential are to be maintained. A simulation of the effect of nitrate which is released from vacuoles showed that in young plants, vacuoles can supply enough nitrogen to sustain growth for several days (Novoa, 1979). But that process does not seem to be adequate to meet the large nitrogen demand during grainfilling. It is then that one observes the breakdown of leaf proteins.

Nitrogen, stored as protein, becomes important after vacuolar nitrogen is used. Ribulose 1,5 bisphosphate carboxylase appears to play a fundamental role not only in photosynthesis but also in storage of reduced nitrogen. Huffaker and Peterson (1974) and Martin and Thimann (1972) observed that protein losses in leaves and seedlings is accounted-for mainly by a decrease in that enzyme. Gregory and Sen (1937) showed that the remobilization of nitrogen is influenced by age and nitrogen supply.

In plants with an adequate supply, protein losses are high only in older leaves; in plants under low supply mature leaves lose protein more or less linearly with age. With extreme deficiency, protein levels of very young leaves may be reduced. Evans *et al.* (1975) estimated that in wheat grown under low nitrogen, virtually all of the grain nitrogen is derived from remobilization. Where nitrogen absorption is still possible, remobilization may account for less than 50 per cent of grain nitrogen. Spiertz (1977) and Spiertz and van der Haar (1978) reported that 60 per cent to 94 per cent of the wheat grain nitrogen is remobilized from vegetative organs under Netherlands conditions.

Field experiments by Austin *et al.* (1976) showed that those wheat varieties which lost the most dry weight from leaves during grainfilling, also took up the least nitrogen in that period, and that variation in leafiness was the main cause of variation in nitrogen content per plant. The high harvest index of modern wheat varieties implies that a larger fraction of the stored nitrogen is remobilized during grainfilling. McNeal *et al.* (1971) also observed that wheat varieties differ in their ability to remobilize, and Huffaker and Rains (1978) have underlined the importance of remobilization for its influence on the protein content of wheat grain.

Remobilization is a major determinant of nitrogen use efficiency by the whole crop. The extreme remobilization common to soybean, lentil and pea that led Sinclair and de Wit (1975) to their self-destruction hypothesis, may simply represent an adaptation for high efficiency in the use of scarce supplies of nitrogen. Leopold (1961) pointed out that remobilization is so efficient in oats that the plant can acquire enough nitrogen during vegetative phase for the entire life cycle. That characteristic can be important in areas of low rainfall where nitrogen uptake is periodically impaired by low moisture supply. Remobilization can be decreased (Spratt and Gasser, 1970a), or increased (Turner, this volume) by water stress. It may be possible to select cereal and legume varieties that are highly efficient in remobilization under conditions of low water supply.

NITROGEN AND YIELD

It seems appropriate to discuss the influence of nitrogen on agronomic yield giving emphasis to cereals, and to differentiate between source-sink and morphological aspects of the problem. The source-sink concept provides a very useful approach but we must keep in mind that other processes such as nutrient uptake, transport and developmental phase can determine whether sink or source is limiting (Evans *et al.*, 1975).

Nitrogen effects on source capacity

The source capacity of plants is primarily determined by leaf area, leaf area duration, and rates of photosynthesis, respiration and animo acid synthesis. Leaf area is the main factor in biomass formation (Watson, 1947), and it varies in amount with plant population and nutrient supply. Leaf area is normally increased by nitrogen (Watson *et al.*, 1958; Langer and Liew, 1973; Pearman *et al.*, 1977; Spiertz and Ellen, 1978; Spiertz and van der Haar, 1978). The greater leaf area can be due to an effect on leaf number or leaf size. Langer and Liew (1973) did not find any effect of nitrogen on the leaf number of the wheat main shoot. But leaf number per plant will increase with nitrogen supply due to an increase in tiller number (Halse *et al.*, 1969; Pearman *et al.*, 1977 with wheat; and Chandler, 1969 with rice). The extra tillers are obtained if nitrogen is applied early (Spratt and Gasser, 1970a) but that has little effect on yield unless they also produce grain. The area of each lamina is also increased by nitrogen due to effects on both cell number and size (Hewitt, 1963). Metivier and Dale (1977) reported that barley grains, rich in nitrogen, produced seedlings with first leaves of larger area than did grains which were low in nitrogen. Pearman *et al.* (1977) observed increases of wheat leaf laminae due to nitrogen. In the opinion of these researchers, the partitioning of the effect of nitrogen on leaf area between cell size and number is of doubtful physiological significance since an inverse relationship between these components is observed with changes in plant competition.

Leaf area duration is also extended by nitrogen (Langer and Liew, 1973; Pearman *et al.*, 1977; Thomas *et al.*, 1978). Thorne's (1974) analysis of Rothamsted experiments before 1965 and experiments in Holland, Australia and Canada, led her to conclude that most of the variations in wheat grain yields due to nitrogen could be related to variation in leaf area duration. However, the correlation did not hold under conditions of intensive agriculture with leaf area indices higher than 7.

If the ratio of grain yield to leaf area duration after anthesis is constant, differences in grain yield can be attributed to differences in leaf area duration. In that case, factors such as nitrogen that affect leaf area duration are likely to be involved. Pearman *et al.* (1977) observed a lower grain: leaf ratio with increasing nitrogen, indicating that the efficiency of leaves for grain production was decreased. The effect of nitrogen was to increase leaf area more than grain yield.

Loomis and Gerakis (1975) pointed out that photosynthetic productivity shows a curvilinear, plateau response to increasing leaf area index. They therefore suggested that leaf area attributes such as duration will not be linearly related to production except at low leaf area indices (less than 2 to 3); and that the integration of per cent cover will correlate with production better than leaf area duration. Excluding the role of leaves in nitrogen and carbohydrate storage for later remobilization, the least area of leaves for the greatest duration of cover would be most efficient.

The rate of photosynthesis also is influenced by nitrogen nutrition. Burström (1943) observed an enhancement in apparent CO_2 assimilation with a rise in the nitrate content of wheat leaf laminae; Osman and Milthorpe (1971) found that the increase in the gross photosynthesis in that species was linear with nitrogen content. Murata (1969) reported a positive correlation between total protein nitrogen content and photosynthetic activity of rice, and Watanabe and Yoshida (1970) observed a similar relationship in that species between chlorophyll content and photosynthesis rate. Medina (1969, 1970) concluded that nitrogen stress reduced photosynthesis of *Atriplex* leaves mainly through changes in carboxylase enzyme activity. Pearman *et al.* (1977) observed only a small effect of nitrogen on the photosynthesis of flag leaves of wheat, but combined with changes in leaf area, photosynthesis per plant was increased by 30 per cent. The effects appear to be dependent on solar radiation; at low light, the change in net photosynthesis of a wheat crop due to nitrogen was very small (Spiertz and van der Haar, 1978).

Thomas *et al.* (1978) did not find any effect of nitrogen on photosynthesis or photorespiration of wheat but suggested that there was an effect on dark respiration. Increased respiration rates have been reported in rice (Tanaka *et al.*, 1966) and wheat (Spiertz and van der Haar, 1978) as a consequence of higher nitrogen levels. As noted earlier, Penning de Vries *et al.* (1974) provide a means for calculating the respiration associated with the biosynthesis of biomass. Protein synthesis is more costly in metabolism than is starch or cellulose synthesis and, as the protein content of new biomass increases, the amount of respiration associated with that synthesis will also increase. More recently, Penning de Vries (1975) estimated that 60 per cent of maintenance respiration is related to protein turnover rate and thus, like biosynthesis respiration, to protein content.

Thus, an adequate supply of nitrogen generally stimulates not only photosynthesis but also amino acid and protein synthesis and respiration. While the greater supply of photosynthate leads to greater growth, the resulting biomass generally has a lower carbon: nitrogen ratio.

Nitrogen effects on yield components that determine the sink capacity of plants

The sink capacity of a cereal plant is determined by the number and size of grains. Their rates of growth, expressed per unit area of land, represent the grain sink capacity. Grain number is dependent on the number of ears per unit area, the number of spikelets per ear and the number of fertile florets per spikelet. The importance of grain number as the main determinant of wheat yield has been pointed out by Thorne *et al.* (1968) and many others (Evans and Wardlaw, 1976; Ledent, 1977; Spiertz and Ellen, 1978; Kolderup, 1978; and by Yoshida, 1972 for rice). In small grains, nitrogen generally increases the number of tillers, resulting in a greater number of ears per unit land area (Halse *et al.*, 1969; Pushman and Bingham, 1976). In cases where ear number is little affected, nitrogen may then cause an increase in ear weight up to 70 per cent over the control (Gasser and Iordanou, 1967; Spratt and Gasser, 1970a). That can result from an increase in the number of spikelets per ear if the nitrogen is applied early (Langer and Liew, 1973; Holmes 1973; McNeal *et al.*, 1971; Pearman *et al.*, 1977). The number of grains per spikelet (fertile florets) was not affected by nitrogen in some experiments (Pearman *et al.*, 1977), but was increased in others when tillers were removed (Langer and Liew, 1973).

Grain weight generally shows much less variation than grain number. Grain weight may be greater with higher nitrogen (Halse *et al.*, 1969; Langer and Liew, 1973; Spiertz and van der Haar, 1978; Spiertz and Ellen, 1978; Pearman *et al.*, 1977) or unchanged (Pushman and Bingham, 1976). It is possible that Pearman's results were due to water stress, although they estimated that this was negligible. In rice, heavy doses of nitrogen promote the sterility of spikelets (Murata, 1969). In wheat, nitrogen promotes the rate of floret development, the number of fertile florets (Holmes, 1973; Langer and Hanif, 1973; Thomas *et al.,* 1978) and the number of grains set (Thomas *et al.*, 1978).

Compensation phenomena observed with yield components in grains (Evans and Wardlaw, 1976) impair the optimization of yields by a simple combination of optimum yield components. As an example, an excess of tillers, as determined early in the crop cycle, is related later to smaller head or grain size. Canvin (1976) suggested that compensation phenomena are due in large part to the relative supplies of carbohydrate and nitrogen during critical stages of development.

The nitrogen effect on protein content of grains and its consequences for yield

It is well established that increasing the supply of nitrogen to cereals results in progressive increments in yield and total protein content of grain (Watson *et al.*,

1958; Halse *et al.*, 1969; Spratt and Gasser, 1970a; Langer and Liew, 1973; Spiertz and Ellen, 1978). Hucklesby *et al.* (1971) reported a 37 per cent increase in the protein content of the wheat variety 'Arthur' that was paralleled with a yield rise, while for the variety 'Blueboy', the increase in protein was associated with a yield reduction. Negative correlations between grain protein and yield were observed by Grant and McCalla (1949), Terman *et al.* (1969), Evans and Wardlaw (1976), Sinclair and de Wit (1975) and Hageman *et al.* (1976). The analysis by Pushman and Bingham (1976) of the ratio of grain protein to grain yield for 10 wheat varieties showed that the slope of the negative regression was not affected by nitrogen supply. The differences in protein content between wheat varieties, at the same level of nitrogen supply, can be up to 14 per cent (Johnson *et al.*, 1963; Pushman and Bingham, 1976), and nitrogen level can change protein content up to 5 per cent (Pushman and Bingham, 1976).

Benzian and Lane (1979) analyzed the relationship between nitrogen supply, grain protein content and yield in numerous experiments at Rothamsted and elsewhere. They found that a greater nitrogen supply increased the grain protein content more or less linearly, while yield showed a dimishing return pattern. When nitrogen is strongly limiting, small additions can result in greater yields but with decreased protein concentrations. That dilution effect comes as endosperm (low protein) is increased relative to embryo (high protein). At higher levels of nitrogen, increases in nitrogen can increase both yield and protein concentration; and even higher levels, if they increase protein levels further, may decrease yield.

Part of the slight negative relationship between protein content and yield at high nitrogen, can be attributed to the higher glucose costs for synthesis of protein than of carbohydrates (1 g glucose yields 0.45 g protein or 0.86 g starch, according to Penning de Vries *et al.*, 1974). The high cost of protein synthesis means that, for the same yield, an increase in protein of 1 per cent would require a 1 per cent increase in assimilate supply (Bhatia and Rabson, 1976).

Donovan *et al.* (1977) observed a lower kernel weight in a high protein variety of wheat as compared to a low protein variety. The difference in weight and protein concentration was due to higher levels of carbohydrates, other than starch, in the low protein variety. Pushman and Bingham (1976) found a positive association between grain weight and protein content while Shokr and Stolen (1979) found a slight negative relationship. Grain weight varies with grain number tending to confuse attempts to correlate grain weight with protein content. Shokr and Stolen also found a positive correlation between grain protein concentration and lodging that could be an additional, although indirect, reason for the negative correlation of grain protein and yield. Kramer (1979), Hageman *et al.* (1976), Evans and Wardlaw (1976) and Benzian and Lane (1979) have expressed the opinion that the negative correlation between grain protein content and yield might be changed with management practices such as supplying nitrogen late in the season.

Morphological characteristics important in the response of small grains to nitrogen

Plant height, stiffness of stems, leaf angles, tillering capacity and ear weight which have important influences on the response of small grains to nitrogen, have been given special attention with rice (Chandler, 1969; Yoshida, 1972). The height of rice plants is increased by nitrogen (Chandler, 1969) and the same is true for wheat (McNeal *et al.*, 1971). Taller plants usually provide a better ventilated canopy, and CO_2 exchange may be improved. But the elongation of basal stem internodes of wheat enhances the danger of lodging. A major advantage of dwarf varieties of small grains is their high resistance to lodging (Chandler, 1969; Porter *et al.*, 1964).

Increased nitrogen inputs can result in extremely dense communities with a large leaf area index, and a more vertical manner of display would then increase productivity (Loomis and Williams, 1969). In rice, small and erect leaves were taken as a breeding objective to achieve a favorable response to nitrogen (Chandler, 1969). But even with adequate nitrogen, rice has seldom been grown at densities sufficient for an erect leaf advantage, and the correlation is more likely due to stiff straw (Loomis and Gerakis, 1975).

The tillering response to nitrogen, and the associated increase in leaf area and ear number can be a central feature of short, erect-leaf varieties. That plasticity tends to maintain a balance in sink-source relationships in the response to nitrogen.

Nitrogen-use efficiency

An efficient use of nitrogen in plant production is an essential goal in crop management. The values for efficiencies in the use of applied nitrogen collected in Table 1 show a wide variation due to differences in experimental conditions and to the hyperbolic nature of the yield response to nitrogen. If we distinguish between biological and economic yields, we can define the efficiency of nitrogen use from the recovery of available or applied nitrogen, or as the ratio kg biomass or kg grain to kg applied nitrogen. Agronomists tend to be more interested in the latter aspect, but the nitrogen content of residues is the input to soil humus and thus becomes a part of the supply for subsequent crops. All of these definitions are global since they involve variations in plant growth due to variations in environ-

Table 1. Some reported values of recorvery fraction (RF) and agronomic and physiological efficiencies in the use of nitrogen.

Crop	RF	Physiological efficiency	Agronomic efficiency	References
Barley	0.58–0.69			Gasser and Iordanou, 1967.
Rice	0.10–0.56	71.4	7.1–40.0	van Keulen, 1977
Wheat	0.44–0.60			Stanford and Hunter, 1973.
Wheat	0.58–0.69			Gasser and Iordanou, 1967
Wheat	0.30–0.90	33.5–39.4	15.0–45.0	Hamid, 1972
Wheat			33.0[a]–88.0[b]	Pushman and Bingham, 1976
Wheat		28.2–46.7	24.1–35.2	Pearman et al., 1977
Wheat	0.28–0.38	42.5–48.3	35.0–[a]–99.0[b]	

[a] High nitrogen supply
[b] Low nitrogen supply

ment (including the availability of water and soil nitrogen), in plant factors such as root uptake efficiency, chemical composition of tissues and nitrogen remobilization, and in management factors including timing, rate, method, and the kind of nitrogen dressing. Leaching, denitrification, mineralization of organic matter and other soil processes are also included.

The ratio kg grain:kg nitrogen absorbed, focuses on physiological efficiency in using the nitrogen actually absorbed, and summarizes the effect of plant factors. Global and physiological efficiencies are related by the recovery fraction (RF, kg nitrogen absorbed:kg applied) so that:

Agronomic efficiency = Physiological efficiency × recovery fraction

or $\quad AE = \dfrac{\text{kg grain}}{\text{kg N applied}} = \left(\dfrac{\text{kg grain}}{\text{kg N absorbed}}\right)\left(\dfrac{\text{kg N absorbed}}{\text{kg N applied}}\right)$

Some reported values of these factors are shown in Table 1.

Agronomic efficiency can be increased by increasing the physiological efficiency, the recovery fraction, or both. The recovery fraction serves as a better index than nitrogen use efficiency for the evaluation of fertilizer practices. It involves two components:

$$RF = \left(\dfrac{\text{kg N absorbed}}{\text{kg N supply}}\right)\left(\dfrac{\text{kg N supply}}{\text{kg N applied}}\right)$$

The first ratio in this equation, represents the uptake efficiency which depends upon root properties such as distribution, surface area and uptake per unit surface area. Surface area of roots is related to the root age and morphology (including branching, length and diameter). Increased nitrogen supply may slow the growth of length and diameter of wheat roots during early spring but later produces larger root systems (Welbank *et al.*, 1974). On the other hand, the uptake per unit area will depend on factors such as concentration of ions around the roots, affinity of the nitrate or ammonium carriers for those ions, and on the number of such carriers (Epstein, 1972; Rao and Rains 1976a; Huffaker and Rains, 1978). The rate of ion uptake changes with time and along the root axes (Milthorpe and Moorby, 1974), but that has received very little study.

Timing is important. Experiments with wheat (Hamid, 1972) show that the maximum recovery of fertilizer nitrogen in the grain was from the nitrogen applied at tillering rather than at earlier or later stages. It may be that greater accumulation, and greater remobilization during grainfilling, occur when nitrogen is applied at tillering. Stanford and Hunter (1973) observed a 8 per cent greater recovery from spring- than from fall-applied nitrogen. In a dry spring, the inverse may be true, since late dressings are less effective under low moisture conditions (Dimitrenko *et al.*, 1977).

The supply of nitrogen is influenced by soil conditions. Leaching, denitrification, immobilization of nitrogen in the biomass of soil microorganisms and then in humus, and mineralization of organic matter are important and also are influenced by water supply. Fernandez and Laird (1959) showed that under optimum soil moisture, wheat yielded 24 kg grain kg^{-1} applied nitrogen but only 11 kg/kg N when water was limiting. The recovered nitrogen in the grain decreased from around 20 per cent to 4 per cent when water supply was restricted in the experiments of Spratt and Gasser (1970b), and from around 30 per cent with water kept af field capacity to 15 per cent at -2.4 bars in the experiments of Thompson *et al.* (1975). In such experiments, water supply would affect the growth rate as well as availability of soil nitrogen.

The rate and kind of fertilizer used and the method of application influence the recovery of nitrogen. The effect of the rate seems variable since Bartholomew and Hiltbold (1952) and Hamid (1972) observed a greater recovery of nitrogen as the doses were increased, while Stanford and Hunter (1973). McNeal *et al.* (1971) and Gasser and Ioardanou (1967) found the reverse. The possibility for nitrogen losses through leaching and denitrification, point to decreased recovery as the general response. Increased recovery can occur where root growth is enhanced or the nitrogen needs of soil bacteria (high carbohydrate residues) are overcome by added fertilizer.

Spratt and Gasser (1970b) found that recovery was greater from nitrate than from ammonium under adequate water supply, but with limited water, the difference was smaller, probably is reflecting the much greater mobility of nitrate in soils. Recovery is improved when the fertilizer is concentrated in the root zone in bands rather than broadcast on to the surface (Daigger and Sander, 1976). Recovery may be greatest with foliar sprays but their practical use is limited by the need for frequent applications.

The physiological efficiency of nitrogen use (kg grain/kg N absorbed) is dependent on the nitrogen efficiency of biomass formation, the effect of nitrogen on carbohydrate partitioning, nitrate reduction efficiency, and remobilization of protein nitrogen from senescent tissues as well as transport and storage functions. Bruetsch and Estes (1976) studied the variation in nutrient efficiency for above-ground dry matter production by 12 corn genotypes. The efficiency of nitrogen use ranged from 59 to 82 kg biomass/kg N absorbed showing the importance of the genetic component. In a field experiment which compared two wheat and three triticale varieties, efficiencies for total dry matter production ranged from 70 to 120 kg biomass/kg N absorbed (Pino, 1979). The efficiency of wheat was greater than triticale. Varietal differences were also detected. Such differences can be due to variations in chemical composition, translocation, nitrogen effects on carbohydrate partitioning and other factors.

Chemical composition, essentially the per cent protein in the biomass or grain, will determine how much biomass or grain can be produced per unit of absorbed nitrogen. Earlier, we saw that 1.6 per cent nitrogen is a minimum content for most small grains. That corresponds to 63 kg biomass/kg N absorbed and to an average protein content of 9 per cent.

Given an estimate of soil nitrogen supply and potential biomass formation, and assuming a certain uptake efficiency, we can extend the example given earlier into an estimate of the supply of available nitrogen required in the soil for maximum production. A graphical method has been proposed for rice by van Keulen (1977) and Rojas (1979). The negative relationship between the protein content of grain and grain yield, is such that increases in nitrogen supplies that generate protein contents above 12 per cent will decrease the agronomic efficiency of nitrogen used for grain production unless a parallel rise in photosynthate supply occurs.

Rather obviously, increasing plant photosynthesis would support a greater assimilation of nitrogen and increased yields. Efficiency will also vary according to the type of protein being made. An increase in the proportion of the amino acid lysine will enhance the glucose requirement for grain growth, by as much as 25 per cent for a high-lysine barley cultivar. Rabson et al. (1978) concluded that 'cultivars with improved protein are likely to lag behind top yielding cultivars in grain yield unless genotypes with more efficient nitrogen and carbon utilization can be identified.' However, their argument is clearly circular since such genotypes could then produce an even greater yield of grain with normal protein!

Nitrogen use efficiency, then, will vary with the chemical composition of biomass. It will also vary with changes in the proportions of various tissues and organs of the plant since these vary in composition and in their contributions to nitrogen assimilation. Carbohydrate (photosynthate) partitioning is strongly involved in such variations. Goodman (1977) explored the possibility of improving the response of *Lolium* to nitrogen. He found strongly heritable variation between species and between populations within species in the response of leaf area and shoot weight to added nitrogen. Root dry weight responses also varied with genotype.

Data presented by Pino (1979) show a lowering in the efficiency of absorbed nitrogen use in biomass and grain formation as a result of increased nitrogen supply. That effect can be due in part to a higher protein content of the biomass, in part to an increased storage of nitrogen and in part to a change in the grain:biomass ratio ('harvest index'). The grain:biomass ratio was increased by nitrogen, in this case, from 0.16 to 0.44 in one of the wheat varieties and from 0.32 to 0.39 in another. This increment explains the reduction in the physiological efficiency of biomass production since grains are richer in protein.

The experiment of Pearman et al. (1977) showed that an enhanced nitrogen supply diminished the physiological efficiency of grain production. They suggested that this was due to stimulation in the use of photosynthates for stem growth, or to increased respiration. In a later experiment, Pearman et al. (1978) found no effect of nitrogen on carbohydrate distribution, so they concluded that the effect of nitrogen was on respiration. Makunga et al. (1978) continued that line of research. They confirmed the negligible effect of nitrogen on carbon distribution, despite the large increase in growth and yield. But their data also show that the grain:straw ratio was decreased by nitrogen one year and increased in another, pointing to variation in the partitioning of assimilates.

McDowall (1972) analyzed the growth kinetics of 'Marquis' wheat, and compared nitrogen use in the growth of the various organs. Roots grew with the highest efficiency and also had the first priority on nitrogen use. Thomas et al. (1978) designed an experiment to study the effect of nitrogen on the efficiency of wheat leaves in producing carbohydrates for the grain. Extra nitrogen increased total photosynthesis, but the extra carbon did not appear in the stem or grain weight and no increase in photorespiration activity was detected. So, they proposed an increase in dark respiration to account for the smaller efficiency in grain production.

Nitrate reduction may depend on the quantity of nitrate reductase present, the activity of the enzyme, or on the availability of nitrate or reductant. The level of nitrate reductase depends on genotype, and it has been suggested that its activity might be useful as a productivity index for grain and grain protein (Hageman *et al.*, 1976). We question the practicality of that approach: enzyme assays are not necessarily simple, and this enzyme is highly inducible by nitrate supply. High levels of nitrate reductase may well be an effect, rather than the cause of, increased protein production. If the enzyme were limiting, one would expect to find that the plant would accumulate nitrate and photosynthate while having a low growth rate. Our analysis fits with the conclusion of Hageman *et al.* (1976) that under general farming conditions, corn seems unable to reach its genetic potential for nitrate reduction and protein production because of lack of nitrate.

Radin (1975), with soybean and cotton, also found that nitrate reduction was more limited by substrate availability than by enzyme levels. Blackwood and Hallam (1979) reported good correlations between yield and nitrate reductase activity in wheat, but suggested that the correlation would vary with nitrate supply and other factors. Dalling *et al.* (1975) found differences among wheat varieties in the yield of reduced nitrogen for the same level of nitrate reductase activity. They suggested various reasons to account for that, including bio-chemical differences of nitrate reductases, nitrate reductase decay rates, competition for reductant by other enzymes, and differences in nitrate pool sizes.

The reduction of nitrite to ammonium requires NADH in nonphotosynthetic tissues but in leaves, nitrite reductase uses reduced ferredoxin as a cofactor which could reduce the cost of reduction. The reduction of ferrodoxin uses light as its source of energy, but the reduction of NAD requires carbohydrate oxidation. Since the use of NADH can place a significant drain on carbohydrate reserves, it is of great interest to estimate how much of the absorbed nitrate is reduced in leaves and how much is reduced in roots. It is possible that root growth is dependent on its own nitrate reduction as was found for cotton (Radin, 1977). If this is true, it serves to limit the amount of nitrate that will be reduced in leaves.

The translocation of reduced nitrogen will influence the efficiency of nitrogen use (Gasser and Iordanou, 1967). Translocation efficiency (estimated from the ratio of grain nitrogen to total organic nitrogen) was found to range between 22 and 78 per cent in various wheat varieties by Dalling *et al.* (1975) and Hageman *et al.* (1976). Pino (1979) observed a greater translocation efficiency for wheat (77 to 80 per cent) than for triticale varieties (53 to 58 per cent). A high translocation efficiency, coupled with large capacities for uptake and assimilation, should improve nitrogen use efficiency (Huffaker and Rains, 1978). Protease activity is required for the breakdown of protein in plant leaves, and that can be important to translocation. The self-destruction hypothesis of Sinclair and de Wit (1975), in which the large amounts of nitrogen required for grainfilling are obtained from vegetative tissues which then lose physiological activity, can be considered an efficient way of using nitrogen. In that case, proteins are made when nitrate is in free supply and used when nitrogen supply is low.

Translocation efficiency can be decreased by lodging (Hageman et al., 1976) and water stress (Spratt and Gasser, 1970a). High nitrogen supply seems to reduce the translocation efficiency in some cases (Gasser and Iordanou, 1967; Hucklesby et al., 1971; McNeal et al., 1971), but not in others (Spiertz and Ellen, 1978).

OVERVIEW

The effects of nitrogen on plant production derive from biochemical, physiological and morphological processes. At the biochemical level, the high costs of nitrogen assimilation and the difference in costs of nitrite reduction between roots and leaves appear as important issues for nitrogen use efficiency. The costs are smaller for leaves in light than for roots. Nitrate reduction and nitrogen fixation costs appear to be similar, indicating the high value in the nitrogen economy of farming of the development of rotations with high yielding cultivars of nitrogen-fixing species. Such cultivars can represent a large energy saving of around 140 l of petroleum equivalents per hectare, assuming a cost of 14 000 kcal/kg fertilizer N (Producers Nitrogen plant, California, using natural gas) and a rate of $100 \, kg \, N \, ha^{-1}$. The problem of developing high-yielding legumes is partially related to the chemical composition of their grains since that accounts for 15 per cent of the large differences in grain yield between cereals and pulses and 50 per cent of the differences between cereals and oil seed legumes (soybean, peanut). Reported yields show that cereals can produce 50 to 100 per cent more than pulses. Net photosynthesis rates per unit leaf area are similar for beans and wheat (Rabson et al., 1978), and harvest indices of legumes are often higher than those of cereals. Clearly there are other important differences which need to be defined. Leaf area development, pH balancing and disease resistance may be among those.

The uptake efficiency of nitrate plants can probably be improved. At the whole-plant level, nitrogen use efficiency is intimately related to carbon metabolism and partitioning. Leaf area and tillering responses to nitrogen are important since they result in increased source and sink capacities. Lodging resistance is an essential characteristic in cereals for tolerance and high response to adequate nitrogen. Remobilization of protein nitrogen also seems to be of great importance, and McNeal *et al.* (1966) suggested that movement of 70 per cent of the nitrogen from leaves and stems to grain would represent good efficiency. At the crop level, the method and timing of the supplies of water and nitrogen have a strong impact on nitrogen use efficiency.

The nitrogen economy of crop production involves considerable complexity, and one must take a systems view in efforts at improvements in management practices. We have attempted to provide a background of understanding for such a systems view.

ACKNOWLEDGEMENTS

Support for a portion of this work was provided by a grant from the National Science Foundation (PFR77–07301 A01). D. K. McDermitt provided critical comments on the manuscript and, with S. Morrison, helped us in conceptualizing the study.

REFERENCES

Aslam, M., Huffaker R. C., Rains, D. W. and Rao, K. P. 1979 Influence of light and ambient CO_2 concentration on nitrate assimilation by intact barley seedlings. Plant Physiol. 63; 1205–1209.

Ashley, D. A., Jackson, W. A. and Volk, R. J. 1975 Nitrate uptake and assimilation to wheat seedlings during initial exposure by nitrate. Plant Physiol. 55, 1102–1106.

Aung, L. H. 1974 Root-shoot relationships. In: Carson, E. W. (ed.), The Plant Root and Its Environment. University Press of Virginia, Charlotteville, p. 29–61.

Austin, R. B., Ford, M. A., Edrich, J. A. and Blackwell, R. D. 1976 The nitrogen economy of winter wheat. J. Agr. Sci., Camb. 88: 159–167.

Barley, K. P. 1970 The configuration of the root system in relation to nutrient uptake. Adv. Agronomy 22: 159–201.

Barta, A. 1976 Transport and distribution of CO_2 assimilate in *Lolium perenne* in response to varying nitrogen supply to halves of a divided root system. Plant Physiol. 38: 48–52.

Bartholomew, W. V. and Hiltbold, A. E. 1952 Recovery of fertilizer N by oats. Soil Sci. 73: 193–201.

Bassioni, N H 1971 Temperature and pH interaction in NO_3^- uptake. Plant and Soil 35: 445–448.

Benzian, B. and Lane, P 1979 Some relationships between grain yield and grain protein of wheat experiments in South-East England and comparisons with such relationships elsewhere. J. Sci. Food. Agr. 30, 59–70.

Bhatia, C. R. and Rabson, R 1976 Bioenergetic considerations in cereal breeding for protein improvement. Science 194, 1418–1421.

Blackwood, G. C. and Hallam, R. 1979 Nitrate reductase activity in wheat (*Triticum aestivum* L.). II. The correlation with yield. New Phytol. 82, 417–425.

Brouwer, R. 1954 The regulating influence of transpiration and suction tension in water and salt uptake by roots of intact *Vicia faba* plants. Acta Bot. Nederland. 3, 264–312.

Brouwer, R. 1962a Distribution of dry matter in the plant. Neth. J. Agr. Sci. 10, 361–376.

Brouwer, R. 1962b Nutritive influences on the distribution of dry matter in the plants. Neth. J. Agr. Sci. 10, 399–408.

Brouwer, R. 1965 Water movement across the root. In: Symp. Soc. Exp. Biology 19, 131–149.

Bruetsch, T. F. and Estes, G. O. 1976 Genotype variation in nutrient uptake efficiency in corn. Agron. J. 68, 521–523.

Burström, H. 1943 Photosynthesis and assimilation of nitrate by wheat leaves. Ann. Royal Agr. Coll. Sweden 11, 1–50.

Canvin, D. T. 1976 Interrelationships between carbohydrate and nitrogen metabolism. In: Genetic Improvement of Seed Protein. Proc. of a Workshop, 1974. p. 172–195. Natl. Acad. Sci., Washington, D.C.

Chandler, R. F. 1969 Plant morphology and stand geometry in relation to nitrogen. In: Eastin, J. D., Haskins, F. A., Sullivan C. Y. and Van Bavel C. H. M., (eds.), Physiological Aspects of Crop Yields. pp. 265–289. Am. Soc. Agron. Madison, Wisconsin.

Chantarotwong, W., Huffaker, R. C., Miller, B. L. and Granstedt, R. C. 1976 *In vivo* nitrate reduction in relation to nitrate uptake, nitrate content and *in vitro* nitrate reductase activity in intact barley seedlings. Plant Physiol. 57, 519–522.

Coic, Y. 1971 Influence du metabolism de NO_3^- dans les racines sur l'etat nutritional de la plant. In: Samish, R. H. (ed.), Recent advances in Plant Nutrition. p. 217–227. Gordon and Breach Sci. Publ., New York.

Cook, M. G. and Evans, L. T. 1976 Effect of size geometry and distance from source on the distribution of assimilates in wheat. In: Wardlaw, I. F. and Passioura J. B. (eds.), Transport and Transfer Process in Plants. p. 393–400. Academic Press, New York.

Daigger, L. A. and Sander, D. H. 1976 Nitrogen availability to wheat as affected by depth of nitrogen placement. Agron, J. 68, 524–526.

Dalling, M. J., Halloran, G. M. and Wilson, J. H. 1975 The relation between nitrate reductase activity and grain nitrogen productivity in wheat. Austr. J. Agr. Res. 26, 1–10.

Dimitrenko, P. A., Thomashevskaya, Y. G., Golovashchuk, Z. T., Insshin, N. A., Semonova, N. K. and Dan'ko, A. Y. 1977 Utilization of fertilizer nitrogen by winter wheat and sugar beet under different fertilizer application conditions. Soviet Soil Sci. 9, 540–551.

Donovan, G. R., Lee, J. W. and Hill, R. D. 1977 Compositional changes in the developing grain of high and low protein wheats. I. Chemical composition. Cereal Chem. 54, 638–645.

Edwards, J. H. and Barber, S. A. 1976 Nitrogen flux into corn roots as influenced by shoot requirement. Agron. J. 689, 471–473.

Epstein, E. 1972 Mineral Nutrition of Plants. Principles and Perspectives. John Wiley and Sons, New York. 412 p.

Evans, L. T., Wardlaw, J. F. and Fischer, R. A. 1975 The physiological basis of crop yield. In: Evans, L. T. (ed.). Crop Physiology. p. 327–355. Cambridge Univ. Press. 374 p.

Evans, L. T. and Wardlaw, I. F. 1976 Aspects if the comparative physiology of grain yield in cereals. Adv. Agron. 28, 301–359.

Ezeta, F. H. and Jackson, W. A. 1975 Nitrate translocation by detopped corn seedlings. Plant Physiol. 56, 48–156.

Fernandez, R. and Laird, R. T. 1959 Yield and protein content of wheat in central Mexico as affected by available soil moisture and nitrogen fertilization. Agron. J. 51, 33–36.

Gasser, J. K. R. and Iordanou, I. G. 1967 Effects of ammonium sulphate and calcium nitrate on the growth, yield and nitrogen uptake of barley, wheat and oats. J. Agr. Sci. 68, 307–316.

Goodman, J. J. 1977 Selection for nitrogen responses in *Lolium*. Ann. Bot. 41, 243–256.

Grant, M. N. and McCalla, A. G. 1949 Yield and protein content of random selections from single crosses. Can. J. Res. 27, 230–240.

Greenwood, E. A. N. 1976 Nitrogen stress in plants. Adv. Agron. 28, 1–33.

Gregory, F. G. and Sen, P. K. 1937 Physiological studies in plant nutrition. VI. The relation of repiration rate to carbohydrate and nitrogen metabolism of barley leaf as determined by nitrogen. Ann. Bot. 1, 521–561.

Hageman, R. H., Lambert, R., Loussaert, D., Dalling, M. and Klepper, L. A. 1976 Nitrate and nitrite reductase as factors limiting protein synthesis. In: Genetic Improvement of Seed Protein. Proc. of a Workshop, 1974. p. 103–131. Natl. Acad. Sci. Washington, D.C.

Hallmark, W. B. and Huffaker, R. C. 1978 The influence of ambient nitrate, temperature and light on nitrate assimilation in sudan grass seedlings. Physiol. Plant. 44, 147–152.

Halse, N. J., Greenwood, E. A. N., Lapins, P. and Boundy, C. A. P. 1969 An analysis of nitrogen deficiency on the growth and yield of Western Australia wheat crop. Aust. J. Agr. Res. 20, 987–998.

Hamid, A. 1972 Efficiency of N uptake by wheat as affected by time and rate of application using N^{15} labelled ammonium sulphate and sodium nitrate. Plant and Soil 37, 389–340.

Harper, J. E. and Hageman, R. H. 1972 Canopy and seasonal profiles of nitrate reductase in soybeans (*Glycine max* L. Merr.). Plant Physiol. 49, 146–154.

Hewitt, E. J. 1963 The essential nutrient elements, requirements and interaction in plants. In: Steward, F. C. (ed.), Plant Physiology, a Treatise. Vol. III, Chapter 2. Academic Press, New York.

Hewitt, E. J., Hucklesby, D. P. and Notton, D. A. 1976 Nitrate metabolism. In: Bonner, J. and Varner, J. E. (eds.). Plant Biochemistry. p. 633–681. Academic Press, New York.

Higinbotham, N. 1973 The mineral absorption process in plants. Bot. Rev. 39, 15–69.

Holmes, D. P. 1973 Inflorescence development of semidwarf and standard height wheat cultivars in different photoperiod and nitrogen treatments. Can. J. Bot. 51. 941–956.

Hucklesby, D. P. C., Brown, M., Howell, S. E. and Hageman, R. H. 1971 Late spring applications of nitrogen for efficient utilization and enhanced production of grain protein of wheat. Agron. J. 63, 274–276.

Huffaker, R. C. and Rains, D. W. 1978 Factors influencing nitrate acquistion by plants; assimilation and fate of reduced nitrogen. In: Nielson, D. R. and McDonald, J. G. (eds.), Nitrogen in the Environment. Vol. 2, pp. 1–43. Academic Press, New York.

Huffaker, R. C. and Peterson, L. W. 1974 Protein turnover in plants and possible means of its regulation. Ann. Rev. Plant Physiol. 25, 363–392.

Hylmo, R. 1953 Transpiration and ion absorption. Physiol. Plant. 6, 333–405.

Jackson, W. A., Flesher, D., and Hageman, R. H. 1973 Nitrate uptake by dark-grown corn seedlings: some characteristics of apparent induction. Plant Physiol. 51, 120–127.

Johnson, V. A., Schmidt, J. W., Hattern, P. J., and Havnold, A. 1963 Agronomic and quality characteristics of high protein F_2 derived families from a Soft Red Winter – Hard Red Winter wheat cross. Crop Sci. 3, 7–10.

Kolderup, F. 1978 Application of different temperatures in three growth phases of wheat. II: Effects on ear size and seed setting. Acta. Agr. Scand. 29, 11–16.

Kramer, J. 1969 The absorption of water by roots. In: Handbuch der Pflanzernahrung und Dungung. pp. 204–234. Springer-Verlag, Berlin.

Kramer, T. 1979 Environmental and genetic variation for protein content in winter wheat (*Triticum aestivum* L.) Euphyt. 28, 209–218.

Langer, R. H. M. and Hanif, M. 1973 A study of floret development in wheat (*Triticum aestivum* L.) Ann. Bot. 37, 743–751.

Langer, R. H. M., and Liew, F. K. Y. 1973 Effect of varying nitrogen supply at different stages of the reproductive phase on spikelet and grain production and on grain nitrogen of wheat. Aust. J. Agr. Res. 24, 647–656.

Ledent, J. F. 1977 Relation entre rendement par epi characters morphologiques a maturité chez diverses variétés de blc d'hiver (*Triticum aestivum* L.) Ann Agron. 28, 391–407.

Lemaire, F. 1975 Action comparée de l'alimentation azotée sur la croissance due système racinaire et des parties aeriennes des vegetaux. Ann. Agron. 26, 59–74.

Leopold, A. C. 1961 Senescene in plant development. Science 134, 1727–1732.

Loomis, R. S., and Gerakis, P. A. 1975 Productivity of agricultural ecosystems. In: Cooper, J. P. (ed.), Photosynthesis and Productivity in Different Environments. Int. Biol. Prog. Vol. 3 p. 145–172. Cambridge University Press, Cambridge, U. K.

Loomis, R. S. and Williams, W. A. 1969 Productivity and the morphology of crop stands: patterns with leaves. In Eastin, J. D., Haskin, F. A., Sullivan, C. Y., and van Bavel, C. H. M. (eds.), Physiological Aspects of Crop Yield. p. 27–47. Am. Soc. Agron., Madison, Wisc.

Luxmoore, R. J., and Millington, R. J. 1971 Growth of perennial ryegrass (*Lolium perenne* L.) in relation to water, nitrogen and light intensity. Plant and Soil 34, 561–574.

Makunga, O. H. D., Pearman, I., Thomas, S. M. and Thorne, G. N. 1978 Distribution of photosynthate produced and after anthesis in tall and semi-dwarf winter wheat, as affected by nitrogen fertilizer. Ann. Appl. Biol. 88, 429–437.

Martin, C. and Thimann. K. V. 1972 The role of protein synthesis in the senescene of leaves. I. The formation of proteases. Plant Physiol. 49, 64–71.

McDowall, F. D. H. 1972 Growth kinetics of Marquis wheat. III: Nitrogen dependence. Can. J. Bot. 50, 1749–1761.

McKee, H. S. 1962 Nitrogen Metabolism in Plants. Clarendon Press, Oxford, 728 pp.

McNeal, F. A., Berg, M. A. and Watson, C. A. 1966 Nitrogen and dry matter in five spring wheat varieties at successive stages of development. Agron. J. 58, 605–608.

McNeal, F. N., Berg, M. A., Brown, P. L. and McGuire, C. F. 1971 Productivity and quality response of five spring wheat genotypes, *Triticum aestivum* L., to nitrogen fertilizer. Agron. J. 63, 908–910.

Medina, E. 1970 Effect of nitrogen supply and light intensity during growth on the photosynthetic capacity and carboxydismutase activity of leaves of *Atriplex patula* sp. *hastata*. Yearbook of the Carnegie Institution. p. 551–559.

Metivier, J. J. and Dale, J. E. 1977 The effects of grain nitrogen and applied nitrate on growth, photosynthesis and protein content of the first leaf of barley cultivars. Ann. Bot. 41, 1287–1296.

Miflin, B. J. and Lea, P. J. 1977 Amino Acid Metabolism. Ann. Rev. Plant Physiol. 28, 299–329.

Milthorpe, F. L. and Moorby, J. 1974 An Introduction to Crop Physiology. Cambridge University Press, Cambridge, U. K. 202 p.

Minotti, P. L., Williams, D. C. and Jackson, A. 1969 The influence of ammonium on nitrate reduction in wheat seedlings. Planta 86, 267–271.

Murata, Y. 1969 Physiological responses to nitrogen in plants. In: Eastin, J. D., Haskins, F. A., Sullivan, C. Y. and Van Bavel, C. H. M. (eds.), Physiological Aspects of Crop Yield. pp. 235–263. Am. Soc. Agron., Madison, Wisconsin.

Neales, T. F., Anderson, M. J. and Wardlaw, J. F. 1963 The role of the leaves in the accumulation of nitrogen by wheat during ear development. Aust. J. Agr. Res. 14, 725–36.

Nicholas, J. C., Harper, J. E. and Hageman, R. H. 1976 Nitrate reductase activity in soybean (*Glycine max.* L. Merr.). I. Effect of light and temperature. Plant Physiol. 58, 731–735.

Novoa, R. 1979 A preliminary dynamic model of nitrogen metabolism in higher plants. Ph. D. Diss. University of California, Davis, 202 pp.

Osman, A. M. and Milthorpe, F. L. 1971 Photosynthesis of wheat leaves in relation to age, illumination and nutrient supply. II: Results. Photosynthetica 5, 61–70.

Pate, J. S. 1966 Photosynthesizing leaves and nodulated roots as donors of carbon to protein of shoot of the field pea (*Pisum arvense* L.) Ann. Bot. 30, 93–109.

Pate, J. S. 1973 Uptake, assimilation and transport of nitrogen compounds by plants. Soil Biol. Biochem. 5, 109–119.

Pate, J. S., Layzell. D. B. and McNeil, D. L. 1979 Modeling the transport and utilization of carbon and nitrogen in a nodulated legume. Plant Physiol. 63, 730–737.

Pearman, I., Thomas, S. M. and Thorne, G. N. 1977 Effects of nitrogen fertilizer on growth and yield of spring wheat. Ann. Bot. 41, 93–108.

Pearman, I., Thomas, S. M. and Thorne, G. N. 1978 Effects of nitrogen fertilizer on the distribution of photosynthate during growth of spring wheat. Ann. Bot. 42, 91–99.

Penning de Vries, F. W. T. 1975 The cost of maintenance processes in plant cells. Ann. Bot. 39, 77–92.

Penning de Vries, F. W. T., Brunsting, A. H. M. and van Laar, H. H. 1974 Products, requirements and efficiency of biosynthesis; A quantitative approach. J. Theor. Biol. 45, 339–377.

Pino, I. 1979 Economia del nitrogeno en cultivares de trigo (*Triticum aestivum* L.) y triticales (*Triticosecale* sp.). Magister Sc. Thesis. Escuela Agronomia, Universidad Catolica de Chile. Santiago. 57 p.

Pitman, M. G. 1977 Ion transport into the xylem. Ann. Rev. Plant Physiol. 28, 71–88.

Porter, K. B., Atkins, I. M. Gilmore, E. C, Lahr, K. A. and Scotting, P. 1964 Evaluation of short stature winter wheats (*Triticum aestivum* L.) for production under Texas conditions. Agron. J. 56, 393–396.

Pushman, F. M. and Bingham, J. 1976 The effect of a granular nitrogen fertilizer and a foliar spray urea on the yield and breadmaking quality of winter wheats. J. Agr. Sci. Camb. 87, 281–292.

Rabson, R., Bhatia, C. R. and Mitra, R. K. 1978 Crop productivity, grain protein and energy. In: Seed Protein Improvement by Nuclear Techniques. Panel Proceedings. p. 3–20. IAEA. Vienna.

Radin, J. 1975 Differential regulation of nitrate reductase induction in roots and shoots of cotton plants. Plant Physiol. 55, 178–179.

Radin, J. 1977 Contribution of the root system to nitrate assimilation in whole cotton plants. Aust. J. Plant. Physiol. 4, 811–819.

Radin, J. 1978 A physiological basis for the division of nitrate assimilation between roots and leaves. Plant Sci. Letters 13, 21–25.

Rains, D. W. 1968 Kinetics and energetics of light-enhanced potassium absorption by corn leaf tissue. Plant Physiol. 43, 394–400.

Raper, C. D., Parsons, L. R. Patterson, D. T. and Kramer, P. 1977 Relationship between growth and nitrogen accumulation for vegetative cotton and soybean plants. Bot. Gaz. 138(2), 129–137.

Rao, K. P. and Rains, D. W. 1976a Nitrate absorption by barley. I: Kinetics and energetics. Plant Physiol. 57, 55–58.

Rao, K. P. and Rains, D. W. 1976b Nitrate absorption by barley. II: Influence of nitrate reductase activity. Plant Physiol. 57, 59–62.

Raven, J. A. and Smith, F. A. 1976 Nitrogen assimilation and transport in vascular land plants in relation to intracellular pH regulation. New Phytol. 76, 415–431.

Rojas, Carlos 1979 Modelo simplifacado para estimar los requerimientos de nitrogeno en arroz. Magister Sc. Thesis. Escuela Agronomia, Universidad Catolica de Chile. Santiago. 62 p.

Shaner, D. L. and Boyer, J. S. 1976a Nitrate reductase activity in maize leaves. I: Regulation by nitrate flux. Plant Physiol. 58, 499–504.

Shaner, D. L. and Boyer, J. S. 1976b. Nitrate reductase activity in maize leaves. II. Regulation by nitrate flux at low leaf water potential. Plant Physiol. 58, 505–509.

Schuurman, J. J. and Knot, L. 1974 Effect of nitrogen on root and shoot of *Lolium multiflorum* var. *westerworldium*. Neth. J. Agr. Sci. 22, 82–88.

Shokr, E. S. and Stolen, O. 1979 Aspects in protein improvement in spring wheat. In: The Royal Veterinary and Agricultural University Yearbook 1979. Denmark Arsskrift. p. 107–122.

Sinclair, T. R. and De Wit, C. T. 1975 Comparative analysis of photosynthate and nitrogen requirements in the production of seed by various crops. Science 89, 565–567.

Spiertz, J. H. J. 1977 The influence of temperature and light intensity on grain growth in relation to carbohydrate and nitrogen economy of the wheat plant. Neth. J. Agr. Sci. 25, 182–197.

Spiertz, J. H. J. and Ellen, J. 1978 Effects of nitrogen on crop development and grain growth of winter wheat in relation to assimilation and utilization of assimilates and nutrients. Neth. J. Agr. Sci. 25, 210–231.

Spiertz, J. H. J. and van der Haar, H. 1978 Differences in grain growth, crop photosynthesis and distribution of assimilates between semi-dwarf and standard cultivars of winter wheat. Neth. J. Agr. Sci. 26, 233–249.

Spratt, E. D. and Gasser, J. K. R. 1970a Effects of fertilizer nitrogen and water supply on the distribution of dry matter and nitrogen between the different parts of wheat. Can. J. Plant Sci. 50, 613–625.

Spratt, E. D. and Gasser, J. K. R. 1970b Effect of ammonium and nitrate form of nitrogen and restricted water supply on growth and nitrogen uptake of wheat. Can. J. Soil Sci. 50, 263–273.

Stanford, G. and Hunter, H. 1973 Nitrogen requirement of winter wheat (*Triticum aestivum* L.) varieties Blue-Boy and Red Coat. Agron. J. 65, 442–447.

204

Tanaka, A. 1958 Studies on the physiological characteristics and significance of rice leaves in relation to their position in the stem. J. Sci. Soil Manure, Japan. 29, 327–333.

Tanaka, K., Kawano, K. and Yamaguchi, J. 1966 Photosynthesis, respiration and plant type of the tropical rice plant. Intl. Rice Research Inst. Tech. Bull. 7, 1–46.

Terman, G. L., Ramig, R. E., Dreier, A. F. and Olson, R. A. 1969 Yield: protein relationships in wheat grain, as affected by nitrogen and water. Agron. J. 61, 755–759.

Thomas, S. M., Thorne, G. N. and Pearman, I. 1978 Effect of nitrogen on growth, yield and photorespiratory activity in spring wheat. Ann. Bot. 42, 827–837.

Thompson, Rex K., Jackson, E. B. and Gebert, J. R. 1975 Irrigated wheat production response to water and nitrogen fertilizer. Univ. Arizona Agr. Exp. Sta. Tech. Bull. 229.

Thorne, G. N. 1974 Physiology of grain yield of wheat and barley. Rep. Rothamsted Exp. Sta. 1973. Part 2. p. 5–25.

Thorne, G. N., Ford, M. A. and Watson, D. J. 1968 Growth, development and yield of spring wheat in artificial climates. Ann. Bot. 32, 425–446.

Tinker, P. B. 1969 The transport of ions in the soil around plant roots. In: Rorison, I. J. (ed.), Ecological aspects of the mineral nutrition of plants. p. 135–147. Blackwell Sci. Publ., Oxford, U. K. 484 p.

Troughton, A. 1977 The rate of growth and partitioning of assimilates in young grass plants. A mathematical model. Ann. Bot. 41, 553–565.

Van den Honert, T. H. and Hooymans, J. J. M. 1955 On the absorption of nitrate by maize in water culture. Acta Bot. Neerl. 4, 376–384.

Van Keulen, H. 1977 Nitrogen requirements of rice with special reference to Java. Contr. Centr. Res. Inst. Agric. Bogor (Indonesia No. 30. 67 p.

Watanabe, H. and Yoshida, S. 1970 Effects of nitrogen, phosphorus and potassium on photo-phosphorylation in rice in relation to the photosynthetic rate of single leaves. Soil Sci. Plant Nutrit. 16, 163–166.

Watson, D. J. 1947 Comparative physiological studies on the growth of field crops. II: The effect of varying nutrient supply on net assimilation rate and leaf area. Ann. Bot. 12, 281–310.

Watson, D. J., Thorne, G. N. and French, S. A. W. 1958 Physiological causes of differences in grain yield in varieties of barley. Ann. Bot. 22, 321–352.

Welbank, P. J., Gibb, M. J. Taylor, P. J. and Williams, E. D. 1973 Root growth of cereal crops. Rothamsted Report for 1973. Part 2. p. 26–66.

Yoshida, S. 1972 Physiological aspects of grain yield. Ann. Rev. Plant Physiol. 23, 437–464.

8. Modelling the interaction of water and nitrogen

H. VAN KEULEN

Center for Agriculture Research, Wageningen, The Netherlands

The amount of useful plant material that a farmer collects from his field at the end of the season, is the integrated result of a large number of processes which interact during the plants' life cycle. In some cases, it may be obvious which growth factor is the major determinant of that final result. Especially in high intensity agriculture as practised in Western Europe, a state of 'potential growth' often exists where the available radiant energy during the growth period determines the yield.

In situations where the growing conditions cannot be controlled, or only to a very limited extent, it is much more difficult to pinpoint the factors responsible for a particular yield. However such knowledge is of prime importance for people who are engaged in agricultural planning and development, because it forms the basis for any attempts to improve the situation. Especially in the developing world, where agricultural research has no long tradition, such knowledge is often lacking.

Intuitive application of principles and knowledge obtained under completely different conditions in more developed countries, rarely leads to the expected improvements since the production system is not understood well enough. Many of the basic processes which govern production may be the same, but the environment may modify the rates of these processes to such an extent that their relative importance changes completely. In each situation, a systematic examination of processes is therefore necessary to asses which are critical. It is often difficult, if not impossible, to make such an analysis in the real world, since the environment is fluctuating and unpredictable. When the relevant processes can be studied separately, they may be combined in models which can be used for simulation experiments. The results obtained may serve as a basis for further experimentation.

WATER USE AND CROP PRODUCTION

The relationship between dry matter production of plants and their use of water has received considerable attention ever since the classical experiments by Briggs

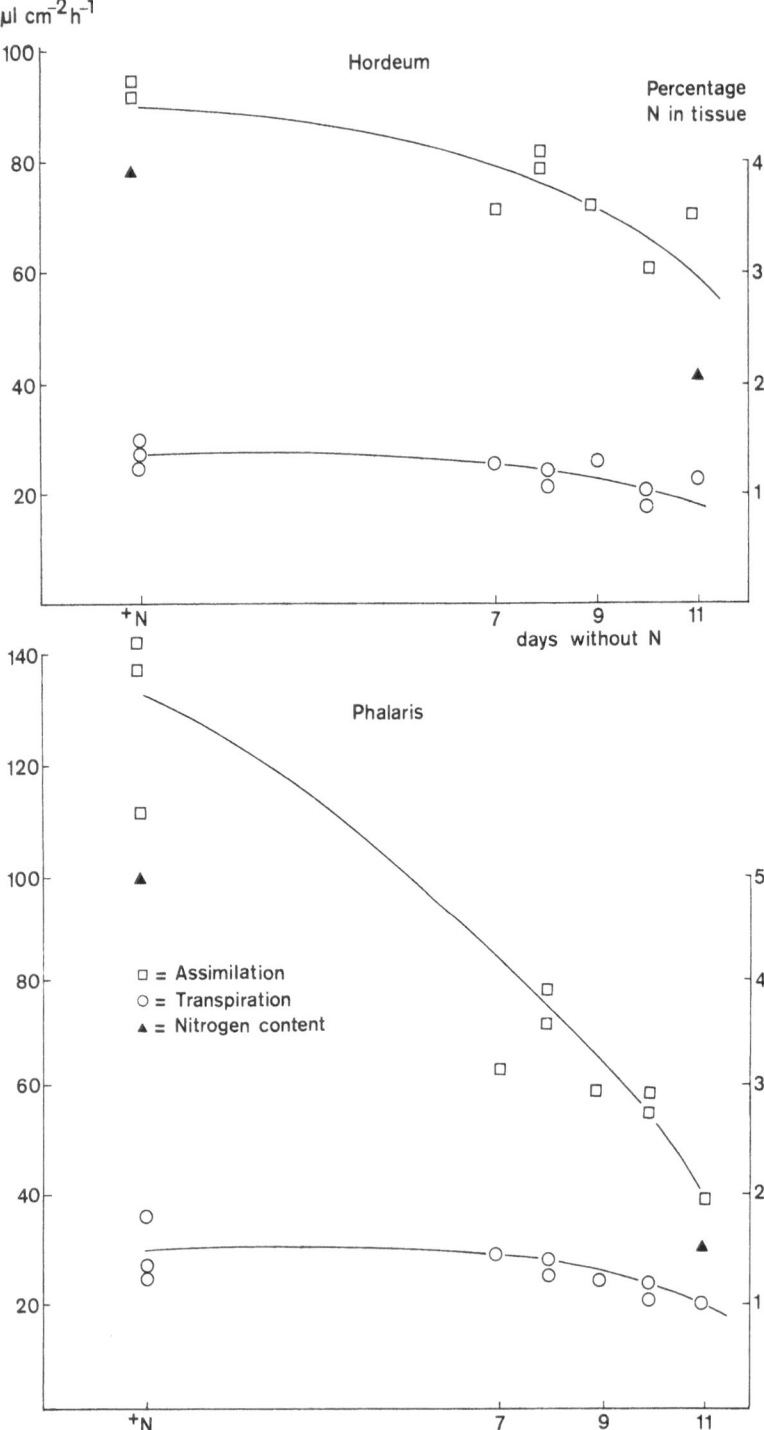

Fig. 1. The influence of the number of growing days without nitrogen to net photosynthesis, transpiration and leaf protein content of two plant species.

and Shantz at the beginning of this century. The analysis of de Wit (1958) showed that the expected relationship depends on the irradiance prevailing during the growing period. In climates with clear skies, photosynthesis is light-saturated; the rate of CO_2-diffusion towards the active site being the rate-limiting factor. Transpiration, on the other hand, is more or less proportional to the irradiance. Hence the ratio between assimilation and transpiration is also proportional to irradiance. Where the level of irradiance is lower and photosynthesis is energy-limited, the ratio of assimilation and transpiration is constant, irrespective of irradiance.

This analysis holds, as de Wit stated, for conditions where 'nutrients are not too low'. What then is the situation when a nutrient, more specifically nitrogen, *is* too low? It seems logical to turn to the basic processes that are involved; photosynthesis and transpiration. In Fig. 1, measurements by Lof (1976) are reproduced. The plants of two grass species were grown in growth chambers on nutrient solution, and measured on subsequent days after transfer to N-depleted solutions. Witholding nitrogen for 11 days results in a considerable decrease in the nitrogen content of the leaves. This decrease is accompanied by a substantial drop in the rate of net photosynthesis, which is more severe in *Phalaris* than in *Hordeum*.

The concurrent drop in the rate of transpiration is about equal for both species, but the decrease is distinctly less than that in photosynthesis. These results suggest that the lower nitrogen concentration in the leaves hardly affected their stomatal conductance. The lower rate of photosynthesis must thus be attributed to increased mesophyll or carboxylation resistance. The ratio between assimilation and transpiration is then much more unfavourable under nitrogen-limited conditions.

Maize plants, grown and measured under similar conditions, behave differently. In Fig. 2, the rate of net CO_2-assimilation is plotted against the conductance for water vapour, i.e. the inverse of the sum of stomatal resistance and boundary layer resistance, obtained from simultaneous measurements of CO_2-exchange and transpiration. Fig. 2a holds for plants which were continuously supplied with nitrogen, while the results in Fig. 2b refer to plants from which nitrogen was withheld for two weeks before the measurements. The regression line calculated from the points in Fig. 2a, is repeated in Fig. 2b for comparison. It is obvious that, despite the much lower values for the maximum rate of CO_2 assimilation of the nitrogen-starved plants, the conductance: net assimilation ratio is not affected. Similar observations were reported recently by Australian research workers (Wong *et al.*, 1979).

The most likely explanation for this proportionality is regulation of stomatal

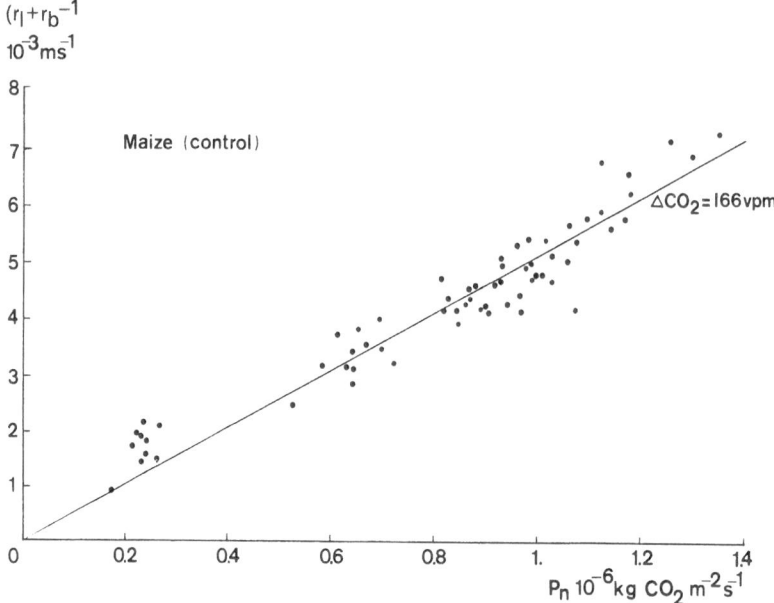

Fig. 2a. The relationship between the rate of net photosynthesis (Pn) and leaf conductance for water vapour $(r_l + r_h)^{-1}$, for maize plants grown under optimum nitrogen supply.

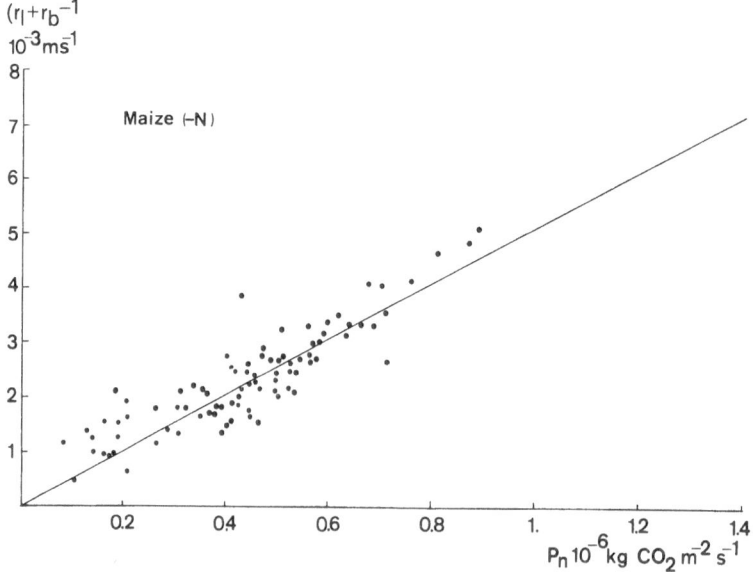

Fig. 2b. The relationship between the rate of net photosynthesis (Pn) and leaf conductance for water vapour $(r_l + r_h)^{-1}$, for maize plants grown for two weeks without nitrogen.

aperture by the CO_2-concentration inside the stomatal cavity, as discussed by Goudriaan and van Laar (1978a). Increased stomatal resistance is here the result of the lower rate of CO_2 assimilation, rather than its cause. Available experimental evidence suggests that stomatal regulation can be present to a greater or lesser extent in the same species, depending on unknown internal or external conditions during its growth. When CO_2-induced regulation is present, the assimilation: transpiration ratio is thus independent of the nutritional status of the plants.

As water use efficiency is generally expressed on the basis of dry matter production, respiratory losses also influence its value. Many quantitative aspects of respiration rate are still uncertain, and it is difficult to predict its influence on water use efficiency. For example, it is expected that lower assimilation rates and lower protein contents in nitrogen-stressed plants be accompanied by lower respiration rates. In general, nitrogenous compounds require relatively large amounts of energy, both for synthesis and for maintenance. However, limited nitrogen supply may lead to continuous remobilization of nitrogenous compounds in older tissue, followed by transport-to and incorporation-into younger tissue which also requires energy. Quantitative assessment of these adjustments can be obtained only from a more thorough study of the respiratory process.

Another phenomenon which distorts the relationships between the assimilation: transpiration ratio at different nutritional conditions and the water use efficiency, is the influence of the nitrogen status on the distribution of assimilates. In general, the top-to-root ratio decreases under nitrogen deficient conditions, in accordance with the functional balance principle. Moreover there is a tendency for a greater reduction in leaf growth than in stems and sheaths. The result of both processes is a less favourable water use efficiency because there is a relative increase in non-photosynthesizing tissue, which unfavourably changes the assimilation: respiration ratio.

So far, water use efficiency has been expressed in terms of water actually used by the plant for transpiration. When the efficiency is expressed in terms of production per unit of water input, either by rain or irrigation, non-productive water losses have to be taken into account also. A major source of such loss is direct soil surface evaporation. Different rainfall patterns may lead to widely varying transpiration: evaporation ratios under otherwise virtually identical conditions (van Keulen, 1975).

Growth under nitrogen deficient conditions implies a slower rate of accumulation of dry matter, which, combined with a different distribution of the material, leads to a prolonged period in which the vegetation does not cover the soil completely. Under such conditions, direct soil surface evaporation is larger than under non-deficient conditions where a closed canopy is reached earlier. The amount of moisture available for transpiration is thus smaller under nitrogen deficient conditions.

In conclusion, management practices that ensure the existence of non-limiting nitrogen supply throughout a plant's life cycle, enhance the efficiency of water use both in terms of input of water and in terms of transpiration. However, in situations where the total moisture supply is limited, application of nitrogenous fertilizers to crops grown for their reproductive organs, may lead to excessive vegetative growth, early use of the available moisture, and hence to water shortage in the economically most important part of the life cycle.

NITROGEN UPTAKE AND ASSIMILATION

A model for mass flow and diffusion

A voluminous literature exist on the transport of ions in the growth medium to the roots of plants. This section does not pretend to give a complete treatment of the subject but illustrates how it can be treated by means of model studies. It also illustrates the hierarchical approach in simulation studies as advocated by several model builders (Goodall, 1975; van Keulen *et al.*, 1976).

A vegetative canopy growing under optimum conditions can increase in weight at a rate of about 200 kg dry matter ha^{-1} day^{-1}. The nitrogen concentration of the tissue may be around 0.025 kg N kg^{-1} (dry weight). When the transpiration rate of such a canopy is 2.5 mm day^{-1}, the concentration of nitrogen in the soil solution must be at least 100 mg kg^{-1} to satisfy the nitrogen demand of the growing canopy by mass flow only. That is a much higher value than normally found in the soil. Therefore it seems, that the difference must be made good by diffusion towards the root surface. In the treatment, it is assumed that nitrogen is present in the soil solution in the form of nitrate, which facilitates the calculations because the complicating processes of exchange with the solid phase of the soil can be neglected.

An individual root is considered, whose distance to its neighbour (a measure of root density) can be changed. The root is considered as a cylinder with a given radius, surrounded by concentric soil compartments. For a given initial salt distribution around this root, and assuming zero concentration at the root surface, the radial flux across each of the compartments can be calculated, and hence the flux into the root.

A numerical model to treat ion transport by mass flow and diffusion in such a system is given by de Wit and van Keulen (1972). When the initial conditions are relatively simple, i.e. a constant initial concentration over the full sphere of influence of the root, an analytical solution for the partial differential equation can be developed (Crank, 1956). The numerical solution obtained with the model was in good agreement with the results obtained from the analytical solution.

In the calculations for this study, a number of different situations have been considered: two rooting densities (average distance between roots, 1 cm and 2 cm); presence and absence of mass flow; and soils with different dispersion coefficients. The results of several runs are shown in Fig. 3. Depletion of the anion store in the soil is slowest with a small root density, in the absence of mass flow (curve 1). Even then, however, 90 per cent of the total amount is removed within five days. Doubling the root density increases transport by diffusion dramatically (curve 5), so that even without mass flow, practically all the anions are removed within two days. Mass flow adds substantially to anion transport only in soils with a relatively high dispersion coefficient (in general loessal-type soils with a rather narrow range of particle sizes).

For the field, these results indicate that virtually all of the anion store is available for uptake by plants when needed, even at very slow transpiration rates.

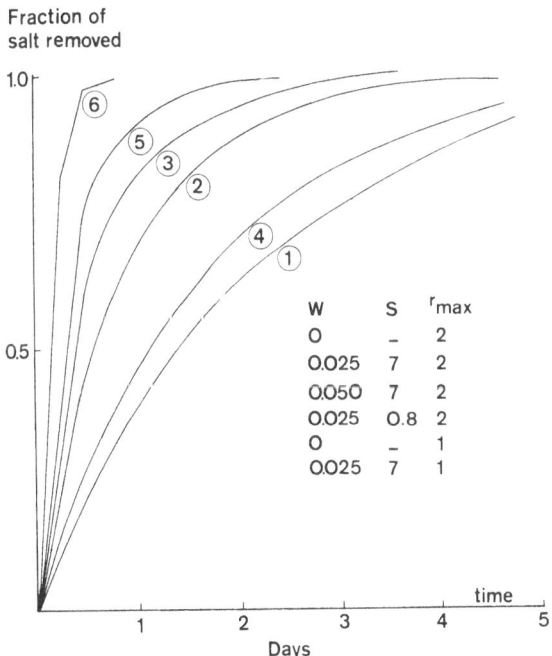

Fig. 3. The amount of anions removed from the soil solution with different root densities (r_{max}), mass flow rates (W) and dispersion factors of the soil (S).

When nitrogen is limiting production, the nitrate released in the soil solution will be taken up almost immediately, and this may explain the very low concentrations of this element found in the soil under growing crops.

The influence of moisture content in the soil on transport is not easy to assess. When moisture content drops, the diffusion coefficient decreases since a smaller surface area of the soil is available for diffusion. The concentration of anions in the soil increases at the same time, thus partly offsetting the smaller diffusion coefficient. There is, of course, a limiting situation when contact between the various moisture 'pockets' in the soil is disrupted and transport ceases altogether. Whether this is a continuous or a discontinuous process is not clear.

Interaction between nitrogen and moisture

In the field, the moisture content in the soil may become so low that nutrient transport and hence uptake are restricted. Even in the temperate zone, Garwood and Williams (1967) observed, during a dry period, depressions in the growth rate of a grass sward which could be remedied, not by irrigation, but by injection of nitrogenous fertilizer in the deeper soil layers. Obviously the plants were able to withdraw water from these deeper layers, where no nitrogen was present. The nitrogen in the upper soil layers was rendered unavailable by their dryness.

The phenomenon has also been observed in the monsoonal tropics where rainfall may cease relatively early in the season. The appearance of rice plants deteriorates soon after the water layer has disappeared from the soil surface. In these paddy soils, it seems unlikely that all of the stored moisture is used up at that time. Here again, it was shown that when nitrogenous fertilizer, applied at the usual time, was placed at a depth of $\pm$ 50 cm, growth rate was not depressed until much later, if at all (van Keulen, unpubl. results).

The effects of nitrogen shortage and water shortage and their interactions were recently illustrated in a series of elegant pot experiments with a system in which nitrogen and water could be applied independently in three compartments (Rehatta et al., 1979). Some of the results have been reproduced in Fig. 4. Moisture shortage with equal availability of nitrogen led to reduced uptake of the element. The uptake must then be governed by the reduced rate of dry matter production, because the concentration of the element in the tissue is at its maximum level.

Where nitrogen is unavailable because there is no moisture for transport to the roots, the concentration of the element in the tissue drops to a much lower value. The total dry matter yield, presumably controlled in these cases by nitrogen supply, is also lower, but the nitrogen in the tissue is diluted to about half of its

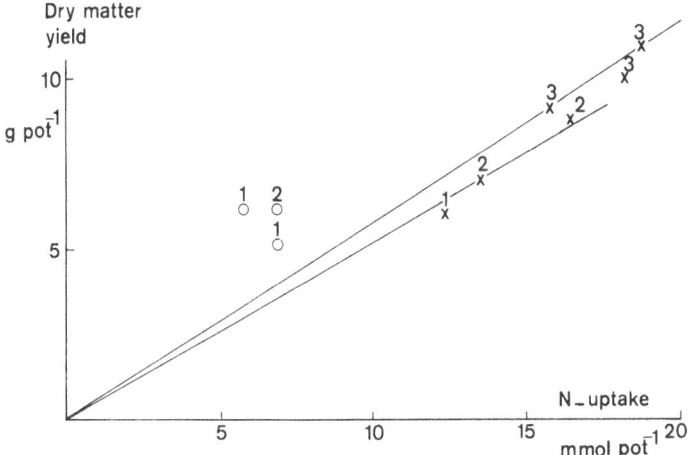

Fig. 4. The relationship between nitrogen uptake and dry matter yield of vegetative rice plants grown in compartmentalized systems. The crosses refer to treatments where nitrogen and water were in the same compartment, the open circles to those where they were separated. The number refers to the number of wet compartments.

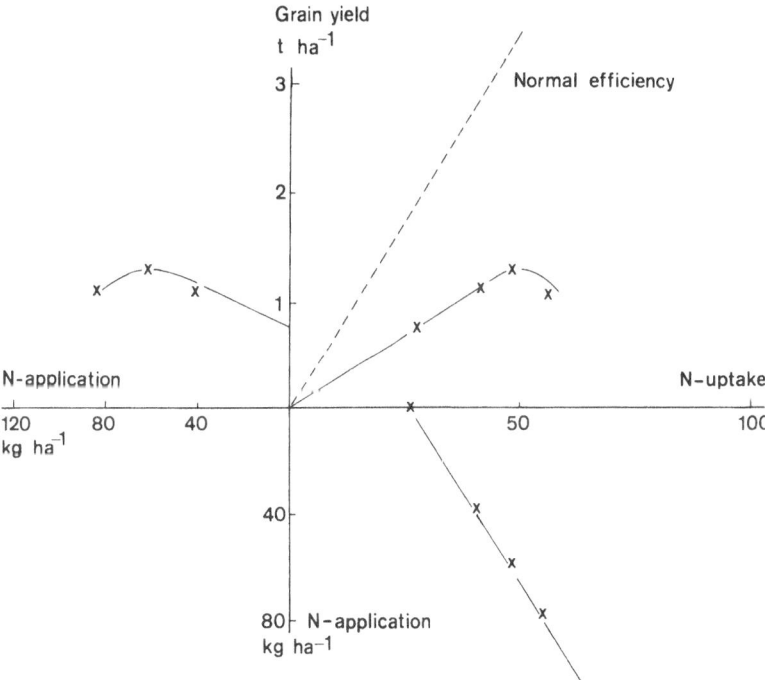

Fig. 5. The relationship between total nitrogen uptake and grain yield, that between nitrogen application and uptake and that between nitrogen application and grain yield for upland rice grown in India.

maximum level. These results illustrate that moisture shortage to the plants may have both a direct and an indirect effect on nitrogen uptake; in the one case governed by physical transport processes in the soil, in the other by metabolic processes in the plant.

The two contrasting effects are again illustrated, in a slightly modified form, in Figs. 5 and 6. For the upland rice, the efficiency of nitrogen utilization in terms of grain yield per unit of N absorbed is about one-third of that normally found (van Keulen, 1977). The reason is that moisture shortage towards the end of the growing period accelerated senescence of the vegetative plant parts. Photosynthesis declined rapidly and the grains could not be filled in the normal way. This resulted in grains with a high protein content and a low grain weight. The effect on the yield-uptake curve was even more conspicuous because translocation of nitrogen from the vegetative tissue to the filling grains was hampered also, so that the straw contained more nitrogen at harvest than under normal conditions. This phenomenon is frequently observed in semi-arid regions; in extreme cases resulting in shrivelled grains.

In Fig. 6, the effect of moisture shortage shows up in a most pronounced way, in the application-uptake curve. At the lower irrigation rate, the soil was desiccated earlier, transport towards the root system was inhibited and uptake ceased at a relatively low application level. It is unlikely that the nitrogen level in the tissue prevented further uptake since the concentration in the abundantly irrigated treatment is about 25 per cent higher than in the dry treatment.

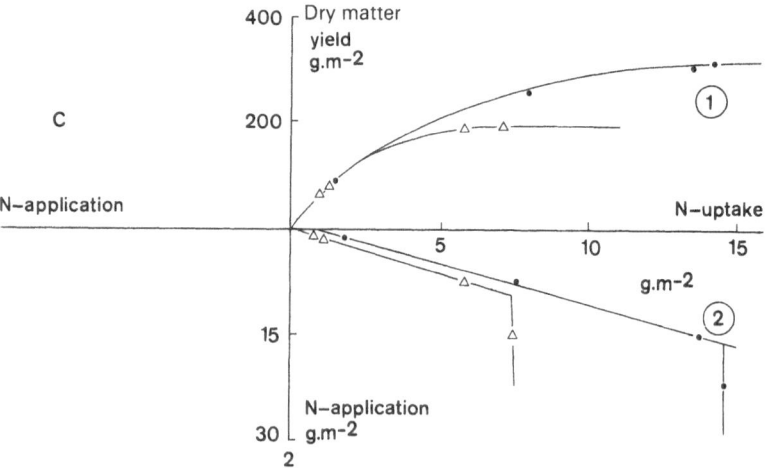

Fig. 6. The relationship between nitrogen uptake and dry matter yield and that between nitrogen application and uptake for ryegrass at two irrigation rates.

THE SIMULATION MODEL PAPRAN

The climate in semi-arid regions is variable and unpredictable. Experimentally, this creates the problem that only a very limited number of possible situations can be studied in any given research period. The records collected during such periods can be exploited, in the first instance, to develop simulation models. When such models have been validated, they may be used both to generate crop responses, for instance, for a series of years long enough to characterize a given region. Or they may be used to extrapolate from the region that has been studied to another one where less detailed information is available.

The model discussed here concentrates on the effects of moisture availability and nitrogen supply on dry matter production.

DESCRIPTION OF THE MODEL

The main elements of the model are given in a simplified relational diagram (Fig. 7). It describes the growth of an annual crop or pasture from the water and the nitrogen balance in the soil below it.

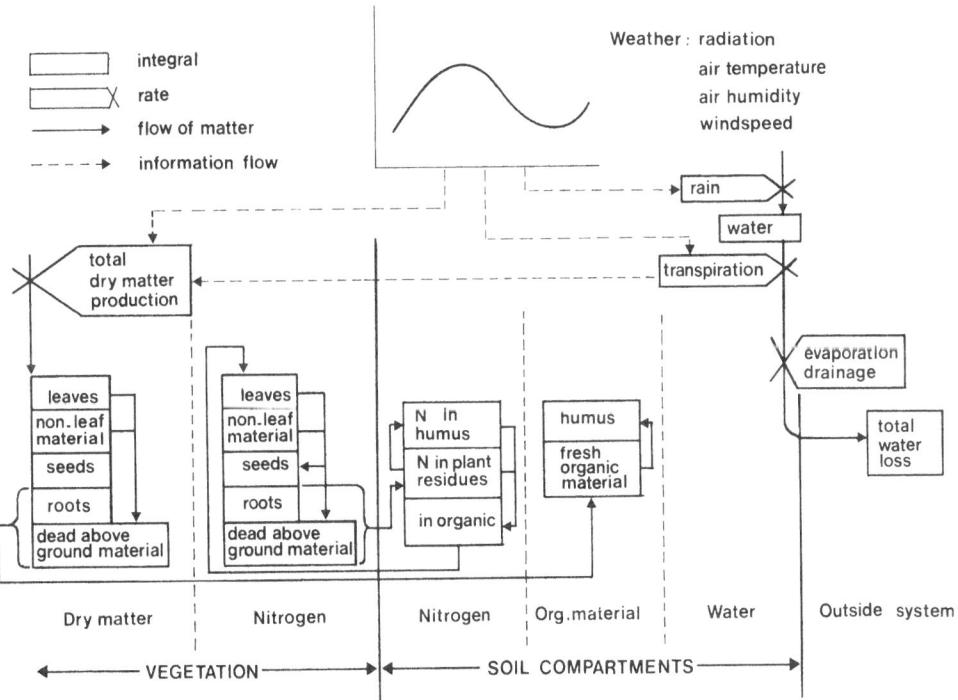

Fig. 7. A schematized relational diagram of the model PAPRAN.

Plant growth

Dry matter accumulation. The calculation of plant production in the absence of nitrogen shortage is based on the observation that a close relationship exists between water use (= transpiration) and dry matter production, when water is the main limiting factor for growth (de Wit, 1958).

First, the potential gross rate of CO_2-assimilation is obtained as a function of leaf area index and irradiance (Goudriaan and van Laar, 1978b) for a given photosynthesis light-response curve of an individual leaf of the species. The rate of maintenance respiration, that is the energy required to continuously rebuild degrading enzymes and to counteract leakage of inorganic ions (Penning de Vries, 1974), is calculated as a function of dry weight present and the protein content of the structural material. The potential rate of increase in dry weight follows from the balance between gross assimilation, maintenance respiration, and growth respiration – the energy lost in the conversion of primary photosynthates into structural plant material, which is expressed as a weight conversion efficiency.

The potential rate of transpiration of the canopy is calculated from the leaf area index and the environmental conditions (level of irradiance, vapour pressure and temperature of the ambient air and wind speed). A descriptive formula is used, (van Keulen, 1975) in which average daily weather variables are applied. The description is based on the results of a detailed model (de Wit *et al.*, 1978).

The ratio of potential daily growth rate and potential transpiration – the water use efficiency – is a central parameter in this model. This ratio is assumed to be constant, irrespective of the water status of the canopy. When plants grow under severe moisture stress, the ratio may change (Lof, 1976) but the actual amounts of water transpired during such periods are so small that only minor errors are introduced in this way.

The actual transpiration rate is derived from the potential rate, taking into account the distribution and activity of the root system and the soil moisture status (see p. 220). The growth rate of the vegetation, as dictated by moisture supply, is obtained by multiplying the actual transpiration rate by the water use efficiency. Nitrogen deficiency of the vegetation leads to a reduction in this growth rate; the reduction factor being a function of the nitrogen concentration in the leaf tissue. Since only the growth rate is affected in the present version, and not the rate of transpiration, nitrogen shortage inevitably leads to a deterioration in overall water use efficiency.

Allocation of dry matter. In the model, the total amount of dry matter is divided into leaves, non-leaf vegetative material (stems, leaf sheaths), roots and seeds. The allocation between the various plant organs is governed by the phenological state of the canopy, represented by its development stage.

First, a part of the available photosynthates is allocated to the roots. Under optimum conditions, the fraction decreases with age from about 0.5 at establishment to 0.025 towards maturity. Moisture shortage of the canopy may modify this fraction. Following the functional balance principle, transpiration deficits favour root growth relative to the shoot. From the above-ground material, seeds have first call on assimilates formed after flowering. The remainder is divided between leaves and non-leaf material, the proportions being governed by the stage of development (progressively less is invested in new leaves with increasing age) and by the nitrogen status of the vegetation (nitrogen deficiency shifts growth in the direction of non-leaf material).

Dying of vegetative tissue. Both leaves and non-leaf material may die as a result of stress due to water or nitrogen deficiency, or to senescence when the season draws to a close and the plant's life cycle is completed.

The rate of dying due to water shortage is governed by the balance between the rate of root water uptake and the rate of transpiration. When the latter exceeds the rate of replenishment, the tissue is dehydrated. This leads to partially irreparable damage. The buffering capacity of the vegetation is taken into account by assuming a time constant for dying of five days. Under severe moisture stress, the stomata of the plants are closed and water loss is restricted to cuticular transpiration only.

The rate of water loss through the cuticle is a function of evaporative demand of the atmosphere and, consequently, dying of the tissue is dependent on both the moisture status of the soil and the environmental conditions. The death rate acts on both the leaves and the non-leaf material in proportion to their respective weights. This description of tissue dying has to be verified experimentally, but it simulates actual situations during drying reasonably well (van Keulen *et al.*, 1980). Dying of the tissue due to nitrogen deficiency is a function of the nitrogen status of the vegetation. Dying begins when the nitrogen content drops below the threshold value for unrestricted growth: the relative death rate gradually increasing until a final value of 0.3 day^{-1} at the absolute minimum nitrogen content.

Senescence is disregarded during the early stages of development of the vegetation. Leaves have a limited life-span and those formed first die early. However, since the leaves are not kept in age classes, application of a relative death rate to the whole existing leaf mass will overestimate death. Leaves are assumed to start dying after flowering, when translocation to the developing seeds accelerates their deterioration. At the end of the plant's life cycle, the vegetation dies from senescence at a relative rate of 0.1 day^{-1}. That causes almost complete drying up of the crop in two weeks, which is in agreement with field observations.

Nitrogen in the crop

Inorganic nitrogen is taken up by the plants from the soil. In principle, it would not make much difference for the model whether it is available as ammonium or as nitrate, but some of the soil nitrogen processes have been described in such a way that nitrate is assumed to be the major component in the soil. As explained before (page 210), the root system of the vegetation is generally dense and efficient enough to explore the total rooted volume. All of the inorganic nitrogen in that volume is assumed to be available for immediate uptake by the vegetation. This is either a result of mass transport with the transpiration stream, or of diffusion along a concentration gradient, created by low concentrations of N at the root surface.

The demand for nitrogen in the vegetation is assumed to be the difference between the current concentration in the tissue and a possible maximum concentration. The latter value is different for the various organs. For each organ, it is a function of the development stage of the vegetation and decreases as the plant approaches maturity. Under conditions of limited supply, the actual amount taken up is divided between shoot and root in proportion to their relative demands. It may be argued that the demand of the root system will be satisfied first, before nitrogen is transferred to the above-ground plant parts since the root is close to the source. However, experimental evidence (van Dobben, 1961), as well as simulation results, does not support that hypothesis.

Nitrogen transferred to the aerial plant parts, is again distributed between leaves and non-leaf vegetative tissue in proportion to the relative deficiencies in the respective organs. After flowering, the nitrogen demand created by the developing seed, is met by translocation from the vegetative tissue. It is assumed that all nitrogenous compounds accumulated in the seed, have passed through the vegetative tissue first.

This process of translocation leads to depletion of the vegetative mass when the supply from the soil at this stage of crop development does not meet the seed requirement. It has often been observed, especially in small grains (Spiertz, 1978) but also in natural vegetation (Penning de Vries *et al.* 1979), that the uptake of nitrogen after flowering is negligible. It is not yet known whether this is due to exhaustion of the soil nitrogen store or to a spatial separation between moisture and nitrogen in the soil, or to decreased root activity, or to a combination of these factors.

When the vegetative becomes increasingly deficient in nitrogen, translocation to the seeds is hampered, resulting in seeds with a smaller nitrogen content. A lower limit for the protein content of the seeds was introduced, forcing cessation of the carbohydrate supply to the seeds when this limit is approached. This description is based on the observation that, under conditions of limited supply, the N-content of seeds is species-dependent only, irrespective of growing conditions (van Keulen, 1977). This descriptive formulation has been applied, since the underlying processes are unknown.

Dying of vegetative tissue also leads to loss of nitrogen from the vegetation. The amount of nitrogen lost with the dying material depends on the cause of death. When death occurs from water shortage or from senescence, the concentration in the dead tissue is equal to that of the live material. When nitrogen deficiency leads to death, translocation of nitrogen is assumed and only the irreversibly incorporated components are withdrawn. This mimics the real-world situation where continuous breakdown and transfer of nitrogenous compounds enables the growth of young tissue at the expense of older ones.

Soil moisture balance

Infiltration. For the description of the water balance in the soil, the total soil depth is divided into homogeneous compartments. Both the number and the thickness of the compartments can easily be adapted, making the model versatile.

Infiltration into the soil, which may result either from rain or from irrigation, is obtained from the rate of moisture supply, taking into account the influence of run-off. In the model, a value of run-off is postulated, so that the rate of infiltration follows. In reality, run-off occurs when the intensity of rain, or of irrigation, exceeds the maximum infiltration capacity of the soil. This effect cannot be ignored, especially in soils which are sensitive to crust formation. An exact calculation of these effects requires a detailed description of the processes of crust formation and the build-up of above-atmospheric pressures in the soil and their influence on the infiltration capacity (Rietveld, 1978). Little attention has been paid as yet to the incorporation of the results of such detailed models into lower resolution models.

The rate of change of moisture content in a compartment is set equal to the difference between the current moisture content and that at field capacity. It is thus assumed that an 'equilibrium' situation develops instantaneously. The compartments are filled up from the top down, until the total amount of water is dissipated or the remainder is drained below the potential rooting zone.

Evaporation Evaporation directly from the soil surface is one of the most important sources of non-productive water loss in semi-arid regions. The major determinant for its magnitude is the distribution of the rainfall. A given amount of precipitation distributed in many small showers leads to extended periods of soil surface wetness and hence to considerable losses through direct soil evaporation. The potential rate of evaporation follows from a Penman-type equation and the distribution of energy between the vegetative cover and the bare soil. The actual rate of water loss depends then on the moisture content of the top compartment. The total water loss is proportioned over the various soil compartments by means of a mimicking procedure (van Keulen, 1975). This procedure gives reasonable results in areas with winter rainfall, but may lead to erroneous results in conditions where the soil surface temperatures reach high values, such as summer rainfall areas. An accurate alternative for such conditions has not been worked out, but is the subject of continuing research (Wösten and van Loon, 1979).

Water uptake by the roots From the potential rate of transpiration and the total root length (that is the vertical extension of the root system), the potential rate of moisture uptake per unit root length is calculated. In each soil compartment, the potential rate may be reduced due to low moisture content or to low soil temperatures. The relationship between soil moisture content and root water uptake is a Viehmeyer-type curve: soil moisture remains freely available until about 30 per cent of available moisture is left, after which there is a sharp reduction until wilting point. Allowance is made for partial compensation when a portion of the root system is in dry soil.

The effect of soil temperature, which is additive to that of soil moisture, accounts for changes both in the viscosity of the water and in the activity of the root system. Finally the uptake of moisture by the roots may be affected by the concentration of solutes in the soil, creating high osmotic pressures. The total amount of moisture taken up from the various rooted soil layers equals crop transpiration.

Nitrogen in the soil

The importance of soil nitrogen transformations has resulted in a voluminous literature on the subject (Bartholomew and Clark, 1965; Tandon, 1974; van Veen,

1977). However, attemps to use that information for the development of detailed models of the soil-nitrogen system have demonstrated the complexity of that system and the limits of our understanding (Beek and Frissel, 1973; Hagin and Amberger, 1974; van Veen, 1977).

In principle, the approach used by van Veen, which is microbiologically based, is theoretically sound since the transformations in the soil are governed by microbial activities. In that and similar models, the microbial biomass is considered as a separate pool. In practice, it is difficult (if not impossible) to properly initialize the microbial pool for specific field situations, or to validate the results that are obtained. In the present approach, the total soil nitrogen store is therefore separated into three states only: (i) mineral nitrogen including NH_4^+, NO_2^- and NO_3^-, (ii) nitrogen in 'fresh' organic material including plant residues which have not yet passed through the microbial tissue and the microbes themselves, and (iii) nitrogen in 'stable' organic material ('soil humus', which has at least once undergone a transformation through the microbes).

Organic matter transformations In each soil compartment, two organic matter fractions are distinguished: (i) the fresh organic material and (ii) the stable organic material.

The rate of decomposition is based on first-order kinetics, i.e. under optimum conditions a constant relative rate is assumed. The specific rate of decomposition is different for various compounds (Hagin and Amberger, 1974): it is of the order of one day^{-1} for easily decomposable proteins and sugars, $\pm$ 0.05 day^{-1} for cellulose and hemi-cellulose, and $\pm$ 0.01 day^{-1} for lignine. In the model, these different rates are introduced in a step-wise manner, as the original amount of fresh organic material is reduced. Different compositions of the added material can be accounted for by changing the switch-values for the rate constants. These rates of decomposition may be modified by conditions of soil moisture, soil temperature and the $C:N$ ratio of the decomposing material. It is implicitly assumed in this description that the number of decomposing bacteria is never a factor which is limiting decomposition. Moreover, instantaneous adaption to different substrates is assumed.

In the calculation of the $C:N$ ratio, the amount of available mineral nitrogen is included. The functional relationship is adapted from the treatment by Parnas (1975). The stable organic material, which has a constant $C:N$ ratio of $\pm$ 10, is decomposed at a much slower rate, again depending on temperature and moisture using the rate constant calculated by Harpaz (1975) for semi-arid conditions. Accretion of stable organic matter results from stable compounds from the fresh organic material and it is assumed that 'humus formation' takes place only when the overall C : N ratio of the decomposing material drops below 25. At that stage, mineral nitrogen is released of which about 20 per cent is incorporated in the stable fraction. Application of the constant C : N ratio of that material also yields the rate of humus formation.

Soil nitrogen processes During the decomposition of the fresh organic material, mineral nitrogen is being released when the carbon of the substrate is used for the build-up of microbial tissue and the supply of energy for the functioning of the microorganisms. Hence the rate of release is a direct function of the decomposition rate and the nitrogen content of the material which is being decomposed. At the same time, the build up of the microbial biomass needs nitrogen for the formation of proteins. The rate at which nitrogen is incorporated is again related to the decomposition rate of the fresh organic material.

The basic assumption is that all carbon released during decomposition is used for the build up of microbial tissue with an average biosynthetic efficiency (Sörensen, 1975). That assumption yields the N-requirement during decomposition, and the composition of the decomposing material dictates whether or not mineral nitrogen is immobilized. Gross release of nitrogen is thus dependent on the nitrogen content of the fresh organic matter originally added. Net release must be considered as originating from the microbial component of that fresh organic material. The present description leads to realistic simulations of the switch from net immobilization to net release, at overall C : N ratios between 20 and 30. The rate of change of the nitrogen in the stable organic material is the balance between its rate of mineralization, directly dependent on the decomposition rate and the rate of immobilization coupled to the rate of release of N from the fresh organic material.

The amount of mineral nitrogen in each soil compartment changes by release-from or immobilization-into the fresh and the stable organic compounds and by uptake by the plants. Furthermore, transport between soil compartments is taken into account. The integration interval of one day used in this model is much larger than the time constant of solute transport (de Wit and van Keulen, 1972; de Wit and Goudriaan, 1974). Again a mimicking procedure is applied. Transport takes place only with movement of water, that is with infiltration.

The concentration of mineral nitrogen transported over the lower boundary of a compartment is obtained by 'mixing' the solutes present in the compartment and those transported into it with water present and that flowing through. This description takes into account mass transport and some of the effects of mathematical dispersion inherently present in such compartmentalized models (Goudriaan, 1973). Upward transport of nitrogen with the soil evaporation stream, or diffusion along developing concentration gradients, is not taken into account. This description is satisfactory in the present situation where the main interest is in the availability of nitrogen to the plants. However, when the exact distribution of solutes in the soil is important, more detailed models of salt transport have to be applied (de Wit and van Keulen, 1972).

When ammonium fertilizers are applied, especially in soils with a relatively high pH, considerable losses of nitrogen may occur through volatilization. The time constants of the chemical processes involved, are again much lower than the time resolution of the present model, so that a detailed description cannot be given. Therefore, it is assumed in the model that ammonium given as a fertilizer, volatilizes with a constant relative rate form the upper soil compartment. At the same time, nitrification takes place at a constant relative rate. The time constants used for these processes are of the order of 5 to 10 days, so that appreciable losses occur only during the first week after the application of fertilizers, when there is no precipitation during that period. The influence of pH and soil temperature have not been taken into account in the present version. A more realistic description of this process would increase the generality of the model.

Processes not considered Some of the processes taking place in the soil-plant system have been ignored. In the first place, denitrification is disregarded for the moment, since in the semi-arid regions for which this model is mainly developed, anaerobic conditions are unlikely to occur. The possibility is recognised that locally anaerobic pockets, (e.g. around plant roots or inside structural elements) can develop as a result of oxygen depletion through intense biological activity, but the problem of simulation is extremely complex (Leffelaar, 1979). In the second place, adsorption of NH_4^+ onto the exchange complex and fixation of these ions into the lattice of clay minerals has not been taken into account. In most situations, these processes play only a minor role in the annual nitrogen balance. However, in specific cases, especially where nitrification is hampered, they could be important and would have to be incorporated.

RESULTS AND DISCUSSION

The performance of the model was studied by analyzing its behaviour in semi-arid conditions. Actual weather data were used for situations where dry matter yields were available both with and without fertilizer. The experimental conditions are described in detail by van Keulen (1975).

The initial conditions assumed in the model were identical for all seasons: at the onset of the rainy season a total amount of 3000 kg ha^{-1} of fresh organic

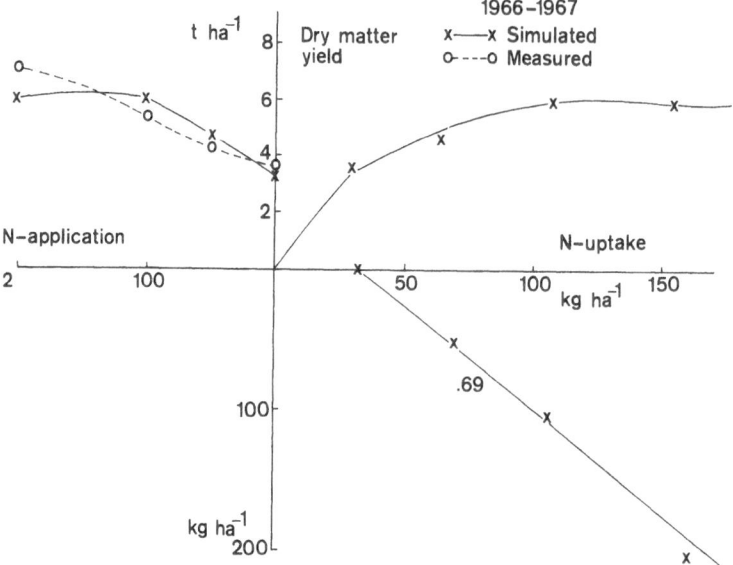

Fig. 8. Simulated and measured results of fertilizer experiments on natural vegetation in the northern Negev in the 1966–67 growing season.

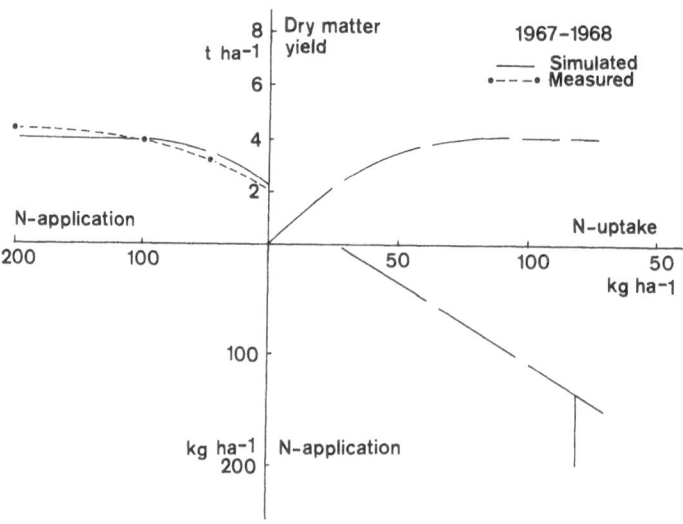

Fig. 9. Simulated and measured results of fertilizer experiments on natural vegetation in the northern Negev in the 1967-68 growing season.

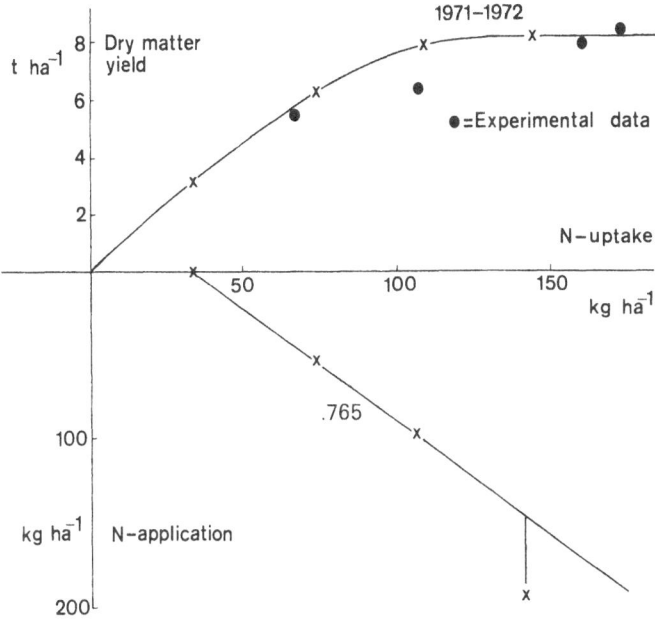

Fig. 10. Simulated and measured results of fertilizer experiments on natural vegetation in the northern Negev in the 1971-72 growing season.

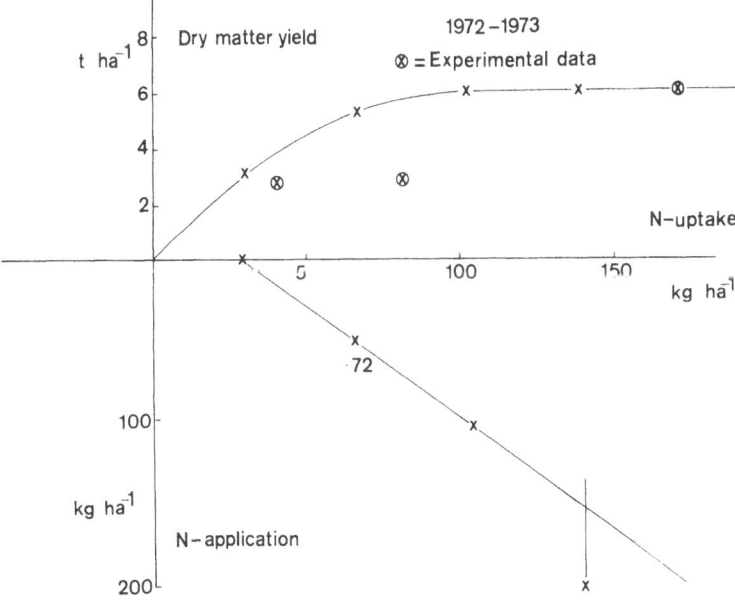

Fig. 11. Simulated and measured results of fertilizer experiments on natural vegetation in the northern Negev in the 1972-73 growing season.

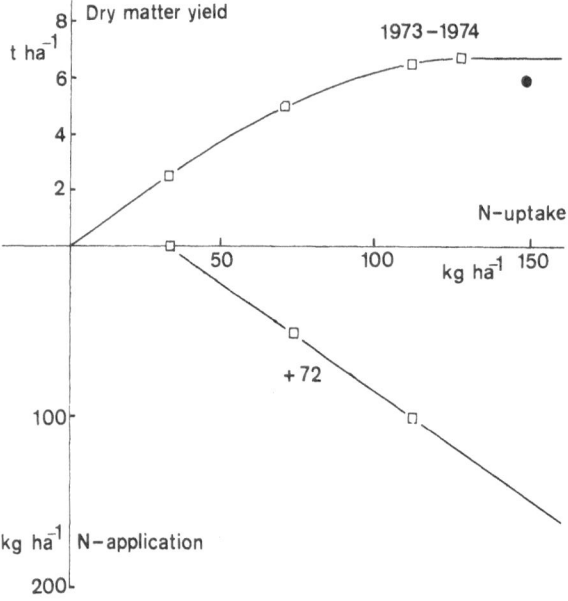

Fig. 12. Simulated and measured results of fertilizer experiments on natural vegetation in the northern Negev in the 1973-74 growing season.

material is assumed to have been added to the upper 60 cm of the profile. The average nitrogen content of that material is set at 0.01 kg N kg^{-1} (dry matter). Fertilizer added is assumed to be mixed evenly through the upper 10 cm of the soil at the onset of the simulation. The model results, presented in Figs. 8 to 12 for the various seasons, hardly show a consistent picture:

In 1966–67, where only application-yield data were available, the yield without fertilizer application is reasonably well estimated. The simulated nitrogen response curve deviates substantially from the measured one: at low application rates the effect of additional fertilizer is overestimated, whereas the maximum experimental yield exceeds the predicted one. The simulated maximum yield is equal to the one predicted by the model ARID CROP, (van Keulen, 1975) which assumes non-limiting nitrogen conditions throughout. Since no uptake values are available, it is difficult to judge whether the discrepancies in the response curve are the result of inaccuracies in the description of plant response to increased nitrogen availability or of a misrepresentation of the application-uptake relationship.

For 1967–68, a lower rainfall year than the previous one, the simulated yield application curve is reasonably close to the measured one over the full range of applications. The drier conditions retarded the decomposition of the fresh organic material, and consequently about 5 kg N ha^{-1} less is available to the vegetation in the absence of fertilizer.

The various experimental treatments in the 1971–72 growing season resulted in the uptake of varying amounts of nitrogen. Unfortunately no 'zero' treatment was measured since all the experimental fields were disced to obtain uniformity. This action resulted in the incorporation into the soil of various amounts of animal excrement accumulated in preceding seasons. Since neither the quantity nor the composition of these excrements could be estimated, their effect on the nitrogen balance is impossible to assess.

The simulated yield-uptake curve is within the accuracy limits of the measured points. The model predicts a levelling of the application-uptake curve at the highest application level, since the crop is 'nitrogen-saturated' throughout the growing period, the predicted maximum uptake being in reasonable agreement with the observed value which was measured at a higher application rate. The simulated yield-uptake curve for 1972–73 deviates considerably from the measured data points. No obvious explanation is forthcoming.

For the lowest point, part of the reason may be the leguminous species in the dry matter which have an inherently higher nitrogen content. The measured maximum uptake suggests that the maximum concentration values applied in the model are somewhat low, since the model predicts about 25 per cent less. The simulated value was again dictated by the ability of the vegetation to store nitrogen.

For the 1973–74 growing season, insufficient measurements are available for comparison. Again, the predicted maximum uptake is lower than the measured one.

The results presented here indicate that in some cases the model seems to be predicting the behaviour of the real world reasonably well, whereas in other situations considerable deviations occur. The problem is that, for a model of this complexity, using only the gross output for validation is rather frustrating. There are so many relationships and parameters involved that almost any result may be obtained by changing their values. That is a very dangerous procedure which can turn simulation into 'a most cumbersome and obscure way of curve fitting' (de Wit, 1970). On the other hand, there are virtually no experimental records available for validating the various elements of the model. This gap again stresses the point that systems analysis and simulation can never be a substitute for experimental work, but may be used as a framework by which relevant problems can be more clearly recognized and experimentation better directed.

228

The development of PAPRAN and its results have again underlined the need for a more systematic and quantitative investigation of the various processes that play a role in the production process in semi-arid regions, and their relative importance. The model, in its present form, does provide a useful starting point for such an investigation, even though at many points it is descriptive rather than explanatory, and needs a much more thorough validation before it can be widely applied with confidence.

REFERENCES

Bartholomew, W. V. and Clark, F. E. 1965 Soil nitrogen. Agronomy 10, Amer. Soc. Agron. Madison, Wisc.
Beek, J. and Frissel, M. J. 1972 Simulation of nitrogen behaviour in soils. Simulation Monographs, Pudoc, Wageningen.
Crank, J. 1956 The mathematics of diffusion. Oxford Univ. Press.
Dobben, W. H. van 1961 Nitrogen uptake of spring wheat and poppies in relation to growth and development. Jaarb. I.B.S. 1961, 54–60 (Dutch with English summary).
Garwood, E. A. and Williams, T. E. 1967 Growth, water use and nutrient uptake from the subsoil by grass swards. J. Agric. Sci. Camb. 69, 125–130.
Goodall, D. W., 1975 The hierarchical approach to model building. In: Arnold G. W. and de Wit C. T. (eds.), Critical evaluation of systems analysis in ecosystems research and management.
Goudriaan, J. 1973 Dispersion in simulation models of population growth and salt movement in the soil. Neth. J. Agric. Sci. 21, 269–281.
Goudriaan, J. and van Laar, H. H. 1978a Relations between leaf resistance, CO_2 concentration and CO_2 assimilation in maize, beans, lalang grass and sunflower. Photosynthetica 12, 241–249.
Goudriaan, J. and van Laar, H. H., 1978b Calculation of daily totals of the gross CO_2 assimilation of leaf canopies. Neth. J. Agric. Sci. 26, 373–382.
Hagin, J. and Amberger, A., 1974 Contribution of fertilizer and manures to the N- and P-load of waters. A computer simulation. Deutsche Forschungsgemeinschaft, Bonn, Technion Research and Development Foundation Ltd. Haifa.
Harpaz, Y., 1975 Simulation of the nitrogen balance in semi-arid regions. Ph. D. Thesis Hebrew University, Jerusalem.
Keulen, H. van, 1975 Simulation of water use and herbage growth in arid regions. Simulation Monographs, Pudoc, Wageningen.
Keulen, H. van, 1977 Nitrogen requirements of rice, with special emphasis on Java. Contr. Centr. Res. Inst. Agric. Bogor, no. 30.
Keulen, H. van, Seligman, N. G. and Goudriaan, J. 1975 Availability of anions in the growth medium to roots of an actively growing plant. Neth. J. Agric. Sci. 23, 131–138.
Keulen, H. van, Seligman, N. G. and Benjamin, R. W., 1980 Simulation of water use and herbage growth in arid regions – re-evaluation and further development of the model ARID CROP. Agric. Systems (in press).
Keulen, H. van, de Wit, C. T. and Lof, H. 1976 The use of simulation models for productivity studies in arid regions. In: Lange, O. L., Kappen, L. and Schultze, E. D. (eds.), Water and Plant Life. Ecological Studies 12, Springer Verlag, Berlin.
Leffelaar, P. A., 1979 Simulation of partial anaerobiosis in a model soil in respect to denitrification. Soil Sci. 128, 110–120.
Lof, H., 1976 Water use efficiency and competition between arid zone annuals, especially the grasses *Phalaris minor* and *Hordeum murinum*. Agr. Res. Rep. 853, Pudoc, Wageningen.

Parnas, H., 1975 Model for decomposition of organic material by micro-organisms. Soil Biol. Biochem. 7, 161–169.

Penman, H. L., 1948 Natural evaporation from open water, bare soil and grass. Proc. Roy. Soc. A. 193, 120–146.

Penning de Vries, F. W. T., 1974 Substrate utilization and respiration in relation to growth and maintenance in higher plants. Neth. J. Agric. Sci. 22, 40–44.

Penning de Vries, F. W. T., Krul, J. M. and Keulen, H. van, 1979 Productivity of Sahelian rangelands in relation to the availability of nitrogen and phosphorus from the soil Proc. Workshop 'Nitrogen cycling in West African ecosystems'. Ibadan, Nigeria.

Rehatta, S. B., Dijkshoorn, W. and Lampe, J. E. M. 1979 Nitrogen uptake by rice plants from a dry soil at maintained water supply from a greater depth. Neth. J. Agric. Sci. 27, 99–110.

Rietveld, J. 1978 Soil non wettability and its relevance as a contributing factor to surface runoff on sandy soils in Mali. Rep. Dept. of Theor. Prod. Ecol. Agric. University, Wageningen.

Sörensen, L. H. 1975 The influence of clay on the rate of decay of amino acid metabolites synthesized in soils during decomposition of cellulose. Soil Biol. Biochem. 7, 171–177.

Spiertz, J. H. J. 1978 Grain production and assimilate utilization of wheat in relation to cultivar characteristics, climate factors and nitrogen supply. Agric. Res. Rep. 881. Pudoc, Wageningen.

Tandon, H. L. S., 1974 Dynamics of fertilizer nitrogen in Indian soils. I. Usage, transformation and crop removal of N. Fertilizer News 19, 3–11.

Veen, J. van, 1977 The behaviour of nitrogen in soil. A computer simulation model. Ph. D. Thesis, Univ. of Amsterdam.

Wit, C. T. de, 1958 Transpiration and crop yields. Versl. Landbouwk. Onderz. (Agric. Res. Rep.) 64. 6, Pudoc, Wageningen.

Wit, C. T. de, 1970 Dynamics concepts in biology. Proc. IBP/PP Technical Meeting, Trebon, Pudoc, Wageningen.

Wit, C. T. de, and van Keulen, H., 1972 Simulation of transport processes in soils. Simulation Monographs, Pudoc Wageningen.

Wit, C. T. de and Goudriaan, J., 1974 Simulation of ecological processes. Simulation Monographs, Pudoc, Wageningen.

Wit, C. T. de et al., 1978 Simulation of assimilation, respiration and transpiration of crops. Simulation Monographs, Pudoc, Wageningen.

Wong, S. C., Bowan, I. R. and Farquhar, G. D. 1979 Stomatal conductance correlates with photosynthetic capacity. Nature 282, 424–426.

Wösten, H. and van Loon, L., 1979 A model to simulate evaporation of bare soils in arid regions. Rep. no. 10 Dept. of Theor. Prod. Ecol. Agric. Univ. Wageningen.

9. Optimizing the use of water and nitrogen through soil and crop management

FLOYD E. BOLTON

Crop Science Dept., Oregon State University, Corvallis, Oregon, U.S.A.

Many factors limit crop production, e.g. diseases, insects, weeds, poorly adapted varieties, poor stands and nutrient deficiencies, but the major limiting factor is the supply of water. When there is little or no chance that the water supply will increase through irrigation or improved rainfall, one must learn how to use it more efficiently.

Water Use Efficiency (W.U.E.) is defined as the unit of economic yield produced per unit of precipitation received. W.U.E. is calculated by dividing the total yield per unit area, e.g. per hectare, by the total precipitation received during the entire period the crop utilizes the land including both fallow and crop periods. For example, if, in a fallow-crop rotation, a grain yield of 3.0 tonnes (3000 kg) per hectare of wheat was produced in an area which received 300 mm precipitation annually, the W.U.E. would be calculated as follows:

$$\frac{\text{Yield per hectare}}{\text{Total precipitation (fallow-crop)}} = \frac{3000}{600} = 5 \, \text{kg ha}^{-1} \, \text{mm}^{-1}$$

This measure of management skill can be used to compare years and locations, and removes the variability of climate in measuring the actual water use efficiency.

Dryland farming systems revolve around the principle that water is the main limiting factor, and to increase or maintain an adequate level of yield, one must maximize the water use efficiency for crop production.

There are two things that must occur in order to utilize the limited moisture supply to achieve increased crop production. We must (i) increase the amount of water stored in the soil, and reduce losses due to evaporation and transpiration; and (ii) make maximum use of the stored water and subsequent precipitation by management of cropping practices geared to the prevalent climatic conditions. The former is related to tillage and moisture conservation practices, and the latter is concerned with crop management practices specifically tailored to fit the available and expected moisture supply to achieve maximum *economic* crop production.

FALLOW-CROP VS ANNUAL CROPPING SYSTEMS

The use of fallow-crop rotations as a farming system is widely practised throughout the world in low rainfall regions. This method has been criticised because of
the supposed inefficiency of soil moisture storage during the fallow period. The
storage efficiency usually ranges between 15 and 35 per cent of the total precipitation that occurs during this time. The critics point out that this is a gross waste
of much limited resource. However, the fallow system is a time-honored method
of producing grain in the sub-humid and semi-arid areas of the Middle East. The
method seems to have been regularly employed in the Tigris-Euphrates delta as
far back as 2400 B.C.

In more recent times, the fallow-crop rotation for producing cereal grains has
become widespread in the U.S.A. In 1900, there were about two million acres of
fallow land in U.S.A. The area steadily increased, until in the 1970's, there were
more than 40 million acres in fallow; mostly in the 17 Western States (Haas *et al.*,
1974). In the ICARDA region, fallow-crop rotations at present constitute the
major cropping system for cereal production. The practice is still controversial,
mostly among professionals, because of the inefficiency in soil water storage and
some of the wind and water erosion problems which are associated with fallow.
The fallow system in low rainfall zones continues to be used extensively, despite
its critics.

Recent innovations in stubble-mulch tillage, have alleviated some of the
erosion problems, and have increased moisture storage. The widespread and
persistent use of the fallow-crop rotation throughout the world, and particularly
in the ICARDA region, requires that this farming practice receives serious
attention in the future Farming Systems research programs.

Examples of the increase in grain yields, and consequently increased W.U.E.,
are shown in Table 1. The Central Great Plains, U.S.A., region is characterized by
a spring-summer rainfall pattern and moderate summer temperatures, as con-
trasted with the Pacific Northwest climate with a winter-spring rainfall pattern
and moderate summer temperatures. The difference in precipitation patterns and
season temperatures is reflected in the W.U.E. values of the two regions. It should
be noted that the W.U.E. was greater in every location under the fallow-crop
rotation.

Improved soil and crop management techniques and better adapted varieties
have shown even greater increases in W.U.E. during the past 20 years (Table 1).
Greb *et al.* (1974) indicated four probable reasons for the increased yield and
W.U.E. for the Central Great Plains regions: (i) increased soil water storage in
fallow resulting from improved stubble-mulch management and weed control;

Table 1. Yield and water use efficiency of winter wheat grown continuously and after fallow at three locations in the Central Great Plains and in a fallow-crop rotation at three locations in the Pacific Northwest, U.S.A.

Location	Average annual precipitation (mm)	Continuous wheat (kg/ha)	Fallow-wheat (kg/ha)	W.U.E.[a] (kg ha^{-1} mm^{-1})	
				Continuous wheat	Fallow-wheat
Akron, Colorado					
60-yr average	419	498	1426	1.2	1.7
1951–70	377	549	1818	1.5	2.4
Colby, Kansas					
49-yr average	470	626	1318	1.3	1.4
1941–63	512	731	1673	1.4	1.6
North Platte, Nebraska					
56-yr average	495	834	2146	1.7	2.2
1951–67	505	684	2747	1.4	2.7
Pendleton, Oregon					
46-yr average	404	1259	3590	3.1	4.4
1954–76	403	–[b]	4270	–	5.3
Moro, Oregon					
65-yr average	290	781	2006	2.7	3.5
1954–76	283	–	2410	–	4.3
Lind, Washington					
55-yr average	241	875	1760	3.6	3.7
1954–76	235	–	2450	–	5.2

[a] W.U.E. on a harvest-to-harvest precipitation basis.
[b] Data not available.

(ii) improved planting equipment that permits deeper penetration to adequate seedbed water, even under relatively dry surface conditions of 8 to 10 cm depth; (iii) improved wheat varieties; and (iv) increased use of commercial fertilizers on those soils known to be deficient in nitrogen and phosphorus.

Professional agronomists and cereal breeders in this region estimated the credit for the yield increases to be for:

improved water storage in fallow, 40 per cent; improved wheat varieties, 40 per cent; improved planting equipment, 10 per cent; and use of nitrogen and phosphorus, 10 per cent.

One of the advantages in the fallow system is the release and accumulation of nitrate nitrogen which greatly reduces the need for supplemental fertilizers (Olson et al., 1953). The use of nitrogen fertilizers can substantially reduce the

Table 2. Grain yields of winter wheat grown in fallow-crop or annual rotation, with and without nitrogen fertilizer in the Central Great Plains and the Pacific Northwest, U.S.A.

Location	Annual precipitation (mm)	Nitrogen applied[a] (kg/ha)	Grain yield		Per cent yield Fallow vs annual (%)
			Fallow-wheat (kg/ha)	Annual wheat (kg/ha)	
Central Great Plains					
Northeast Colorado	457	0	1886	741	254
1956–70		45	1818	1145	159
Northwest Nebraska					
1956–70	445	0	1886	875	216
		45	2222	1347	165
Pacific Northwest					
Pullman, Washington	516	0	2970	1347	220
1922–45		67	3037	2094	145
Pendleton, Oregon	406	0	2997	815	368
1931–50		67	3219	1710	188
Ritzville, Washington	287	0	1414	741	191
1953–57		22	1616	875	185
Harrington, Washington	295	0	2761	808	342
1953–57		34	3098	1549	200

[a] When nitrogen was applied, the rate reported gave optimum yields.

yield advantage of fallow vs annual cropping systems as shown in Table 2. However, in terms of W.U.E. and the economics of producing cereals in this region, the fallow-crop system still holds the advantage.

It is estimated that a fallow-crop rotation needs only to produce 170 per cent of the annual cropping system to be economically equal in terms of profit margin. In regions with less than 500 mm annual precipitation, the yield difference under a well-managed fallow system is usually greater than 170 per cent of the annual cropping system.

Another important factor to be considered is the stability of a cropping system. Small, poor farmers cannot afford a crop failure. Smika (1970), in a study comparing 27 years (1940–1967) of fallow-wheat with continuous wheat under semi-arid conditions, showed the fallow system did not have a crop failure, whereas continuous wheat failed more than 30 per cent of the time. The average W.U.E. was 80 per cent greater for the fallow system when the annual precipitation was between 246 and 430 mm. At least 580 mm of annual precipitation was required before annual wheat, with nitrogen fertilizer, used water as efficiently as fallow-wheat without added nitrogen.

Smika (1970) also showed that in this semi-arid region, average fallow-wheat yields without added nitrogen were 2.6 times higher than annual wheat yields with nitrogen. He concluded from this long term study that, in all aspects, the fallow-winter wheat cropping system was much superior to an annual winter wheat cropping system and was necessary for stable winter wheat production under semi-arid conditions in the Great Plains of U.S.A.

ADVANTAGES AND DISADVANTAGES OF FALLOW

The advantages and disadvantages of fallow vs annual cereals in U.S.A. may not always apply to other wheatgrowing areas of the world. In areas with relatively mild winters and long periods of high temperatures during the spring and summer, a crop may not respond as readily as described above. In the spring wheatgrowing areas of the Northern Great Plains of U.S.A., the fallow system does not show as great an advantage. For example, the winter wheat fallow period is only 14 months, the stubble is upright during the single winter, and by the next spring, rains tend to keep fallow fields from wind-eroding, although some water erosion may occur. However for spring wheat, the fallow period is 18 to 21 months, with two winters and little or no soil protection during the second winter. It is during the second winter and spring of fallow that the erosion potential of spring wheat reaches its maximum.

The advantages of winter wheat fallow are: (i) it increases soil water storage; (ii) has a stabilizing influence on grain yields; (iii) increases W.U.E.; (iv) releases nutrients (especially nitrogen) by mineralization, hence requires less fertilizer; (v) allows for more timely cultivation; (vi) needs fewer field operations per unit of grain produced, and (vii) allows a higher net profit per unit area.

There are some disadvantages: (i) fallow may enhance erosion; (ii) tend to deplete organic matter and reduce the overall level of fertility; (iii) intensify leaching of some nutrients to below the root zone (only in higher rainfall sites).

The advantages far outweigh the disadvantages in areas where fallow for winter wheat production is used properly.

SOIL MANAGEMENT FOR MOISTURE CONSERVATION

Even though fallow-crop rotations have been used in the Middle East since ancient times, the grain yields in many areas are well below the potential. The adoption of a particular cropping system, whether it be an annual or fallow-crop rotation, does not guarantee success. The conservation of soil moisture requires close attention to management of the land from harvest to harvest. The traditional use of fallow land as 'weedy' pasture during the non-crop period is widespread. The value of the crop aftermath, volunteer cereals and weeds as forage for animals is difficult to assess. However, if the farmer's goal is increased grain production, then changes are required in the traditional use of the fallow.

In the Anatolian Plateau of Turkey during the past decade, soil management trials with fallow (Bolton, 1973; Hepworth et al. 1975; Guler, et al. 1978) have shown that it is possible to increase yields of winter cereals substantially by improved methods. The trials included several types of implements such as the mouldboard plow, disk plough, and sweep-chisel plough for initial tillage, and the sweep plough and harrow and rodweeders for secondary tillage operations. None of the tillage implement combinations showed an advantage in yield when used in a proper and timely manner. However, those implements which leave most of the crop aftermath on the soil surface reduced wind and water erosion hazards. The main effect on grain yields compared to farmers' fields, was the timeliness of tillage operations, particularly initial spring tillage, weed control, and seedbed preparation.

In adaptive research trials on farmers' fields at several locations for four years, the grain yields were 80 to 90 per cent higher than on the adjacent fields. The methods used in adaptive research trials were not radically different from the former methods with regard to the implement used or number of tillage operations, except in time and manner of operations. This fact led to relatively rapid adoption of the improved tillage methods.

Improved soil management that increases soil water storage and reduces evaporation will not itself increases grain yields. If it is not followed by good crop management techniques, the advantage of increased moisture supplies is often lost. In the trials mentioned above, the best available information gained from crop management trials was used to make maximum use of the moisture supply.

CROP MANAGEMENT AND WATER USE EFFICIENCY

Yield potential in dryland areas is limited by the moisture supply, but the *actual yield* obtained is limited by the production techniques. Yields in the dryland cereal regions of the world, in both the developed and developing nations, can be substantially increased if the admittedly limited moisture supply is used more efficiently. Some of the more critical crop management factors that affect the efficient use of limited moisture supplies are: Weed control, stand establishment, plant nutrient-moisture balance, adapted varieties, and plant populations and spacings.

WEED CONTROL AND WATER USE EFFICIENCY

Although control of weeds is not directly related to management factors which affect the water use efficiency of the crop plant, such as variety, stand establishment, and moisture-fertilizer balance, it certainly has an important role in water availability. Any living organism that uses water (and all do!) and is under conditions of limited moisture supply, will reduce the economic yield in direct proportion to the amount of water used. Under adequate moisture levels or irrigation where water is not the major factor limiting the yield potential, some weeds may be tolerated without loss of economic yield.

Generally speaking, the more limited the moisture supply, the more critical adequate weed control becomes. Hepworth *et al.* (1975) showed that, even under poor soil and crop management, timely chemical weed control resulted in large increases of yield. With improved crop management and good weed control practices, the increase in grain yield averaged about 33 per cent (Fig. 1).

Weed control alone increased grain yields by about 64 per cent, but at a yield of only 1.54 t ha^{-1} at a low level of management. Under good soil and crop management, a certain amount of cultural weed control had been achieved so that the addition of chemical control during the crop growth period did not show as great a yield increase. However, in total yield, increases due to weed control under improved management averaged 0.8 t ha^{-1} as compared to about 0.5 t ha^{-1} yield increases under poor management.

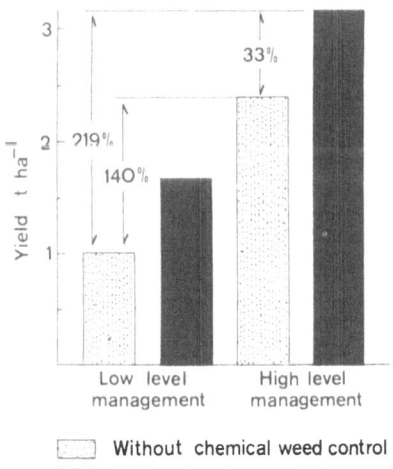

Fig. 1. Effect of weed control and management on yield of winter wheat in Turkey, 1974-75.

238

STAND ESTABLISHMENT AND WATER USE EFFICIENCY

The term 'stand establishment' is used in this chapter in preference to the commonly used terms such as seeding date, planting date, or date of seeding. Such terms describe the time the seeds are placed in the soil, but do not necessarily ensure that the following events result in an established stand. Too often, a date of seeding trial is presented in the literature with no reference to time of emergence, crown root development or the appearance of tillers. All the above events must occur before one can state that a stand has been 'established'.

Time of stand establishment

In dryland cereal-producing areas with a winter precipitation pattern, the optimum time of stand establishment may have a considerable effect on water use efficiency (in relation to economic yield) by ensuring that the growth of the crop is adjusted to the available soil moisture. The data in Fig. 2, which are averages of many trials over a period of years, show the optimum range for stand establishment in eastern Oregon. A number of the early trials reviewed showed no differences in yield due to planting data. Closer inspection of the data showed that in many cases the seeds were placed in dry soil for several dates before a significant amount of precipitation occurred. In these cases, the date of seeding may have differed by two to four weeks, but the time of emergence was the same. Since the introduction of deep-furrow drills and the placement of seeds into moist soil from the previous fallow, the data have been more consistent and have represented different dates of emergence.

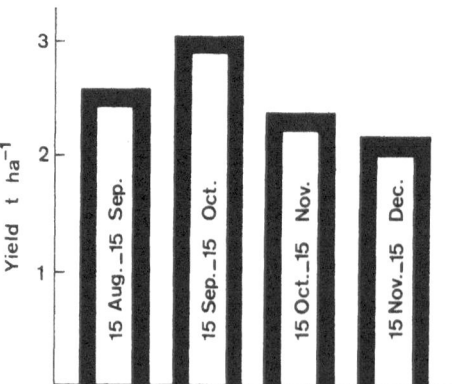

Fig. 2.   Effect of stand establishment and grain yield in Eastern Oregon, USA. (Long term average).

Long-term observations of precipitation in relation to grain yields at three locations in the Pacific Northwest (Bolton, 1976, unpublished data), have shown that whenever significant rainfall occurred during the period of optimum stand, established (September 15 to October 15) grain yields were significantly higher. This apparently was the result of adequate soil moisture for emergence and subsequent plant development before low winter temperatures occurred.

As soil moisture conservation practices improved and deep-furrow drills were introduced, the seeding into residual moisture became a standard practice. However, there are still seasons when the moisture level for seeding is too deep even for the deep-furrow drill. Often, the time of seeding is delayed beyond the optimum range while waiting for adequate rainfall. In most instances, this results in reduced yields because the plants are not adequately developed before the cold weather begins.

Pehlivanturk (1975) showed that time from emergence (after adequate moisture was present) to crown root development and tillering was greatly increased after soil temperatures reached a certain range. This study indicated that when the mean soil temperature at seeding depth dropped below 10 °C, the length of time for seedling development of crown roots and/or tillers was dramatically increased. The number of days from emergence to tillering were 28, 51, and 125 from October 1, October 15, and November 1, respectively. The mean soil temperatures at seeding depth for these planting dates were 16, 9, and 5 °C, respectively.

Seeding into dry soil is risky, and presents problems with crusting after rains and sometimes weed control. However, reductions in yield and late stand establishment are well documented and consistent. The risk in reduced yields from crusting or weeds is not predictable. It would seem that timely seeding under dry soil conditions outweighs any advantages in waiting for adequate moisture to wet the seed zone.

MOISTURE: FERTILIZER BALANCE AND WATER USE EFFICIENCY

In areas of limited precipitation, a balanced nutrient supply enables the crop to make more efficient use of the available moisture (Arnon, 1975, 1972; Olson *et al.*, 1964). The recommended use of fertilizers is the most economical way of increasing crop water use efficiency (Viets, 1962, 1967, 1971, 1972). It has been demonstrated that a nutrient-deficient plant, even though it is not growing or is growing slowly, is using water at about the same rate as a nutritionally balanced plant, yet will produce a considerably lower yield (Asana, 1962; Aspinall *et al.*, 1964; Brown, 1972; Leggett, 1959). Some recent information by Aktan (1976) has

240

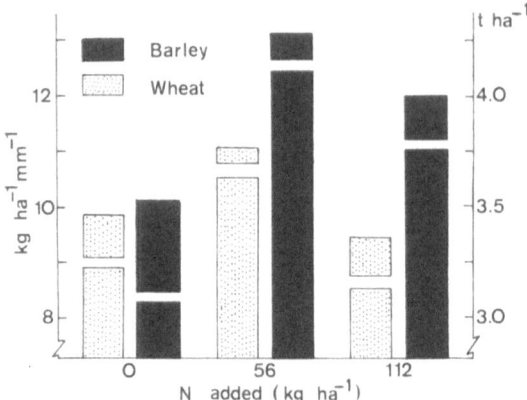

Fig. 3. Effect of nitrogen fertilizer on W.U.E. and grain yield (cross bar) of winter wheat and winter barley at Moro, Oregon, USA, 1975.

confirmed these findings, and indicates that different levels of soil moisture at seeding require different levels of applied nitrogen in order to balance the moisture-nitrogen supply for maximum water use efficiency. An example of this study is shown in Fig. 3. These data show that water use efficiency can be decreased markedly either by a deficiency or by an excess of nitrogen. It is also shown that a barley crop uses both water and nitrogen more efficiently than wheat, at least in this region.

The highly variable moisture supply, typical of dryland regions, requires that fertilizer requirements be tailored to the season if the best efficient use of water is to be accomplished. In addition, different soil management methods, such as bare fallow, stubble mulch minimum tillage or no-till (chemical fallow) systems may require different levels of nitrogen to reach the proper moisture-fertilizer balance (Brown et al., 1960; Greb et al., 1967; Koehler et al., 1967; Oveson and Appleby, 1971).

The amount of plant residues left on the soil surface or mixed in the upper few centimeters can have a large influence on the amount of nitrogen available to the crop plant (Leggett et al., 1974; Smika et al., 1969; Smika, 1970). Generally, the more surface residues present, the more nitrogen is required to balance fertilizer needs with the moisture supply.

Even though nitrogenous fertilizers receive the major emphasis, other fertilizer elements play an important role in increasing the water use efficiency. Phosphorus, potassium, sulphur, and the other minor elements required for plant growth and development, must be present in the proper available amounts or yield levels are reduced (Brown *et al.*, 1960; Brown, 1972; Gardner, 1964). These other fertilizer elements are not as dependent on the moisture supply as nitrogen, but unless they are present in the proper available form and amount, they will have a profound effect on the water use efficiency.

ADAPTED VARIETIES FOR DRYLAND CEREAL PRODUCTION

A key element in increasing the W.U.E., thereby increasing economic yield, is the use of the best adapted variety which is presently available. All cereal breeding programs have similar goals, viz., increasing yield potential and quality, maintaining resistance to pests (insects and diseases), lodging, shattering, and other agronomic characters. Resistance or tolerance to drouth and heat is sometimes included among the breeding goals, but these are much more difficult to select for.

Selection for varieties with high yielding potential and resistance to the prevalent crop pests has been very successful in the past two decades. In many

Table 3. Grain yield, precipitation, and water use efficiency at two locations in the fallow-crop rotation area of the Pacific Northwest, U.S.A.

Dates	Span years	Yield (t ha^{-1})	Total precipitation (mm)	Water use efficiency (kg ha^{-1} mm^{-1})
Lind, Washington				
1922–1974	54			
Kharkov		1.58	477	3.31
Improved variety		1.71	477	3.58
1931–1952	22			
Kharkov		1.35	518	2.60
Improved variety		1.35	518	2.60
1953–1974	22			
Kharkov		2.01	457	4.39
Improved variety		2.36	457	5.16
Pendleton, Oregon				
1931–1974	44			
Kharkov		3.03	805	3.76
Improved variety		3.51	805	4.36
1931–1952	22			
Kharkov		2.53	820	3.08
Improved variety		2.84	820	3.46
1953–1974	22			
Kharkov		3.15	800	3.93
Improved variety		4.16	800	5.20

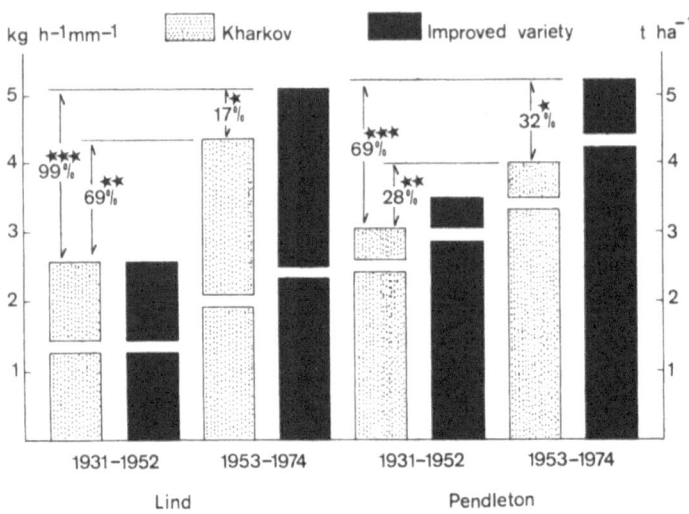

Fig. 4.   Effect of improved varieties and crop management of water use efficiency and grain yield (cross bar) of winter wheat in two locations in the Pacific Northwest, USA.
Per cent increase due to: *improved varieties, **improved management. ***interaction of improved varieties and management
Kharkov: long-term standard check variety.
Improved variety: variety most widely grown during the indicated period.

cases, the amount of water needed for maximum yield greatly exceeds its availability in the dryland cereal regions. Many of the varieties with high yield potential under adequate moisture have also produced higher yields under limited moisture.

Breeders have long recognized that selection for drought tolerance or heat resistance is extremely difficult because no suitable selection criteria are available at present. Selection in early generation material under dryland conditions is very risky because of the extreme variability from season to season (Hurd, 1971). As a consequence, many breeders have resorted to testing larger numbers of later generation lines in the dryland areas, hoping to isolate types that are better adapted to drouth. This approach has generally been quite successful in the Pacific Northwest of U.S.A.

An example of the progress made in varietal improvement for dryland conditions is shown in Table 3 and Fig. 4. These data represent grain yields from standard variety trials grown under a fallow-crop rotation in which the variety Kharkov was present as the long-term check variety. The trials were conducted on the Lind Dryland Research Unit, Lind, Washington, and the Columbia Basin Agricultural Research Center, where the annual rainfall averages about 240 and

400 mm, respectively. A standard meteorological station is located at each experimental site. Since improved varieties are constantly changing over a period of years, the grain yields represent those varieties which were the most widely grown commercially during each period. For example, at Lind, Washington, between 1931 and 1952, the varieties grown commercially in this region were Turkey, Rio, Rex, and Elgin. From 1953 to 1974, the most widely grown varieties were Omar, Moro, Gaines, and Nugaines.

At the Lind site, the increases in W.U.E. for grain yield averaged about 99 per cent for the past 22 years when compared to the previous 22-year period. This increase can be separated into three categories: (i) increase due to improved varieties (17 per cent), (ii) increase due to improved soil and crop management (69 per cent), and (iii) the increase due to the interaction of improved varieties and improved soil and crop management (13 per cent). This region is the lowest rainfall region for cereal production in U.S.A.

The value of improved varieties does not show as great an influence on grain yields as improved crop management, when considered alone. However, when the influence of improved varieties is added to improved crop management (17 per cent + 13 per cent), an overall increase of 30 per cent in W.U.E. is achieved. In the more arid cereal-producing regions, it appears that crop management plays a more dominant role than improved varieties. However, to gain maximum benefit, both elements must be incorporated in the production system.

As the precipitation increases, the roles of improved varieties and improved crop management shift in emphasis as shown by the data in Table 3 and Fig. 4 for the Pendleton location. The increase in W.U.E. due to improved varieties is almost twice that shown for the Lind location (32 compared with 17 per cent). In addition, the effect of improved crop management is less than half that of the lower rainfall zone at Lind (69 compared with 28 per cent). This indicates that varietal improvement plays a larger role in W.U.E. under more favorable climatic conditions. The interaction between improved varieties and improved crop management shows about a 9 per cent increase in W.U.E. as compared to a 13 per cent in W.U.E. at the Lind location.

The overall W.U.E. in terms of grain produced per unit volume of water received was about the same for both locations, i.e., 5.20 and 5.16 kg ha^{-1} mm^{-1} for Pendleton and Lind, respectively. This indicates that both locations are producing at about the same level of W.U.E. but at different yield levels because of the large difference in precipitation. These data clearly indicate the advantage of using the best adapted varieties for maximum W.U.E. and increased grain yield level.

Plant population and stand establishment

The optimum number of plants per unit area to obtain maximum economic production varies widely in different environments. Under well-watered or irrigated conditions, the plant population per unit area above a certain minimum number seems to have little or no effect on the subsequent yield (Donald, 1963). With limited moisture conditions, the plant population and distribution can exhibit substantial influence on the water use efficiency and the subsequent grain yield (Harper, 1961).

In cereal crops, adjustments in plant populations and distribution, because of the methods of seeding and nature of the crops, must be made either within rows or between rows. Increasing the distance between plants within the rows and reducing the distance between rows to achieve a more uniform distribution over a unit area may defeat its purpose when moisture is limited.

Brown and Shrader (1959) showed that plants with little competition in the seedling stage tend to produce excessive vegetative growth and deplete soil moisture that could be used for grain production at a later date. Larson and Willis (1957) concluded that to gain maximum benefit from limited moisture, the intra-row population should be at a level great enough to provide adequate competition in the seedling stages to prevent excessive growth and moisture use and that the inter-row spaces be adjusted to fit the moisture supply. This proposal is generally contrary to the usual practice in many dryland areas of the world, including both developed and developing countries.

Seeding rates and row spacings have been traditionally used for two purposes which are not actually related to the best plant population for maximum yield: (i) seeding rates within the row are usually greater than necessary in order to provide ground cover for wind and water erosion control. (ii) close row spacings are often preferred to provide greater competition for weeds that may grow between the rows. These concepts may have been partially successful in their purposes, but undoubtedly contribute to reduced grain yields in many seasons. The work of Arnon (1972) and Arnon and Blum (1964) suggests that better use of a limited moisture supply is possible with more careful attention to the plant population and distribution in dryland regions.

Many studies in several different dryland regions of the world have shown that seeding rates can be reduced and row spacings increased without reducing even increasing yields many times (Alessi *et al.,* 1971; Bolton, 1973; Guler, 1975; Wilson, 1969). Reasons for not using more carefully adjusted seeding rates and row spacings are many and varied but none seems to be related to more efficient use of moisture. There is considerable evidence that varieties differ in their response to seeding rates and row spacing in different climatic zones (Guler, 1975). Guler found that some varieties had different optimum plant population numbers at each location, while other varieties responded in about the same way at a given seeding rate over several locations.

SUMMARY

To optimize the use of water and nitrogen under rainfed conditions with a limited moisture supply, primary attention must be given to the efficient use of water. This may be accomplished in two ways: (i) by storing the maximum amount in the soil profile by improved tillage and moisture conservation practices, (ii) making maximum use of the stored water and subsequent precipitation through improved crop management practices.

As the water use efficiency increases, the response to nitrogen fertilizers will improve and be reflected in higher crop yields. Variation in seasonal precipitation typical of semi-arid, rainfed regions requires that fertilizer applications be based on current moisture supplies rather than long-term averages.

Rainfed crop production has been most successful when the 'package of practices' approach has been used. This means using the best combinations of tillage for moisture conservation and seedbed preparation, selection of adapted varieties, optimum planting and rates and dates, fertilizers applied at the proper rates and times in relation to moisture supplies, timely weed control in both fallow and crop seasons, and harvesting methods. The timing and application of the various components of the package is very important and may vary from one locality to another.

The proper package of practices must be determined by extensive, applied field research conducted over relatively long periods. Each element in the production package must be done in the right sequence or the advantage of other elements is often lost. For example, if an improved variety is introduced without the addition of weed control or fertilizers or better seedbed preparation, oftentimes the old, local variety will still yield as well or better. If fertilizers are introduced as an improved practice and weeds left uncontrolled, the weeds may respond more to the increased fertility level than the crop, and may in some cases actually reduce yields. If improved initial tillage practices are introduced, but secondary tillage for weed control and seedbed preparation is neglected, the end result may be worse than the traditional method.

In the developing countries the rapid development of a complete production package is not generally possible. Many of the resources (e.g. equipment, herbicides, fertilizers) are not available at present, and require considerable time and

246

investment to develop. However, if certain elements of the production package are properly applied and in the right sequence, substantial yield increases are possible. The key to increase in crop yields is sorting out those practices which give the greatest benefits and which can fit into the local traditional systems. As other resources become available and additional field research is conducted, the other elements of the production package can be applied.

REFERENCES

Aktan, Selman 1976 Nitrate-nitrogen accumulation and distribution in the soil profile during a fallow grain rotation as influenced by different levels of soil profile moisture. M.S. Thesis, Oregon State University, Corvallis.
Alessi, J. and Power, J. F. 1971 Influences of method of seeding and moisture on winter wheat survival and yield. Agron. J. 63, 81–83.
Arnon, I. 1975 Physiological principles of dryland crop production, pp. 3–145. In: Gupta, U. S. (ed.), Physiological Aspects of Dryland Farming, Oxford & IBH, New Delhi.
Arnon, I. 1972 Crop Production in Dry Regions, 2 vols., Leonard Hill, London.
Arnon, I. and Blum, A. 1964 Response of hybrid and self-pollinated sorghum varieties to moisture regime and intra-row competition. Israel J. Agric. Res. 14, 45–53.
Asana, R. D. 1962 Analysis of drought resistance in wheat. Arid Zone Res. 16, 182–190.
Aspinall, D., Nicholls, P. B. and May, C. H. 1964 The effect of soil moisture stress on the growth of barley. I. Vegetative development and grain yield. Aust. J. Agric. Res. 15, 729.
Bolton, F. E. 1973 Soil and crop management research at the Wheat Research and Training Center: October 1970 to August 1973, Ankara, Turkey. Termination Report to Rockefeller Foundation and Oregon State University under RF-OSU (RF70011).
Brown, D. A., Place, G. A. and Pettiet, J. V. 1960 The effect of soil moisture upon cation exchange in soils and nutrient uptake by plants. Trans. 7th Int. Congr. Soil Sci. 3, 443–449.
Brown, P. L. 1972 Water use and soil water depletion by dryland wheat as affected by nitrogen fertilization. Trans. 7th Int. Congr. Soil Sci. 63, 43–46.
Donald, C. M. 1963 Competition among crop and pasture plants. Adv. Agron. 15, 1–118.
Gardner, W. H. 1964 Research for more efficient water use: soil physics, pp. 85–94. In: Research on Water, Soil Sci. Soc. Am., Madison, Wisconsin.
Greb, B. W., Smika, D. E., Woodruff, N. P. and Whitfield, C. J. 1974 Chapter 4 – Summer fallow in the Central Great Plains. In: Summer fallow in the Western United States. Cons. Res. Rept. No. 17, ARS-USDA, April 1974.
Greb, B. W., Smika, D. E. and Black, A. L. 1967 Effect of straw mulch rates on soil water storage during summer fallow in the Great Plains. Soil Sci. Soc. Amer. Proc. 31, 556–559.
Güler, M., Ünver, I., Pala, M., Durutan, N. and Karca, M. 1978 Orta Anadolu'da 1972–1977. Nadas Toprak Hazirliği ve Bugdam Yetiştirme Tekniği Araştirmalari. Ankara 1978.
Güler, Mengu 1975 Yield and other agronomic characters of winter wheat as affected by five seeding rates and three different environmental conditions. M.S. Thesis, Oregon State University, Corvallis.
Haas, H. J., Willis, W. O. and Bond, J. J. 1974 Chapter 2. Summer fallow in the Northern Great Plains (spring wheat). In: Summer Fallow in the Western United States. Cons. Res. Rpt. No. 17. ARS-USDA, April 1974.
Harper, J. L. 1961 Approaches to the study of plant competition. In: Milthorpe, F. L. (ed.). Mechanism in Biological Competition, pp. 1–39, 5th Symp. Soc. Exp. Biol.
Hepworth, H. M., Zinn, T. G. and Andersen, W. L. 1975 Improved techniques for dryland mean more wheat from fallow farming. 2nd Ed. USAID/Oregon State University Team. Ankara, Turkey. 112 p.

Hurd, E. A. 1971 Can we breed for drought resistance? pp. 77–88. In: Drought Injury and Resistance in Crops. CSSA Special Publ. No. 2, Madison, Wisconsin.

Koehler, F. E. and Guettinger, D. L. 1967 Moisture-nitrógen relationships for wheat production in Eastern Washington. Proc. 18th Annual Pacific Northwest Fertilizer Conf., Twin Falls, Idaho. pp. 93–94.

Larson, W. E. and Willis, W. O. 1957 Light, soil temperature, soil moisture and alfalfa red clover distribution between corn rows of various spacings and row directions. Agron. J. 49, 422–426.

Leggett, G. E., Ramig, R. E., Johnson, L. C. and Masse, T. W. 1974 Summer fallow in the Western United States. U.S. Dept. Agr., Agr. Res. Serv., Conserv. Res. Rpt. 17, 111–135.

Leggett, G. E., Ramig, R. E., Johnson, L. C. and Masse, T. W. 1959 Relationships between wheat yield, available moisture and available nitrogen in Eastern Washington dryland areas. Washington Agr. Exp. Sta. Bull. 609.

Olson, R. A., Thompson, C. A., Grabouski, P. E., Stukenholtz, K. D., Frank, K. D. and Dreier, A. F. 1964 Water requirement of grain crops as modified by fertilizer use. Agron. J. 56, 427–432.

Olson, R. A. and Rhoades, H. T. 1953 Commercial fertilizer for winter wheat in relation to the properties of Nebraska soils. Nebr. Agr. Exp. Sta. Bul. 172.

Oveson, M. M. and Appleby, A. P. 1971 Influence of tillage management in a stubble mulch fallow-winter wheat rotation with herbicide weed control. Agron. J. 63, 19–20.

Pehlivanturk, Alpaslan 1976 Effect of soil temperature, seeding date, and straw mulch of plant development and grain yield of two winter wheat and two winter barley cultivars. M.S. Thesis, Oregon State University, Corvallis.

Smika, D. E., Black, A. L. and Greb, B. W. 1969 Soil nitrate, soil water, and grain yields in a wheat-fallow rotation in the Great Plains as influenced by straw mulch. Agron. J. 61, 785–787.

Smika, D. E. 1970 Summer fallow for dryland winter wheat in the semi-arid Great Plains. Agron. J. 62, 15–17.

Viets, F. G., Jr. 1962 Fertilizers and the efficient use of water. Adv. Agron. 14, 223–264.

Viets, F. G., Jr. 1967 Nutrient availability in relation to soil water, pp. 458–467. In: Hagan, R. M., Haise, R. H. and Edminster, W. (eds.), Irrigation of Agricultural Lands, Agron. Series. Am. Soc. Agron., Madison, Wisconsin.

Viets, F. G., Jr. 1971 Effective drought control for successful dryland agriculture. In: Drought Injury and Resistance in Crop Plants. CSSA Spec. Pub. No. 2, pp. 57–75.

Viets, F. G., Jr. 1972 Water deficits and nutrient availability, pp. 217–239. In: Kozlowski, T. T. (ed.), Water Deficits and Plant Growth, Vol. III. Academic Press, New York.

10. Optimizing the use of water and nitrogen through breeding of crops

R. A. FISCHER

Division of Plant Industry, CSIRO, Canberra City, Australia

Although many crops are grown in ICARDA's region, wheat and barley dominate. This paper concentrates on these two crops. Much breeding effort and physiological research has been expended on wheat and barley and many of the principles which have emerged from this work can be applied to less important crops.

Breeding of both wheat and barley over the past 20 years has lifted dryland yields modestly in certain Mediterranean-type environments (e.g. southern Australia, Israel). The methodology of this program has been traditional, with heavy reliance upon empirical yield testing. The physiological basis of this genetic progress is not understood, although at least two of the factors involved are better phenological adaptations (i.e. a more appropriate flowering date) and increased yield potential (i.e. greater yield in the absence of water stress).

Many other traits have been identified as important for increasing dryland yield but, with the possible exception of the presence of awns and a conservative or low tillering growth habit, claims are not supported either by good experimental evidence or unanimity of opinion. Hence we must concentrate in our discussions on the experimental approaches needed to provide evidence to justify trait screening or to improve breeding strategies in any other way. I will review these approaches, as they have been used for wheat and barley, and attempt some suggestions.

Elucidation of the physiology of performance under drought takes two general forms; this separation probably reflecting the philosophy of the investigator as much as anything else. Pragmatists attempt to proceed from yield differences to underlying process differences. I call this the *black box* approach. At the same time, theorists wish to predict differences in yield from an understanding of processes. This procedure is exemplified in the *ideotype* strategy.

BLACK BOX APPROACH TO DROUGHT RESISTANCE

One of the hardest aspects of work on drought, and a particular problem with the black box approach, is the demonstration of *consistent* differences among geno-

250

types in yield. Droughts vary, interactions abound, and more seriously, experimental error is usually large. Pooling records from many experiments may help, and an example of this, relevant also in other respects, is now given.

Laing and Fischer (1977) took yields for 33 wheat cultivars, tested at 44 dryland sites, over a number of countries, mostly between latitudes 20 °N and 40 °N. Site yield was correlated with growing season rainfall (for maximum yield vs rainfall, $r^2 = 0.68$; slope 7 kg ha^{-1} mm^{-1}) and hence it was assumed that moisture was the major environmental factor which was influencing yield. Individual cultivar yields were regressed against site mean yield as in Eberhardt and Russell (1966).

Regression slopes and mean yields differed significantly between cultivars, suggesting different responses to drought. We related these criteria of yield performance to the presence or absence of dwarfing genes, an important consideration then and today. The mean regression slope for 18 Norin 10-derived short wheats (1.151 ± 0.023) was significantly greater than that for 10 tall wheats of similar flowering date (0.879 ± 0.033) (Fig. 1).

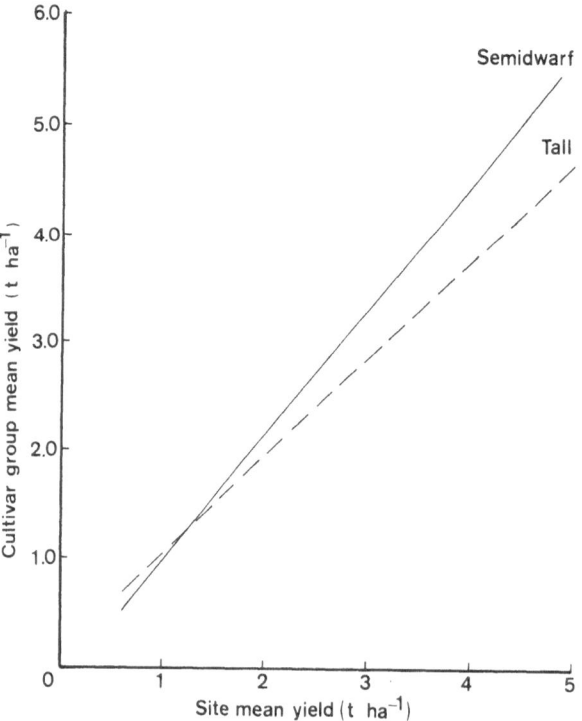

Fig. 1. The mean grain yield of 18 semidwarf cultivars and 10 tall cultivars as a function of site mean yield (mean of 33 cultivars); 44 dryland sites of the 6th and 7th ISWYN (Laing and Fischer, 1977).

Nevertheless the mean yield, or more appropriately the yield potential, of the former group was so much greater than that of the latter group (+ 20 per cent for yield potential, i.e. yield at the wettest sites), that the short wheats outyielded the tall ones not only in wet but also in moderately dry environments, and were no worse than them even in the driest environments. We went on to identify, within the short group, two cultivars of identical flowering date and mean yield but

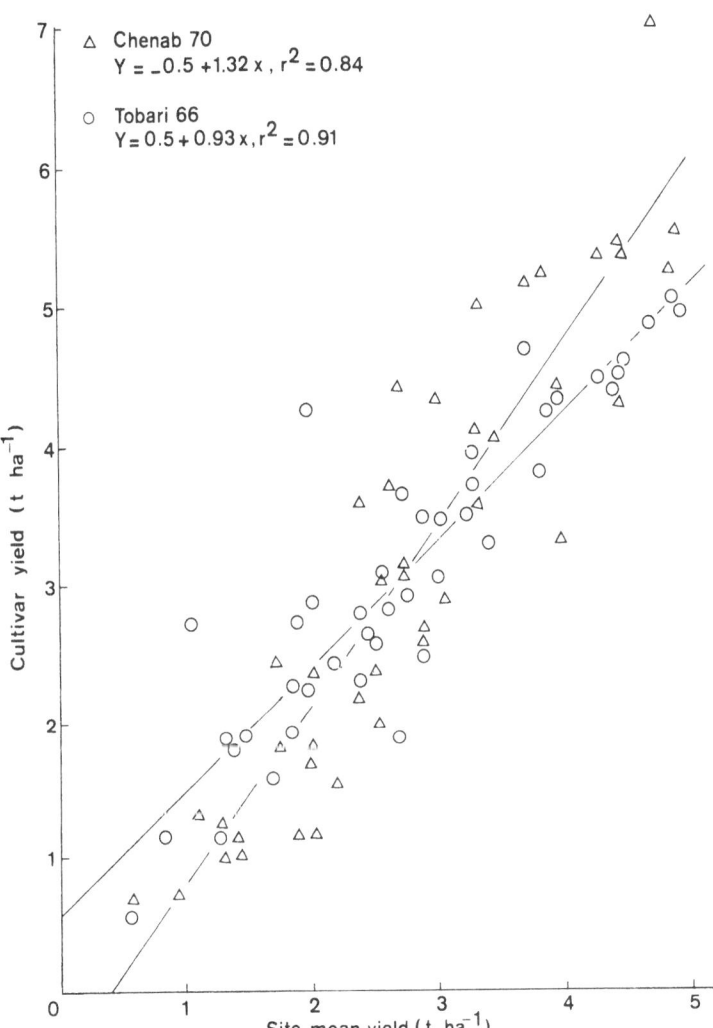

Fig. 2. The mean yields of two semidwarf cultivars (Tobari 66, circles, and Chenab 70, triangles) as a function of site mean yield (same data set as Fig. 1); regression slopes 1.32 (Chenab 70) and 0.93 (Tobari 66).

differing significantly in regression slope and hence in response to drought (Fig. 2). This pair, we suggested, would be appropriate for detailed physiological study.

Relative yield under drought

The positive correlation in the above study between cultivar yield potential and regression slope has been seen in similar studies by others (Eberhardt and Russell, 1966; Johnson *et al.,* 1968; Brennan 1974). It does not seem surprising that cultivars with higher yield potential suffer greatest absolute losses in yield when water is short. This highlights another question inherent in the quantification of performance under drought: is yield the best criterion of drought resistance? Some have suggested that relative yield under drought (i.e. Y_D/Y_P where Y_D = yield under drought and Y_P = yield potential) may be a better indicator of underlying drought resistance traits. Following such reasoning, Fischer and Maurer (1978) defined a drought susceptibility factor (S) such that for each cultivar:

$$Y_D = Y_P(1 - S.D)$$

where D is a measure of drought intensity, conveniently given by 1.0 minus the mean relative yield of all cultivars under the drought.

In a field experiment in which 53 cultivars were exposed to controlled Mediterranean-type or terminal drought (termination of irrigation at or before flowering in the rainless climate of northwest Mexico), Fischer and Wood (1979) attempted to relate the significant cultivar variation revealed in S to differences in growth, morphology, and water relations (midday leaf water potential and leaf permeability to viscous air flow). These quantities were measured both in the absence and presence of the drought. Before analysis of the records, all traits measured under drought and including S, were corrected for drought escape as indicated by the differences in anthesis date (drought yield was negatively related to days to anthesis, $r = -0.39$).

For the 34 bread wheat cultivars tested, and in the case of relationships with traits measured in the absence of drought (wet traits), drought susceptibility (S) was positively related to yield potential and to many of the yield components. Two orthogonal principal components, derived from these wet traits, explained much of the variation in S. These components were dubbed the 'high kernel number/m²' component ($r = 0.64$) and the 'high harvest index – low stature' component ($r = 0.51$). The increase in drought susceptibility, as defined here, with increase in yield potential was not expected. In fact I had argued elsewhere that high harvest index ought to be a desirable trait under all conditions of moisture supply (Laing and Fischer, 1977). However, the above results point to a trade-off between yield potential and stress resistance, the physiological nature of which is not clear at present. This suggests also that the weight given to selection for yield potential, compared to selection against drought susceptibility, will ideally depend on the expected drought intensity of the target environment. Of all the wet traits measured, selection for increased yield potential via increased total dry weight (i.e. grain plus straw) had the least undesirable effect on S.

For people who are uneasy with this new term 'drought susceptibility', we can also look at the relationships between yield under drought (40 per cent of potential yield on the average) and the same wet traits. Again the best single wet trait for predicting yield under drought was total dry weight (genotypic $r = 0.70$). Also yield under drought was positively related to plant height in the wet ($r = -0.33$), whereas yield potential was negatively related to height (genotypic $r = -0.47$).

Physiologists may be more interested in relationships in the above experiment between drought susceptibility and drought yield, on the one hand, and other traits measured under drought, in particular plant water status, on the other. However, despite significant and consistent cultivar differences in plant water status (Fischer and Sanchez, 1979), there were few significant correlations with S, yield or yield components under drought. At best, 10 per cent of the genotypic variations in yield under drought could be explained by variation in plant water status potential, leaf permeability) under the same conditions. Again, total dry weight was the trait most closely related to drought yield (genotypic $r = 0.69$); harvest index variation was less important (genotypic $r = 0.42$).

The results of this massive exercise in data collection and analysis refer to one set of genotypes and one drought occurrence. Given the representative nature of the cultivars and, hopefully, of the drought, the absence of clear-cut results is disappointing. The experience of others working with more limited sets of cultivars (e. g. Kaul and Crowle, 1971; Shimshi and Ephrat, 1972; Jones, 1977) has been essentially similar. Experimental errors are inevitably high (26 per cent and 30 per cent of the variation, respectively, in S and drought yield between cultivar means could be attributed to experimental error in our nine-replicate study). Also there may be no unique pathway to superior performance under drought. It was pointed out that had a better soil type been chosen for the study (e.g. deeper coarser-textured profile), relationships with plant water status under drought might have been clearer.

The study of Fischer and Wood (1979) did provide some useful information on differences between wheat, barley and triticale. Six barley cultivars on average yielded no better than 34 wheat cultivars under drought and, after correcting for

254

differences in anthesis date (barley was several days earlier), barley yielded less and was no less drought susceptible.

Barley was characterized by superior early growth giving greater green ground cover before and during the drought and consequently greater final total dry weight. Barley appears to escape late drought not only by flowering earlier, but also by making more of its growth early in the season. On the other hand seven triticale cultivars had the lowest average yield under drought, and highest drought susceptibility. Growth under drought was not inferior, however, nor was plant water status; but kernel number per spike was more sensitive to drought.

Small plots and spaced plants

The above studies all refer to yield measurements in plots which were large enough to be free from edge effects (Fig. 3). In many published studies, relationships have been sought with yield determined in small or narrow plots, single rows, hill plots or even spaced plants. Because of interference due to inter-genotypic competition (for soil moisture) and/or edge effects which may be

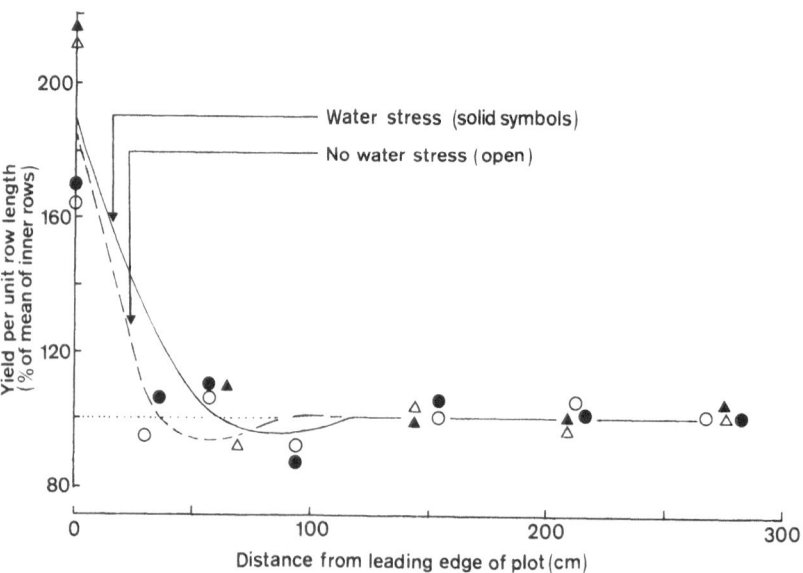

Fig. 3. Grain yield per unit row length (% of mean yield of inner rows) as a function of perpendicular distance from the leading edge of an exposed plot; open symbols = no water stress, closed symbols = post-flowering water stress which lowered yield 40 per cent; circles – samplings from western edge inwards, triangles = samplings from southern edge inwards. CIANO (lat. 27 °N), 1971–72.

Table 1. The effect of plot size on the yield of two wheat cultivars in the absence and presence of drought. Results of two adjacent experiments with differing plot sizes but the same set of 10 cultivars and the same drought treatments. Ciano 1973–74.

Cultivar	Grain yield, g m^{-2} (as % of all cultivars)	
	Large plots[a]	Small plots[b]
	No drought	
Mean 10 cultivars	520 (100)	485 (100)
Cocorit 71	700 (135)	724 (149)
Gabo	438 (84)	375 (77)
	Drought	
Mean 10 cultivars	281 (100)	301 (100)
Cocorit 71	244 (87)	363 (121)
Gabo	315 (112)	265 (88)
SE cultivar mean	10	14

[a] Plots were: 9 rows × 3 m long and 30 cm apart; the central 1.5 m of each of the central 5 rows was harvested (= 2.25 m²).
[b] Plots were 2 rows × 3 m long and 30 cm apart, separated by a 30 cm path; the central 2 m of each row was harvested giving an area, including path, of 1.80 m²

responsible for an atypical availability of resources (e.g. a large soil volume per plant), yields in such conditions must be regarded with great suspicion, not only in the absence of drought (Fischer, 1979b), but especially in its presence (e.g. Table 1, Fig. 4).

A different and more relevant question is whether such plantings, often typical for early generation material in breeding programs where the seed supply and/or space may be limited, provide appropriate conditions, either in the presence or absence of drought, for the assessment of other traits possibly related to drought resistance. Obviously genotypic differences in phenology and, to a large extent, morphology are reliably expressed under these conditions. One might even expect spaced planting under drought to magnify the effects of differences in the rate and extent of root exploration, differences which ought to be reflected faithfully in large plot yields under some conditions. The experiment of Fischer and Wood (1979) also contained spaced plantings (30 × 60 cm) of the genotypes and some of these possibilities are examined (Table 2).

The first two columns of Table 2 do show that across 34 wheat cultivars, height and kernel weight in spaced plants, both in the presence and absence of drought, were closely related to the same traits in plots under drought, and some other similarly paired traits showed significant correlations. However, no trait of non-

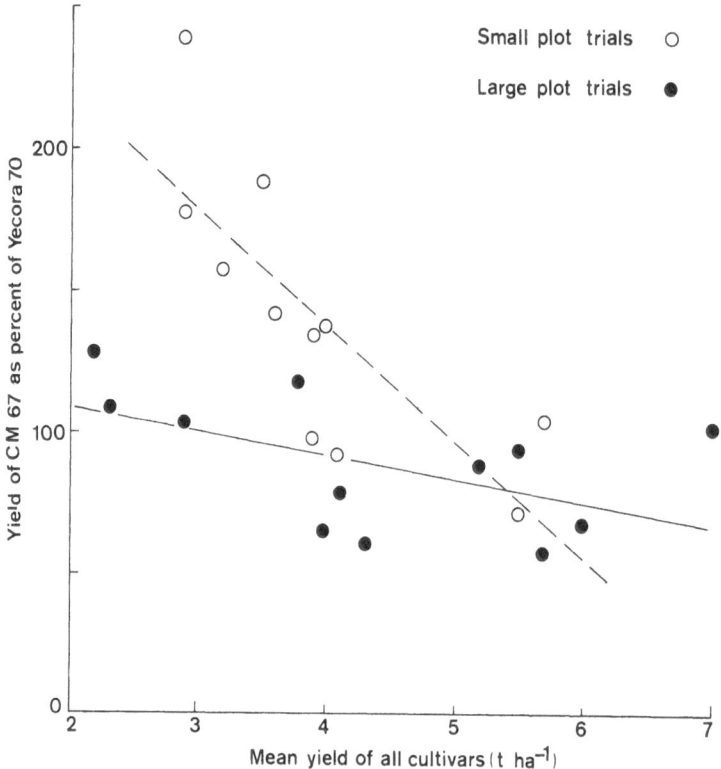

Fig. 4. Yield of barley cultivar CM67 (percentage of the yield of wheat cultivar Yecora 70) as a function of the mean yield of all cultivars; closed circles = experiments with large plots (no edge effect), open circles = experiments with small plots (2 × 30 cm rows × 3.0 m long with one unsown row between adjacent plots, central 2.0 m harvested). Experiments comprising 12 to 54 genotypes with a range of drought treatments, these treatments causing most of the variation in mean yield. CIANO, 1973–75.

droughted spaced plants was significantly correlated with plot yield under drought (third column, Table 2), while for traits of droughted spaced plants, only yield and total dry weight were significantly correlated (fourth column).

Of the spaced plant traits, the best correlation with yield in droughted plots was obtained with drought susceptibility, calculated in a manner analogous to the calculation of S outlined earlier. Even so, the proportion of variation explained ($r^2 = 0.18$) was quite small. These poor relationships between spaced plant traits and yield under drought, contrast with the case of yield in the absence of drought or yield potential when, for example, spaced harvest index (Fischer and Kertesz, 1976) or leaf permeability (CIMMYT, 1977) may be useful selection criteria.

Table 2. Phenotypic (and genotypic) correlations between traits measured on spaced plants in the absence (first column) and the presence (second column) of drought and the same traits[a] measured in plots under similar drought conditions[b]. Phenotypic correlations between grain yield under drought [b] and traits measured on spaced plants in the absence (third column) and presence (fourth column) of drought; genotypic correlations in parenthesis. Thirty four breadwheat cultivars, CIANO 1974–75.

Trait	Trait in drought plot vs same trait in		Grain yield in drought plot vs	
	non-drought spaced plant	drought spaced plant	non-drought spaced plant trait	drought spaced plant trait
Grain yield	0.01	0.38*	0.01	0.38 (0.68)
Total dry weight	0.18	0.39*	−0.01	0.36* (0.63)
Harvest index	0.41*	0.30	−0.03	0.08 (0.11)
Kernel weight	0.80**	0.83**	0.32	0.27 (0.41) (0.35)
Kernel number	0.28	0.56**	−0.17	0.20 (0.33)
Spike number	0.41*	0.42*	−0.21	0.23 (0.33)
Kernel/spike	0.55**	0.50**	0.04	0.04 (0.06)
Plant height	0.92**	0.96**	0.14	0.15 (0.19)
Drought susceptibility (S)	0.18		−0.42*	

[a] Where appropriate traits are expressed on a per plant basis for the spaced planting, and a per m² basis for the plots.
[b] All plot traits under drought corrected for differences in anthesis date.
*, ** Significant at 5% and 7% levels respectively.

Conclusion

What is the future of this so-called black box approach? I believe that it should be part of any substantial physiological attack on drought resistance. In any case, at centres such as ICARDA, the genetic material is often already in the field and yields are being recorded in breeding programs. What is required is much more attention to quantification and/or control of the drought environment, and elimination of obvious extraneous factors such as disease, bird and pest damage and plot edge effects. Irrigation control in rainless environments, rain covers and fungicides are among the tools of such research.

Simple measurements or observations of the genotypes being compared, such as ground cover, dry matter production, flowering date and leaf water potential, should permit useful conclusions underlying causes of yield differences. Also, exposure to the difficulties of eliminating extraneous factors and obtaining consistent yield differences in field experiments under drought is valuable and sobering experience for physiologists. At the same time, studies which propose to examine drought resistance but have as their end point, yield performance in spaced plants or unbordered small plots, ought to be recognized as generally useless for our purposes.

IDEOTYPE APPROACH TO DROUGHT RESISTANCE

The formulation of complete ideotypes and of ideotype traits for dry conditions seems to appeal to the physiologist, if one is to judge by the frequency with which physiological papers conclude with recommendations of selection criteria for the plant breeder. At the heart of this appeal, is a desire to predict useful change through scientific understanding. The intention may be good but usually insufficient consideration is given to the full ramifications of recommended traits in the context of dryland crops, and little or no attention is given to verification of the recommendations. While ideotype formulation will receive most attention here, ideotype verification must also be mentioned.

IDEOTYPE AND IDEOTYPE TRAIT FORMULATION

Essential to the prediction of genetic improvement in dryland crop plants is an understanding of the type of water limitation encountered (timing, duration, and so on) and of the mechanisms by which water limitation affects yield. An appreciation of the likely genetic variability to be found is also important, as is an explanation of why natural evolution and/or artificial selection have not already produced the recommended ideotype.

Environmental framework

We have already heard much about the nature of the water limitation likely to be seen by winter cereal crops in Mediterranean-type environments. Most useful is the quantification of the pattern of water shortage in terms of probabilities so as to permit risk analysis (Anderson, 1974) and allowance for different soil depths and agronomic strategies. In my discussion of ideotypes, I will assume very generally that we have some chance of temporary seedling water stress, combined

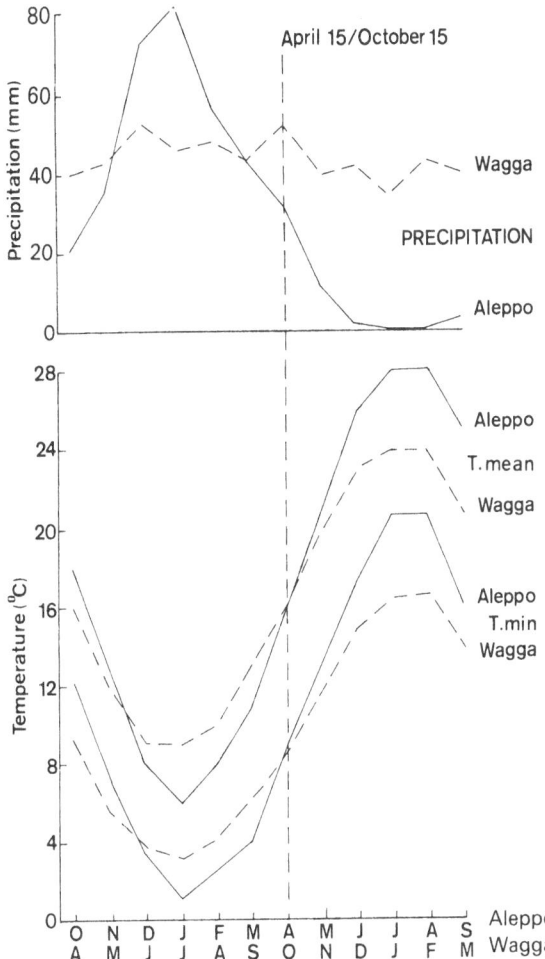

Fig. 5. Mean monthly rainfall, mean temperature and minimum temperature at a pair of Mediterranean-type stations (Aleppo and Wagga Wagga) with approximately matching thermal regimes and total annual rainfall (see also Table 3).

with a high probability of stress beginning after, or sometimes even before, flowering and continuing with increasing intensity, except for occasional partial interruptions, until maturity is reached.

Independent growth limitations during winter due to low temperatures, low solar radiation and waterlogging, and the possibly of meteorological hazards of frost and hot winds in the spring, are other physical factors which ought to be considered. Fig. 5 shows the key climate considerations at Aleppo and at Wagga Wagga, an approximately matching station in Australia.

Nature of the effects of water limitation

We have already discussed the general effects of water deficit on crop yield. For winter cereals, I use the simple conceptual model of Passioura (1977) and Fischer and Turner (1978) as a framework for further analysis. Passioura (1977) suggested that water-limited yield is the product of three largely independent variables: water use, water use efficiency in dry matter production, and harvest index (grain yield as a fraction of total dry matter production). In turn, harvest index is, to the first approximation at least, related to the proportion of the water use which occurs after flowering in determinate crops such as wheat or barley (Passioura 1977). Elaborating this framework for dryland wheat crops in south eastern Australia (Fischer, 1979a), I divided water use or evapotranspiration into components of transpiration and soil evaporation, and defined transpiration efficiency as dry matter production per unit of transpiration (reciprocal of transpiration ratio). Yield is therefore expressed as the product of total transpiration, transpiration efficiency and harvest index. Yield improvement will be considered in these terms.

Maximizing crop transpiration

The first question is how might total crop transpiration as a fraction of water supply (soil store at sowing plus crop season rainfall) be increased by genetic change? This fraction is less than 1.0 because of (i) soil evaporation losses in the crop, and sometimes because of (ii) run-off and through drainage, or (iii) unused available water in the soil profile at maturity. Soil evaporation losses have been estimated (but not measured) to be 20 to 50 per cent of total crop evapotranspiration with wheat (Fischer and Turner, 1978; Doyle and Fischer, 1979). A quick approach to full ground cover reduces these losses (see Fig. 6) but carries other risks (see later). Many superficial roots, rapidly reactivated after any rain, could be desirable in reducing the evaporation component, but without measurements for various cultivars of the partitioning of water loss from the surface layers of the soil between evaporation and root uptake, this must be considered to be a very tentative suggestion.

Run-off and deep drainage ought to be lessened by early sowing, giving greater water use before the wettest period of mid- and late winter is reached; for this, cultivars need to be suited to early sowing, particularly in terms of their phenology.

Water available in the profile (say the top 150 cm) at maturity, provided it occurs frequently and is not fossil water, strongly suggests that a deeper and more

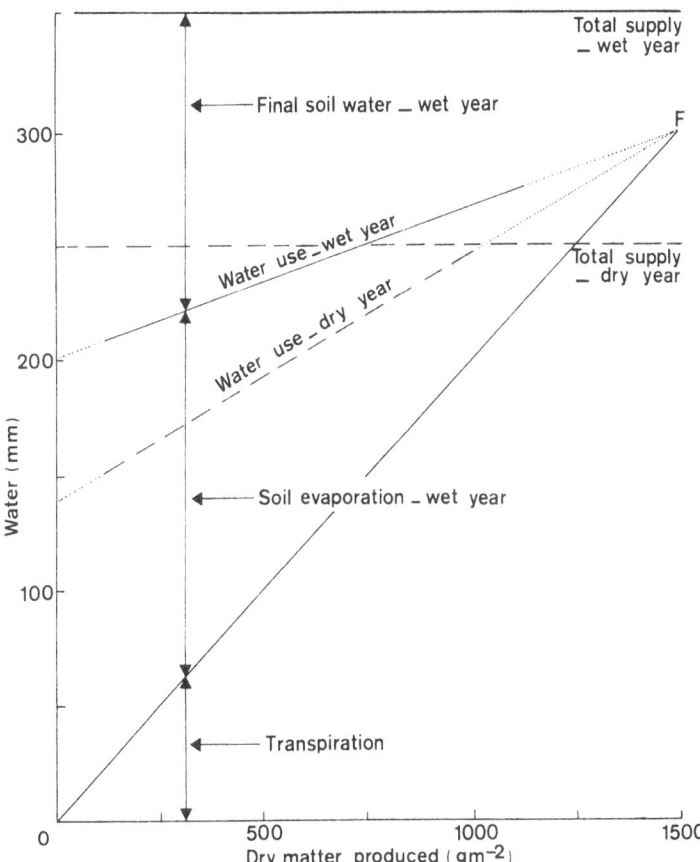

Fig. 6. Relationship between the dry matter production and evatranspiration of wheat crops to a given date in the spring approximating flowering. Two hypothetical seasons at Wagga Wagga, Australia, are illustrated, the wetter one having greater seasonal rainfall and a greater soil evaporation (bare soil evaporation for the period being indicated by the y-axis intercepts). Average transpiration efficiency is assumed to be the same for each season, and is given by the reciprocal of the slope of the line from the origin to F, where F represents the growth and evapotranspiration of a crop with full ground cover (zero soil evaporation) throughout the period (See also Fischer, 1979a and page 264 for further explanation).

extensive root system would be desirable. A longer developing (later) cultivar may be called for, since this is likely to have more extensive rooting (e.g. Asana and Singh, 1968). This involves either earlier sowing or later flowering; the latter poses other problems which will be considered later. Genetic variation in rooting may also be sought within the desirable maturity class and would seem to be one of the least uncertain ways to higher yield through greater water use (provided there is water to use), even when allowance is made for the extra dry matter investment required for the extra roots (Taylor, 1980).

However there are no clear cut precedents for the 'more roots' strategy for improvement of dryland yield. The often-quoted Canadian work of Hurd (e.g. Hurd, 1974) provides, at best, circumstantial evidence for greater yield of durum wheat varieties having deeper roots. Sojka (1974) found that the tall cultivar Gabo had a greater mass of roots and greater water extraction at depth (below 75 cm) than the semidwarf cultivars Yecora 70 and Cocorit 71, and had a higher leaf water potential under drought. Nevertheless a number of other field studies (Lupton *et al.*, 1974; Cholic *et al.*, 1977) indicate similar depths and extents of rooting for semidwarf and tall wheats, or similar leaf water potentials during drought (Fischer and Sanchez, 1979).

While rooting depth may not be related to the semidwarf habit, the search for variation in root systems has barely begun and should be pursued, at least for situations in which current varieties are not generally utilizing all the soil water available. In some soils, it seems possible that winter waterlogging followed by sudden drought, as is common in Mediterranean-type environments, may aggravate the problem of incomplete exploitation of subsoil moisture. Also, selection to increase crop growth rate at suboptimal winter temperatures is likely to lead indirectly to increases in rooting.

Maximizing transpiration efficiency

The second aspect of water-limited yield, water use efficiency or transpiration efficiency, has proved most intriguing. Initially it held the fascination of the pragmatists, since it seemed such an obvious candidate for screening, using for example, growth or gas exchange measurements on single plants in pots. However I know of no reported differences between cereal cultivars, provided care was taken to include foots in any growth sample, to exclude soil evaporation from the water use measurement, and to maintain other environmental factors equal.

These disappointing results for the 'screeners' are proving now to be of special interest to other physiologists. Not only are cereal cultivars not different, but herbaceous C_3 plants in general don't vary much and the only clear cut genotypic difference in transpiration efficiency is an approximate twofold difference between C_3 and C_4 plants (Fischer and Turner, 1978). A further puzzle is the general observation that transpiration efficiency does not increase much, if at all, with water stress. This surprising stability probably results from an even greater uniformity in intercellular CO_2 concentration in illuminated leaves of C_3 and of C_4 plants (Wong *et al.*, 1979). Why evolution has allowed so little diversity here is not altogether clear, but this suggests that chances of genetic variation in transpiration efficiency are limited.

Theoretically, factors which reduce the leaf to air vapour pressure gradient, for example reductions in the radiation load upon the leaf hence in leaf temperature resulting from altered leaf colour or orientation, ought to increase transpiration efficiency. But stomatal closure in response to lowered leaf water potential, as I mentioned before, does not seem to increase efficiency despite the theory, and this is because internal or residual resistance to CO_2 rises in parallel with stomatal resistance during a gradual (normal) onset of stress. However stomatal closing in response to reduced atmospheric humidity will increase average transpiration efficiency by reducing transpiration during those periods of the day when transpiration efficiency is lowest (Cowan and Farquhar, 1977). Studies of species differences of possible adaptive significance in this stomatal response have barely begun.

The mention of atmospheric humidity brings us to the final interesting aspect of transpiration efficiency: the degree to which environmental variation in transpiration efficiency is dominated by variation in vapour pressure deficit of the air. Several workers have pointed to relatively simple empirical relationships between transpiration efficiency (or its reciprocal) and air saturation vapour deficit (Bierhuizen and Slatyer, 1965) or potential evaporation (de Wit, 1958). Fischer (1979a) used unpublished data of Warren and Lill from Wagga Wagga in south eastern Australia relating transpiration efficiency to Class A pan evaporation in order to calculate expected transpiration efficiency (TE) for wheat in each month of the year (Table 3). Growth during the winter months is cheap in terms of

Table 3. Monthly climatic averages for Aleppo (A), Syria (36 °N, 400 m) and Wagga Wagga (W), Australia (35 °S, 200 m); months arranged to correspond seasonally (i.e. October at Aleppo = April at Wagga Wagga).

Month	Radiation[a] $MJ\,m^{-2}\,d^{-1}$		Pan Evap[b] $mm\,d^{-1}$		TE[c] $m^{-2}\,mm^{-1}$	
A	A	W	A[d]	W	A	W
Oct	—	13.9	4.1	3.3	6	6
Nov	—	9.8	1.3	1.8	9	8
Dec	—	8.1	0.7	1.2	?	9
Jan	—	8.3	0.7	1.3	?	9
Feb	—	10.8	1.4	1.9	9	8
Mar	—	15.9	2.9	2.8	7	7
Apl	—	20.4	5.0	4.9	5	5
May	—	25.5	6.6	6.9	4	4
Jun	—	28.0	9.5	9.1	3	3

[a] Total solar radiation;
[b] USWB Class = E_p;
[c] Transpiration efficiency calculated from E_p: TE = $10.2 - 1.30\,E_p + 0.053\,E_p^2$ (See Fischer 1979a);
[d] Estimated from Penman E_o data (Perrin de Brichambant and Wallen 1963) × 1.4.

transpirational cost (9 gm^{-2} mm^{-1} or 11 mm per t ha^{-1}), but this cost rises sharply from early spring to mid summer when it reaches 40 mm per t ha^{-1}.

Strategies to maximize crop growth in months for which TE is highest, will maximize total growth on a limited water supply. Full ground cover during the winter is thus desirable on these grounds also; again early sowing appears to be advantageous. There has been one exception to the above relationship derived at Wagga Wagga. TE was lower than predicted for one winter-early spring growth period when temperatures were low and there were frequent frosts; crop growth rate was also low (Doyle and Fischer, 1979). This suggests that selection for greater crop growth rate at sub-optimal winter temperatures, a very feasible selection goal in view of experience with other crops (e.g. maize), ought to give greater overall dry matter production even although total water supply is limiting. In my terms, transpiration efficiency may be increased and, as well, water loss through soil evaporation during winter replaced by extra transpiration due to greater ground cover. I would give this area high priority.

Maximizing harvest index

I have talked about strategies to maximize total dry matter production without regard to its distribution, the third of the physiological bases of water limited yield which I mentioned at the outset. Distribution in terms of harvest index is, in fact, largely related to the temporal pattern of DM production (and hence of T and TE) and it is the distribution of DM production between the pre- and postanthesis periods which is the key issue. For example, conditions favouring much DM production before flowering and little afterwards will tend to produce much dry matter in stems and leaves but little in grain and hence a low harvest index. This need for a balance between dry matter production (and hence water use) before and after flowering means that generally there will be an optimum amount of preflowering growth which will be less the lower the total water supply. Given a fixed flowering date (see later), the optimum arises because (i) more dry matter at anthesis usually means less water (and hence growth) after anthesis; and (ii), even without this effect, more dry matter at anthesis means more kernels to be filled which, should the number seriously exceed postanthesis water or dry matter supply, will lead to shrivelling.

Effect (i) is illustrated in Fig. 6 for hypothetical seasons with low and high amounts of pre-flowering rain at Wagga Wagga in south-eastern Australia. Linear relationships with positive ET intercepts at zero DM have always been found when DM at a fixed date in the spring, at around flowering, has been varied by sowing rate and date and fertilizer treatments (Fischer, 1979a; Doyle and

Fischer, 1979). These relationships have not, I believe, been examined for situations in which genotypic effects cause the variation in dry matter production, but I assume that they will, in general, remain unchanged. The slope of the ET v DM line is a function of the prevailing TE for the period and the magnitude of soil evaporation, indicated by bare soil evaporation at the extrapolated y-intercept. The wet and dry years are assumed to vary only in total preflowering water supply and soil evaporation; the latter being greater in the wet year. The point F

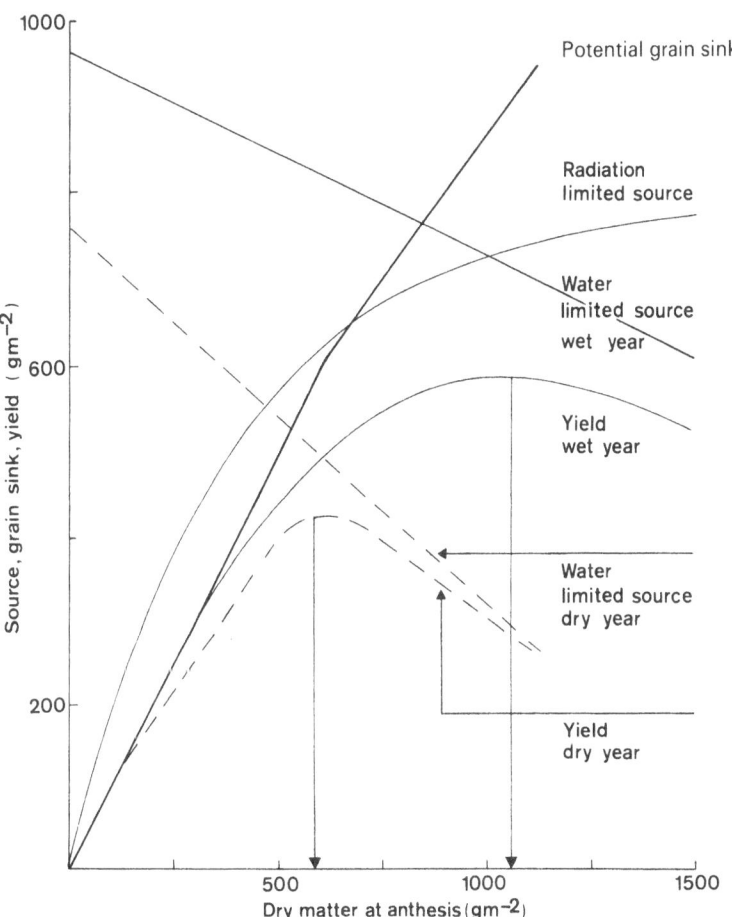

Fig. 7. Post-flowering assimilate supply, potential sink, and actual grain yield as a function of dry matter at flowering for the two seasons illustrated in Fig. 6. Flowering on October 1 (April 1 in northern hemisphere) is assumed as well as similar post-flowering weather for both seasons, this being average weather for October at Wagga Wagga (see Table 3). See text for explanation of assimilate supply and sink.

represents a crop with full ground cover throughout, hence the reciprocal of the slope of the line joining F to the origin is the prevailing TE.

The following figure (Fig. 7) illustrates the post-flowering situation for the two crops of Fig. 6, and has been discussed at length by Fischer (1979a). Again I refer to Wagga Wagga, assuming flowering on October 1 and average October weather. The potential sink curve is assumed to be a function of dry matter at flowering $(DM)_a$ to which kernel number is related, and of potential kernel weight (kernel weight without source limitation during grainfilling, a characteristic largely controlled by genotype). The water-limited source curve comprises two components: preanthesis source fixed at 10 per cent of DM_a (Bidinger et al., 1977 notwithstanding Gallagher et al., 1976), and postanthesis source given by likely postanthesis transpiration multiplied by the appropriate weather-determined TE.

Transpiration after anthesis is assumed to be the soil moisture available at anthesis plus October rainfall, discounted for likely soil evaporation. This assumes that the crop uses all available water. The upper limit on yield is given by whichever is lower, sink or source. Actual yield may be somewhat lower, but observations suggest that optimal DM_a will be close to, probably slightly higher than, that quantity of DM_a which gives potential sink equal to potential source. To the left of this line, crops are sink limited and not fully utilizing postanthesis moisture, although they should achieve maximum harvest index; to the right, they are source or water limited and the harvest index will fall. The optimal DM at flowering will obviously vary from year to year.

Figs. 6 and 7 facilitate the discussion of many agronomic and genetic strategies for increasing yield. For example it has been proposed that the root resistance of the wheat crop be increased by selection for narrow xylem vessels in seminal roots so that, in dry seasons, pre-flowering growth and transpiration are restricted in favour of post-flowering growth (Passioura, 1972). In terms of Fig. 7, this assumes that dry matter production at flowering is superoptimal in dry seasons, which in turn implies that a saving in preanthesis transpiration will give a saving in total preanthesis water use; because of soil evaporation, the saving in water use may be considerably less than that in transpiration.

Optimal flowering date

In considering flowering date strategies in terms of Figure 7, the decline in expected TE as the spring advances, combined with the decline in precipitation, means that the potential production of dry matter drops drastically from early spring onwards. Moreover, the ratio of potential sink to dry matter at flowering decreases as the photothermal quotient decreases. (This is the mean daily solar radiation divided by the difference between mean and basal temperature, see Nix, 1976 and Fischer and Maurer, 1976). For these reasons, it is concluded that over much of southern Australia, the flowering of winter cereals should be as early as permitted by the risk of spring frost (Nix, 1975; Fischer, 1979a). Any consequent tendency towards an insufficient production of dry matter by the time of flowering (and hence insufficient yield potential or sink size) is best countered by improving winter growth and seeding early, which is usually possible in that environment. Both genotypic changes and improved agronomic practices will assist this trend e.g. vigorous establishment in difficult seed bed moisture regimes, good growth at low temperatures, direct drilling, good soil N status, and high seeding density. Also the amount of dry matter needed at anthesis to generate a sink representing a given potential yield is less with modern short wheats, being only about 1.5 times this yield (Fischer, 1979a) compared with a factor of 2 or more for tall wheats.

As I have suggested for southern Australia, achieving a given optimal flowering date for a region requires cultivars with a special phenological pattern in which flowering date is insensitive to sowing date (or alternatively several cultivars of differing maturity class). The lowest sensitivity I have noted in wheat is of the order of 0.2 day delay in flowering date per day delay in sowing date over the atumn-winter (see also Syme, 1973).

In south-eastern Australia, it is probable that selection of genotypes which are better able to tolerate spring frost, thereby permitting even earlier flowering, would further improve yields. This is a difficult prospect with complexities due to subtle interaction of frost effect with stage of development of the spike, variable occurrence of frost and of plant freezing in the field, and so on, but some progress has been made recently with controlled chamber screening (Single and Fletcher, 1979). Such research is most appropriate in view of the potential gains and likelihood of useful genetic variation.

Are flowering date considerations as straightforward in other more typical Mediterranean climates as I have suggested for southern Australia? The deterioration of expected post-flowering assimilate production is likely to be even greater with flowering after, say, April 1 at Aleppo than for the Wagga Wagga example discussed earlier. Similarly the relationship of sink to dry matter at flowering could be expected to deteriorate. On the other hand, the colder winter temperatures suggest that it may be more difficult to achieve satisfactory total dry matter production at flowering (i.e. sink) if flowering is as early as permitted by frost risk (April 1). The salient difference of climate is that temperatures rise very rapidly from later winter to early summer at places such as Aleppo (i.e. 4.5 °C per month, cf 3.5 °C per month at Wagga Wagga). Also the apparently sharp onset of winter rains probably implies that sowing is often delayed by excessive wet, further reducing the chances of achieving a respectable yield potential.

One might conclude from the frequency analysis of water availability of Smith and Harris (this volume) that early flowering would not take advantage of the 25 per cent or so very wet years when the potential for production into the late spring (May) would seem great. Without going into risk analysis, the simple answer to this question can be gained from considering irrigated wheat. My data from northwest Mexico (lat 27 °N) suggest that quite early flowering is best (late February) even under irrigation, consistent with the steady decline in the photo-thermal quotient after early spring. It may be that what is needed to take advantage of occasional wet springs is a crop species or cultivars which are strongly indeterminate or at least can continue to produce flowers when conditions remain wet. Some desert annuals appear to adopt this developmental strategy (Cohen, 1971; Mulroy and Rundel, 1977). Wheat at very low seeding density also tends to behave in this manner.

Obviously we need better biological data and more computer modelling to give better guidelines on the question of flowering date and phenological strategies for particular regions. However it must be borne in mind that disease and its control are other major factors in the complex interaction between the timing of crop growth and the environment. For example, in the presence of *Septoria* and susceptible cultivars, later flowering is favoured because it facilitates disease escape.

Preanthesis plant water stress

Preanthesis drought was not explicitly mentioned in my framework which places all the emphasis on the more common postanthesis drought. There are probably two types to consider: seedling drought, and stress during early spring just before flowering. Neither is very common in my experience, but the former would become more frequent if early seeding was sought, or long fallowing eliminated. For such situations, there may be useful genetic variation in ability to germinate and emerge at low soil moisture, as there appears to be in rangeland species, and/or in ability to survive seedling drought. Both traits appear to be amendable to rapid screening.

A separate problem arises with early seeding on well-fallowed land in the absence of rain. This technique involves deep placement of the seeds to reach the stored moisture, and even though this may be facilitated by use of deep furrow drills, seedlings with short coleoptiles (e.g. semidwarf cultivars) have difficulty in emerging (Allan *et al.*, 1962). For this situation, selection of semidwarf genotypes with long coleoptiles has been attempted with some success in the Pacific Northwest of USA.

Finally, with regard to early moisture stress, it is sometimes implied in the literature (e.g. the number of studies of stress effects on floral initiation in cereals) than an early stress experience imprints some limitation on final yield. One relevant observation is that, commonly, yield of late cultivars is less affected by an early stress period than that of early cultivars whose development appears to be accelerated by stress, unless it is very severe, and yield potential accordingly limited (Fischer *et al.*, 1977; R. A. Fischer and P. C. Wall, unpublished CIMMYT results; Indian subcontinent experience with dryland cultivars).

The explanation of this phenomenon lies, I believe, not so much in special effects of stress upon the initiating spike, but rather in reduced leaf area development and hence reduced dry matter production during the critical 25 days or so before anthesis. Growth reductions induced by early water stress did not lead to lower yield when enough water became available after the stress period to allow the canopy to develop until full ground cover was achieved. The chance of meeting this condition is less with an early cultivar than with a late one, since the critical period arises sooner. The effect of seedling stress on concurrent and subsequent root development may, however, complicate the situation if later stress is expected.

The critical preanthesis period just mentioned needs further consideration because it has been suggested that early spring drought in some years will lead to plant stress at this time, and additionally because many workers point to the heightened sensitivity of yield processes, in particular seed number, to water stress during this period, which lasts approximately from emergence of the leaf preceding the flag leaf until anthesis. In fact, it is tempting to suggest the existence of a clear bottleneck to yield, with reduced plant water potential reducing seed number more than leaf area (Fischer, 1973), and thereby limiting yield, particularly if water supply subsequently improves. It is even teleologically plausible that natural selection has favoured plants which for cropping situations are too conservative in adjusting seed number to water stress. Along these lines, I have attempted to compare wheat species and cultivars for the sensitivity of kernel number per spike to lowered plant water potential. A valid comparison requires that genotypes be at the same stage of spike development. When this condition was satisfied, there appeared to be no marked differences between ten genotypes, six of which are shown in Fig. 8. However the value of this study is limited by the rapid rate at which the stress was imposed.

A reassessment of the critical period philosophy suggests that it may not

270

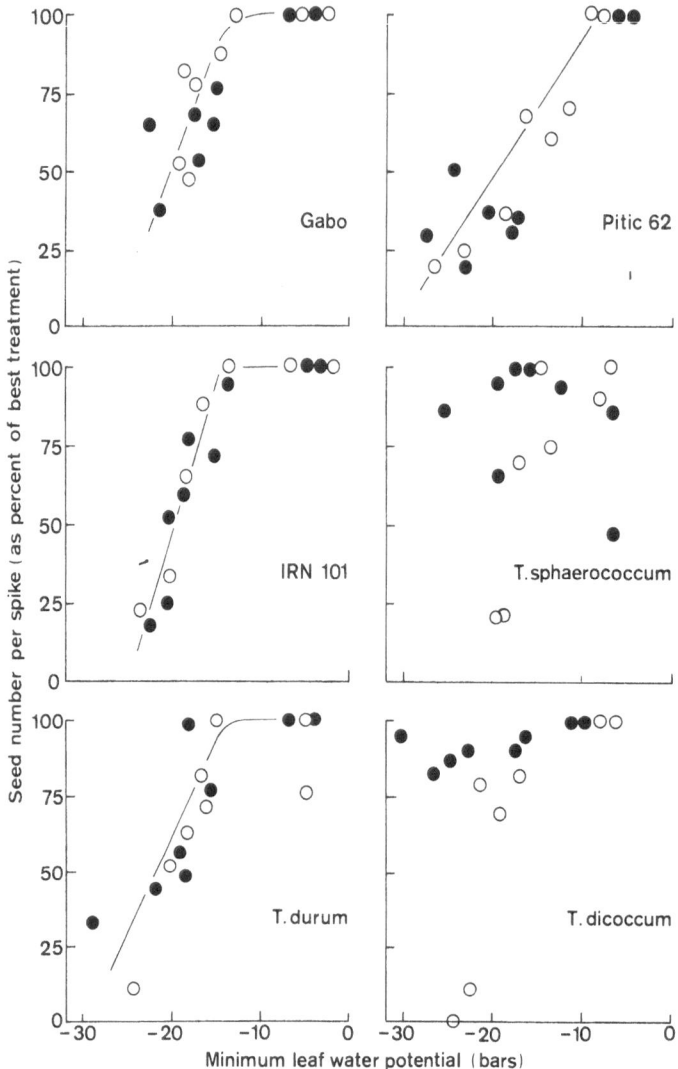

Fig. 8. Kernel number per spike (as a percentage of the best treatment) for 6 wheat genotypes and the effect of the minimum daytime plant water potential reached with rapid soil drying during the critical preanthesis phase. The genotypes were *Triticum aestivum* cv. Gabo, Pitic 62, and IRN101, *T. Durum* 'dwarf', *T. dicoccum* cv W12, and *T. spaerococcum* cv (5). The procedure was similar to that used in Fischer (1973). There were two stress dates (one and 8 days after flag leaf emergence, open and closed symbols respectively) four stress levels and two replicates making a total of 16 pots per cultivar. For each pot, shoot water potential was determined just before rewatering and kernel number per spike measured at maturity. (R. A. Fischer, unpublished data).

represent the type of bottleneck to dryland yield outlined above. There is growing evidence that, with gradual stress more typical of field situations, the reduction in seed number is proportioned to the reduction in dry matter production or assimilation during the critical period, or, less precisely, the reduction in dry matter production at anthesis (Day *et al.*, 1978; Fischer, 1980).

For example, in the experiment of Bidinger *et al.* (1977), Mediterranean-type drought led to significant depressions in leaf water potential, relative to control, beginning some 14 days before anthesis and steadily increasing until maturity. This stress reduced dry matter production by the wheat crop by 31 per cent over the 24 days preceding anthesis; kernel number m^{-2} at maturity, was similarly reduced. The data of Day *et al.* (1978) provide good circumstantial evidence that reduced water use and reduced dry matter production represent the key to reduced seed number in water-stressed barley crops.

Fischer and Turner (1978) have suggested that a guide to the critical period mechanism of stress damage may be provided by separating seed number into two components, namely (i) inflorescence dry weight at anthesis and (ii) seeds per unit inflorescence weight (reproductive efficiency). Stress effects which operate solely upon inflorescence weight (normally in wheat 15 to 25 per cent of total dry weight at anthesis depending on genotype) are more likely to be assimilate mediated. In a controlled environment study, rapid stress at 22, 16, 9 or 2 days before anthesis, always reduced inflorescence weight, but rapid stress at 16 and 9 days also reduced reproductive efficiency (Fischer, 1980).

If it is found to be generally true that stress in the field before flowering reduces kernel number via reduced overall assimilate supply, then it would seem more difficult to select for resistance to this effect. However, if part of the effect is, say, hormone mediated, as some suggest, then progress may possible. Either way, the occurrence of preanthesis stress complicates, but does not invalidate, the approach to the dryland grain yield represented in Fig. 7 i.e. kernel number will be more closely related to dry matter production during the critical ontogenetic period just before flowering than to simply total dry matter at flowering.

Hot spell damage

One other phenomenon related to water stress must at least be mentioned to complete the picture. In various countries, including some with Mediterranean climates, hot winds of low humidity are reputed to damage cereals during grain-filling even when soil moisture appears to be adequate; in other words, comparatively brief periods of atmospheric stress have a disproportionately large effect on yield. This has been discussed recently by Fischer (1980). Both high plant temperature and low plant water potential appear to be involved, and some evidence suggests that permanent loss of photosynthetic tissue, especially leaf laminae, during the stress period, may be the primary cause of reduced kernel weight.

There is also evidence that high tissue nitrogen, which occurs in crops grown after a long period of leguminous pasture, predisposes plants to such hot spell damage. However so little definitive information exists on the precise causes of hot spell damage that it is not possible to suggest ways by which resistance to such damage can be increased. On the other hand, it should be possible to predict the frequency of damaging atmospheric stress, even allowing for interactions with soil moisture, and hence to examine the possible advantages of, for example, stress escape through earliness. Simple climatic analysis suggests, as one might have expected, that these atmospheric stress days are associated with years of lower-than-average rainfall (e.g. Lomas and Shashoua, 1974).

IDEOTYPE VERIFICATION

In order to verify predictions about desirable traits, variation in the traits in question must be generated and its consequences for performance measured. Variation may be created by artificial manipulation such as cutting off awns, or in the case of the hypothesis of Passioura (1972), forcing plants to grow upon only one (instead of three or more) seminal root. Such experiments can be performed fairly rapidly and their results are helpful, but in final analysis testing of traits using genetic sources of variation is essential. This is a slower procedure, for such variation must be located and then incorporated into a given genetic background, or preferably many backgrounds, via isogenic line or isogenic population techniques combined with efficient screening for the trait.

Before effects on performance can be measured, a reasonable degree of homo-zygosity is needed, implying comparisons at the F4 generation or later. By that stage too, enough seed of each line can be produced to enable the multi-site testing in plots which is essential if the results are to convince any plant breeder. Overall, this is a large project for any trait, as was recognized in an earlier symposium on this subject (Moss et al., 1974), and, understandably, there are few examples where traits for dryland performance have been thus tested.

Centres for the improvement of temperate cereals such as ICARDA or CIMMYT have the germplasm and resources, the continuity and the goal orientation necessary to undertake such ideotype testing. To date, however, little has been done, probably because breeders are too busy with more obvious problems e.g. quality and disease resistance; and because physiologists are too few. One or two full-time physiologist/breeders need to be assigned to this area.

In many cases, ideotype testing can be part of the regular breeding system; so many crosses are made and F_2 populations generated that variation in potentially useful traits is probably already present in the breeding nurseries. The variation needs to be identified, recorded, and sometimes protected from the loss which results from normal selection by breeders as the material is advanced by any pedigree system to the yield-testing stage. I have discussed more fully elsewhere the advantage of such a research component in the regular breeding program (Fischer, 1977). It forces the breeders to think more clearly about their selection criteria and strategies, and injects an element of realism into the physiologists' speculations.

One of the first points which becomes obvious is that ideotype testing is more limited by lack of techniques for rapid screening, than by lack of plausible ideotype traits to test. While some morphological traits can be screened rapidly, many other traits cannot. There is great scope for physiological input here, as revealed by examples such as seedling insensitivity to giberellic acid as a screen for Norin 10 dwarfing genes (Gale and Law, 1977), or non-destructive measurement of xylem vessel diameter in seminal roots of seedlings as a source of water-conservative rooting systems (Richards and Passioura, 1980).

OPTIMIZATION OF NITROGEN USE

I have left this subject until last since I do not believe that the breeder of non-leguminous crops needs to do much explicit selection in this area. Optimization of nitrogen use implies maximizing grain produced per unit of nitrogen supplied to the crop either as fertilizer or through leguminous plants grown beforehand. Regardless of the actual level of nitrogen supply concerned (it may be the economic optimum for maximum profit or something less), recent genetic improvement in the yield potential of cultivars has led to greater nitrogen efficiency in terms of yield per unit N supply. The high yielding cultivars are not only more N responsive, but also have equal or superior yields at all levels of nitrogen input. This is illustrated by curve B (high yield potential) compared to curve A (low yield potential) in Fig. 9 (see also Bhardwaj et al., 1975). I know of no data supporting the existence of response like curve C. There may be cross-overs in response curves, but they seem to occur at such low N input levels ($< 1 \, t \, ha^{-1}$ yield) that they are not relevant to improved agricultural practice.

The improved N efficiency of higher yielding cultivars seems to be related to more grain per unit of nitrogen taken up by the crop, rather than more nitrogen uptake (Fischer and Wall, 1976). This, in turn, appears to be associated with

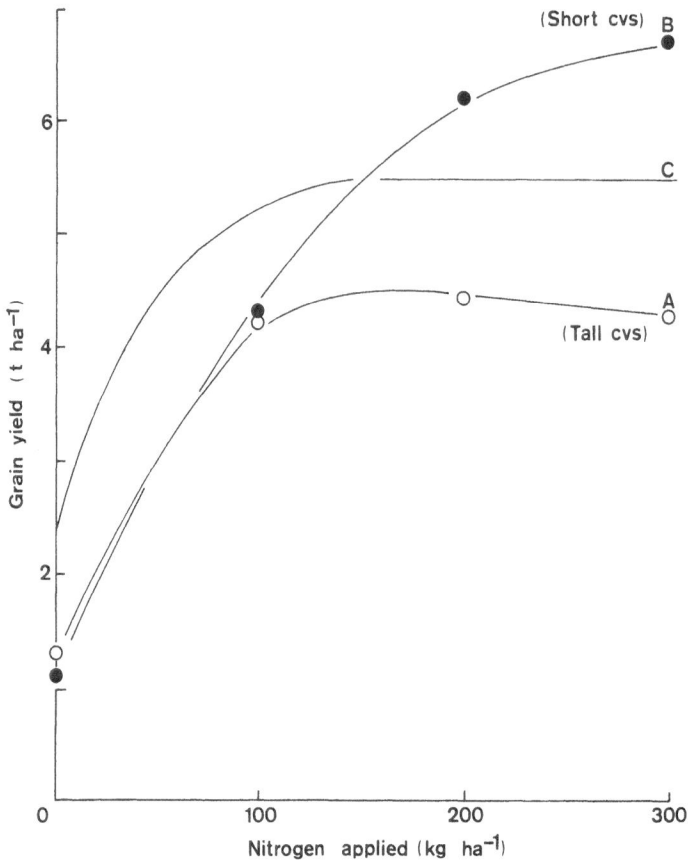

Fig. 9. Average response of grain yield to applied fertilizer nitrogen for tall wheat cultivars of low yield potential (*T. aestivum* cvs Mentana, Yaqui 50 and Nainari 60, curve A) and short cultivars of high yield potential (*T. aestivum* cvs Siete Cerros 66, Yecora 70 and Cajeme 71, curve B). CIANO, 1972–73, full irrigation, low initial soil fertility (R. A. Fischer, unpublished) Curve C represents an hypothetical cultivar showing high N efficiency only at low levels of N supply.

higher harvest indices, partly because of better lodging resistance. It is also associated with a greater proportion of total nitrogen uptake in the grain at maturity (higher nitrogen harvest index); this reaches 75 per cent in modern wheats (Fischer and Wall, 1976).

With harvest index approaching 50 per cent there may be little scope for further improvement of yield per unit N uptake via increased harvest index. Also, if removal of produce is the major drain on nitrogen in the system, even this improvement in efficiency is illusory since the extra grain implies more N removed as produce and this must eventually be replaced. If, on the other hand, other sources of N loss are significant, such as denitrification and leaching, there may be possibilities of genetic change to reduce these losses i.e. increase N uptake as a proportion of N supply.

Both situations are associated with wet conditions which are most likely to occur in mid to late winter; thus rapid early growth, perhaps aided by early sowing and early absorption of soil nitrogen, would reduce the available soil nitrogen exposed to such loss; also, by increasing early water use, it may reduce the maximum soil water contents encountered. Under wet conditions, considerable quantities of nitrogen may be available lower in the profile (Storrier, 1965), but not beyond the depth at which roots might be expected; increased root activity at depth may be another way of reducing leaching losses. But we need more information on N losses in dryland soils since, on present accounts, they might be too small to justify explicit breeding to reduce losses.

Can the cereal crop be a source of nitrogen gain through *in situ* symbiotic or non symbiotic fixation? The latter process could occur either during the crop cycle powered by root exudates, or after the cycle when the energy of the crop residue could conceivably drive microbial N fixation (?) These possibilities might involve the breeder, or genetic engineer, but because of their long term and fundamental nature, do not seem appropriate research for ICARDA at present.

Finally, and although another chapter is dealing with N × water interactions, it is worthwhile mentioning the general reduction in yield response to nitrogen with water limitation, and the existence of negative responses to nitrogen in extreme drought situations (e.g. Fischer and Kohn, 1966). The latter might provide some scope for cultivar selection since it seems to be only partly explained by excessive early growth exhausting limited water supplies before grainfilling (Fischer and Kohn, 1966; Dann, 1969). High plant nitrogen content may predispose the plant to water stress injury (Fischer, 1980); there may also be an interaction with low molybdenum content (Lipsett and Simpson, 1973). Such responses can be expected to be modified by genotype.

ACKNOWLEDGEMENT

I am most grateful to Mrs. A. Stafford who prepared the figures.

REFERENCES

Anderson, J. R. 1974 Risk efficiency in the interpretation of agricultural production research. Rev. Marketing Agric. Econ. 42, 131–184.
Allan, R. E., Vogel, O. A. and Peterson, C. J. 1962 Seedling emergence rate of fall-sown wheat and its association with plant height and coleoptile length. Agron. J. 54, 347–350.

Asana, R. D. and Singh, D. N. 1968 On the relation between flowering time, root-growth and soil-moisture extraction in wheat under non-irrigated cultivation. Ind. J. Pl. Physiol. 10, 154–69.

Bhardwaj, R. B. L., Jain, N. K., Wright, Bill, C., Sharma, K. C., Gill, G. S. and Krantz, B. A. 1975 The Agronomy of Dwarf Wheats. New Delhi: Indian Council Agric. Research.

Bidinger, F., Musgrave, R. B. and Fischer, R. A. 1977 Contribution of stored pre-anthesis assimilate to grain yield in wheat and barley. Nature 270, 431–33.

Bierhuizen, J. F. and Slatyer, R. O. 1965 Effect of atmospheric concentration of water vapour and CO_2 in determining transpiration-photosynthesis relationships of cotton leaves. Agric. Meteorol. 2, 259–70.

Brennan, P. S. 1974 Analysis of genotype x environmental interactions for several characters of wheat. M. Agric. Sc. Thesis, Univ. of Queensland.

Cholic, F. A., Welsh, J. R. and Cole, C. V. 1977 Rooting patterns of semi-dwarf and tall winter wheat cultivars under dryland field conditions. Crop Sci. 17, 637–9.

CIMMYT 1977 CIMMYT Report on Wheat Improvement 1977. El. Batan. Mexico.

Cohen, D. 1971 Maximizing final yield when growth is limited by time or limiting resources. J. Theor. Biol. 33, 299–307.

Cowan, I. R. and Farquhar, G. D. 1977 Stomatal function in relation to leaf metabolism and environment. Symp. Soc. Exp. Biol. 31, 471–505.

Dann, P. R. 1969 Response by wheat to phosphorus and nitrogen with particular reference to 'haying-off'. Aust. J. Exp. Agric. Anim. Husb. 9, 625–29.

Day, W., Legg, B. J., French, B. K., Johnston, A. E., Lawlor, D. W. and Jeffers, W. de C. 1978 A drought experiment using mobile shelters: the effect of drought on barley yield, water use and nutrient uptake. J. agric. Sci. 91, 599–623.

Doyle, A. D. and Fischer, R. A. 1979 Dry matter accumulation and water use relationships in wheat crops. Aust. J. Agric. Res. 30, 815–29.

Eberhardt, S. A. and Russell, W. A. 1966 Stability parameters for comparing varieties. Crop Sci. 6, 36–40.

Fischer, R. A. 1973 The effect of water stress of various stages of development on yield processes in wheat. UNESCO: Plant Response to Climate Factors. Proc. Uppsala Symp. 1970, 233–241.

Fischer, R. A. 1977 The physiology of yield improvement: past and future. Proc. SABRAO Conference, Canberra, February 1977.

Fischer, R. A. 1979a Growth and water limitation to dryland wheat yield in Australia: a physiological framework. J. Aust. Inst. Agric. Sci. 45, 83–94.

Fischer, R. A. 1979b Are your results confounded by interogenotypic competition? In 'Proc. 5th International Wheat Genetics Symp'. Vol. II. Publ. Indian Soc. Genetics and Plant Breeding pp. 767–77.

Fischer, R. A. 1980 Influence of water stress on crop yield in semiarid regions. In: Turner, N. C. and Kramer Wiley, P. J. (eds.), Adaptation of Plants to Water and High Temperature Stress. Wiley Interscience, New York, pp. 323–39.

Fischer, R. A. and Kohn, G. D. 1966 The relationship of grain yield to vegetative growth and post-flowering leaf area in the wheat crop under conditions of limited soil moisture. Aust. J. Agric. Res. 17, 281–95.

Fischer, R. A. and Kertesz, Z. 1976 Harvest index in spaced populations and grain weight in microplots as indicators of yielding ability in spring wheat. Crop Sci. 16, 55–59.

Fischer, R. A. and Maurer, R. 1976 Crop temperature modification and yield potential in a dwarf spring wheat. Crop Sci. 16, 855–59.

Fischer, R. A. and Wall, P. C. 1976 Wheat breeding in Mexico and yield increases. J. Aust. Inst. Agric. Sci. 42, 139–148.

Fischer, R. A., Lindt, J. L. and Glave, A. 1977 Irrigation of dwarf wheats in the Yaqui Valley of Mexico. Expl. Agric. 13, 353–69.

Fischer, R. A. and Turner, N. C. 1978 Plants productivity in the arid and semiarid zones. Ann. Rev. Plant. Physiol. 29, 277–317.

Fischer, R. A. and Maurer, R. 1978 Drought resistance in spring wheat cultivars. I. Grain yield response. Aust. J. Agric. Res. 29, 897–912.

Fischer, R. A. and Sanchez, M. 1979 Drought resistance in spring wheat cultivars II. Effect on plant water relations. Aust. J. Agric. Res. 30, 801–14.

Fischer, R. A. and Wood, J. T. 1979 Drought resistance in spring wheat cultivars III Yield associations with morpho-physiological traits. Aust. J. Agric. Res. 30, 1000–1020.

Gale, M. D. and Law, C. N. 1977 The identification and exploitation of Norin 10 semi-dwarfing genes. Plant Breeding Institute Ann. Report 1976, Cambridge, pp. 21–35.

Gallagher, J. N., Biscoe, P. V. and Hunter, B. 1976 Effect drought on grain growth. Nature 264, 541–42.

Hurd, E. A. 1974 Phenotype and drought tolerance in wheat. Agric. Meteorol. 14, 39–55.

Johnson, V. A., Shafer, S. L. and Schmidt, J. W. 1968 Regression analysis of general adaptation in hard red winter wheat (Triticum aestivum L.) Crop. Sci. 8, 187–91.

Jones, H. G. 1977 Aspects of the water relations of spring wheat (Triticum aestivum L.) in response to induced drought. J. Agric. Sci. 88, 267–82.

Kaul, R. and Crowle, W. L. 1971 Relations between water status, leaf temperature, stomatal aperture, and productivity in some wheat varieties. Z. Pflanzenzücht 65, 233–43.

Laing, D. R. and Fischer, R. A. 1977 Adaptation of semidwarf wheat cultivars to rainfed conditions. Euphytica 26, 129–39.

Lipsett, J. and Simpson, J. R. 1973 Analysis of the response by wheat to application of molybdenum in relation to nitrogen status. Aust. J. Exp. Agric. Anim. Husb. 13, 563–6.

Lomas, J. and Shashoua, Y. 1974 The dependence of wheat yields and grain weight in a semi-arid region on rainfall and on the number of hot, dry days. Israel J. Agric. Res. 23, 113–121.

Lupton, F. G. H., Oliver, R. H., Ellis, F. B., Barnes, B. T., House, K. R., Welbank, P. J. and Taylor, P. J. 1974 Root and shoot growth of semidwarf and taller winter wheats. Ann. Appl. Biol. 77, 129–44.

Moss, D. N., Woolley, J. T. and Stone, J. F. 1974 Plant modification for more efficient water use: the challenge. Agric. Meteorol. 14, 311–20.

Mulroy, T. W. and Rundel, P. W. 1977 Annual plants: adaptations to desert environments. Bioscience 27, 109–14.

Nix, H. A. 1975 The Australian climate and its effects on grain yield and quality. In: Lazenby, A. and Matheson, E. M. (eds.) Australian field crops. Wheat and other temperate cereals. Vol. 1. Angus and Robertson, Sydney. p. 183–226.

Nix, H. A. 1976 Climate and crop productivity in Australia. In: Proc. Symp. Climate and Rice, Int. Rice Res. Inst., Los Banos, Philippines, p. 495–508.

Passioura, J. B. 1972 The effect of root geometry on the yield of wheat growing on stored water. Aust. J. Agric. Res. 23, 745–52.

Passioura, J. B. 1977 Grain yield, harvest index, and water use of wheat. J. Aust. Inst. Agric. Sci. 43, 117–20.

Perrin de Brichambaut, G. and Wallén, C. C. 1963 A study of agroclimatology in semi-arid and arid zones of the Near East Tech. note 56 W.M.O. Geneva.

Richards, R. R. and Passioura, J. B. 1980 Seminal root morphology and water use of wheat. I Environmental effects. Crop Sci. 20 (in press).

Single, W. W. and Fletcher, R. J. 1979 Resistance of wheat to freezing in the heading stages. In Proc. 5th International Wheat Genetics Symp. Vol. I. Pub. Ind. Soc. of Genet. and Plant Breeding. pp. 188–191.

Shimshi, D. and Ephrat, J. 1972 A study of inter-varietal differences of stomatal behaviour in wheat, in relation to transpiration and potential yield. Final Rep. to Ford Foundation, April 1972, Project AII/6, Publ. Agric. Res. Org. Volcani Center, Bet Dagan, Israel.

Sojka, R. W. 1974 Comparative drought response of selected wheat varieties. Ph. D. Diss., Uni. of Calif., Riverside, Dep. Soil Sci. Agric. Eng. (Uni. Microfilms, Ann Arbor, No. 77–14, 412).

Syme, J. R. 1973 Quantitative control of flowering time in wheat cultivars by vernalization and photoperiod sensitivity. Aust. J. Agric. Res. 24, 1–9.

Storrier, R. R. 1965 The leaching of nitrogen and its uptake by wheat in a soil from southern New South Wales. Aust. J. Expl. Agric. Anim. Husb. 5, 323.

Taylor, H. M. 1980 Modifying root systems of cotton and soybeans to increase water absorption. In: Turner, N. C. and Kramer, P. J. (eds.), Adaptation of Plants to Water and High Temperature Stress. Wiley Interscience, New York, pp. 75–84.

de Wit, C. T. 1958 Transpiration and crop yields. Versl. Landbouwk. Onderz. 64(6), 1–88.

Wong, S. C., Cowan, I. R. and Farquhar, G. D. 1979 Stomatal conductance correlates with photosynthetic capacity. Nature 282, 424–26.

11. Plant improvement for semi-arid rangelands: possibilities for drought resistance and nitrogen fixation

D. A. JOHNSON*, M. D. RUMBAUGH and K. H. ASAY

*Crops Research Laboratory, Utah State University, Logan, Utah, U.S.A.

About one-third of the land area of the earth comprises arid and semi-arid climates. Because of aridity and adverse physical factors such as rockiness, rough topography, or shallow soils, many of these lands are used primarily as rangelands rather than as croplands (Stoddart *et al.*, 1975). Rangelands are an important resource: they have produced, and continue to produce, forages for native and domestic grazing animals. As world population expands, rangelands will undoubtedly play an increasingly important role in food production. Demands on rangelands will probably increase as lands used for cultivated forages are diverted to grain production for direct human consumption.

Vast expanses of the world's rangelands are producing far below their potential (McGinnies, 1957). Poor grazing management has seriously depleted many productive and high quality forage species. On many arid and semi-arid rangelands, deterioration has reached a point where succession or natural plant replenishment will not take place within a reasonable time, even with proper grazing management (Moore, 1960). Revegetation offers the only acceptable alternative.

Objectives for introducing desirable range species into semi-arid rangelands include: (i) revegetation of deteriorated rangelands and abandoned croplands; (ii) expansion of the grazing season; and (iii) protection of areas from erosion (Stoddart *et al.*, 1975). To meet these objectives, a combined approach involving both proper management and the reintroduction of desirable forage species is required to return these areas to desirable levels of utilizable production. An integrated strategy, involving both improved plant materials and proper grazing management, can increase forage production on deteriorated rangelands more effectively than either of these methods by itself (McGinnies, 1957).

Improved plants for semi-arid rangelands

Because of the costs associated with the revegetation of rangelands, only the best available plant materials should be used. Unless plant cultivars are developed

279

specifically for range renovation in semi-arid areas, optimum returns from revegetation projects cannot be assured. Just as grain farmers in dryland areas do not plant grain cultivars which have been developed in high rainfall areas, neither should range managers in semi-arid areas plant forage cultivars which have been developed in mesic environments. However, managers of semi-arid ranges face the disadvantage of a lack of commercially available, adapted, and improved forage cultivars.

Because of the distinct wet and dry seasons typical of many semi-arid range-lands, it is difficult to maintain forage throughout the dry season (Naveh, 1972). Plant populations on properly managed semi-arid rangelands usually include grasses, forbs, and shrubs. Perennial grasses are desirable because they generally produce a high proportion of utilizable forage. Because of their tillering capacity, large below ground biomass, and large nutrient reserves, many range grasses are resistant to grazing and, if stocked and used wisely, are highly persistent.

Legumes produce high quality forage, and their nitrogen-fixing ability could maintain or increase soil fertility by providing nitrogen; an often limiting nutrient on many semi-arid rangelands. Shrubs should also be considered in rangeland revegetation projects (McKell, 1975). Shrubs typically tap deep soil moisture reserves not used by grasses and forbs and thus provide palatable forage to carry livestock through the dry season. Combinations of grasses, forbs, and shrubs extend the grazing season by providing a wide base of forage, rather than one period of peak forage quality and quantity. They also provide a broader eco-logical base for increased ecosystem stability, something that a single species could not do.

Although many areas with Mediterranean-type climates typically have an abundance of annual species in their vegetation, perennial species were once an integral component of plant communities in these areas (Rossiter, 1966). Clear-ing, cropping, overgrazing, and fire all helped to reduce or even eliminate many perennial trees, shrubs, forbs, and grasses.

Whether rangelands with Mediterranean-type climates should be managed for annual or perennial species is very controversial. Rossiter (1966) suggested that the inclusion of perennial species under high stocking rates would probably not increase production. The key for maintaining perennial species appears to be moderate stocking rates. Thus, for short term improvements in areas with Mediterranean-type climates, management for annuals may be the best alterna-tive. Nevertheless, in the long term, these areas should probably be managed for perennial plant production. Under proper stocking rates and timing of grazing, perennial species should provide a more reliable grazing resource over a longer portion of the grazing season.

Annual species may have particular applicability in integrating livestock production into rainfed farming systems (Draz, 1977). Traditional cropping systems in Mediterranean-type areas frequently involve rotation between wheat and fallow. In this farming system, millions of hectares are plowed for grain production and left fallow the following year. Oram (1956) explored the possibility of introducing pasture and fodder crops to replace fallow. By planting annual legumes in otherwise fallow land, millions of hectares of land could produce badly needed forage and alleviate grazing pressures on some rangelands. These legumes also could add valuable nitrogen to typically nitrogen-deficient soils. Other possibilities, such as the expansion of irrigated forage crop production, may also provide closer integration between livestock and agricultural production systems (Draz, 1977). In either case, use of improved plant species will probably be critical in achieving optimum returns in these cropping systems.

Native species are a logical starting place for the examination of adapted plant materials, but they might not be the best for revegetation. On many sites grazing pressures may have altered environmental conditions in such a way that other species than natives might be better adapted for revegetation. Additionally, some parts of the world have flora better suited for forage production. Therefore, plant materials from other parts of the world with similar climates should be examined in any plant improvement program. The key consideration should not be country of origin, but rather suitability of the plants to the present range environment and the projected use of the site.

In a program to provide improved forages for the semi-arid western U.S. rangeland, the U.S. Department of Agriculture's Agricultural Research, in cooperation with Utah State University, has assembled a forage and range

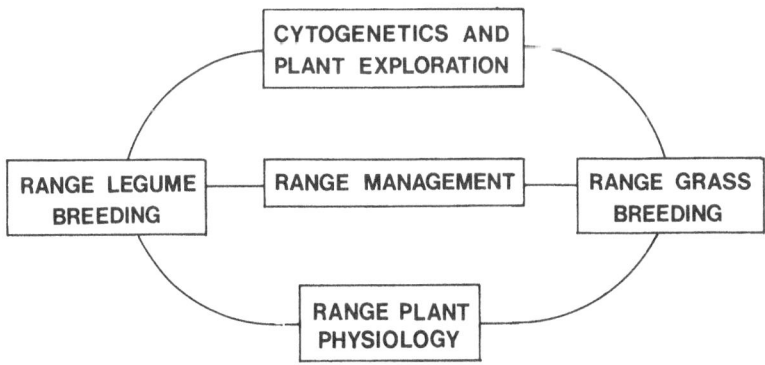

Fig. 1. Diagram depicting research scientists of the Forage and Range Research Project at Logan, Utah. Solid lines identify the primary interactions among the individual researchers.

research team at Logan, Utah (Fig. 1). This team includes a cytogeneticist, plant physiologist, range scientist, and two plant breeders. This chapter deals mainly with the team's approach for improving forages for semi-arid rangeland. Although our program focuses on grasses and forbs, many of the same principles and procedures are applicable to the selection and improvement of shrubs.

APPROACH TO PLANT IMPROVEMENT FOR SEMI-ARID RANGELANDS: INCREASING DROUGHT RESISTANCE

Drought stress is probably the most important environmental factor that affects growth on arid and semi-arid rangelands (Brown, 1977). Consequently, improved range cultivars must have drought resistance mechanisms; adaptations that allow plants to grow and survive in areas subjected to periodic water deficits. Plant adaptations to drought are classically categorized into avoidance and tolerance mechanisms (Shantz, 1927; May and Milthorpe, 1962; Parker, 1968; Levitt, 1972). Avoidance mechanisms allow the plant to escape drought stress, and tolerance mechanisms enable the plant either to postpone or to withstand dehydration. Adaptations to drought stress include both morphological and physiological mechanisms, and are discussed in relation to crop and pasture plants by Turner (1979) and Turner and Begg (1978), respectively, and have been considered by Turner and Begg in the fifth chapter of this volume.

Adaptations for drought, involving tolerance to dehydration, would theoretically be most advantageous in forage plants in which leaf production is important. Adaptations for dehydration tolerance should allow range plants to produce maximum leaf growth at a given water potential. Dehydration postponement mechanisms, such as enhanced root growth or increased stomatal sensitivity, would theoretically be less desirable than tolerance mechanisms because they usually develop at the expense of aboveground growth. In reality, however, dehydration postponement mechanisms are also important for successful adaptation to drought stressed environments. As a result, improved range plants would probably combine both dehydration postponement and tolerance mechanisms enabling them to survive periods of severe drought, as well as continuing active growth during periods of less severe drought.

Genetic improvement of drought resistance in annual crops growing in water-limited areas was reviewed for sorghum (*Sorghum bicolor* [L.] Moench) by Sullivan and Ross (1979) and Blum (1979), rice (*Oryza sativa* L.) by O'Toole and Chang (1979), and wheat (*Triticum aestivum* L.) by Hurd (1976) and Townley-Smith and Hurd (1979). For forages the only documented genetic improvement and selection for drought resistance has been by Wright (1975) with the grass species *Panicum*, *Eragrostis*, and *Bouteloua*.

Johnson (1980) identified five major steps necessary in genetic improvement for drought resistance: (i) drought characterization; (ii) definition of selection criteria; (iii) assemblage of a broad genetic base; (iv) development of screening techniques; and (v) application of the screening procedures. Each step will be examined in relation to the genetic improvement of drought resistance in semi-arid rangeland species.

Drought characterization

Hanson (1972), Reitz (1974), Boyer and McPherson (1975), and Fischer and Turner (1978) emphasized the importance of defining not only the amount of precipitation, but also its distribution. Both intensity and duration of the water deficit are important in determining the particular drought adaptations that are key factors associated with plant responses to drought (Begg and Turner, 1976).

Timing is also important because in environments where drought is un-predictable or sporadic, plants performing better in one season may perform worse in another (Boyer and McPherson, 1975). Consequently the potential for providing improved plant materials is probably greatest in environments where drought occurs predictably during the same part of the growing season each year.

Fig. 2 depicts a climatic diagram (Walter and Lieth, 1960) of a representative range location near Snowville, Utah, which we use as a test site. For comparison, climatic diagrams are also depicted for areas with typical lowland Mediterranean-type climates and environments which are characteristic of the high, dry, cold plateaus in Turkey, Iran, and Afghanistan.

Snowville receives an annual rainfall of 244 mm, mostly during late-fall, winter, and early spring. The low precipitation during the summer is relatively ineffective because of large evaporative losses associated with high temperature and low humidity. Consequently, plant production at Snowville depends heavily on winter-spring moisture, and takes place largely during spring and early summer when temperatures are moderate.

The lowland Mediterranean-type climates (Fig. 2) for Damascus and Aleppo, Syria, are characterized by a much higher temperature regime than that at Snowville. Mean annual temperature is about 18 °C at Damascus and Aleppo, and about 7 °C at Snowville. Additionally, the mean daily minimum temperature of the coldest month is about 3 ° and 1 °C at Damascus and Aleppo, respectively, and about − 12 °C at Snowville. The amount of precipitation is similar for Snowville and Damascus and Aleppo, but the pattern of monthly rainfall differs.

284

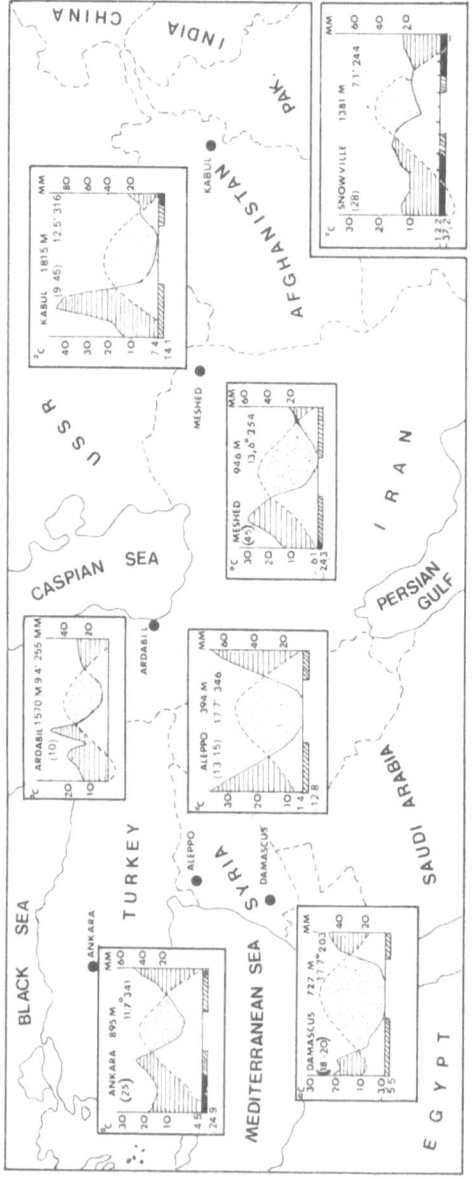

Fig. 2. Climatic diagrams of selected locations in the Middle East and of Snowville, Utah, using the format of Walter and Lieth (1960). The X-axis represents the 12 months of the year beginning in January. The darkened bar along the X-axis indicates months of the year when the mean daily minimum is below 0 °C and the hatched bar depicts months with absolute temperature minima below 0 °C. The dashed line portrays mean temperatures and the solid line indicates precipitation as delineated on the Y-axis. Numbers to the left of the Y-axis beginning at the top are the mean daily minimum temperature of the coldest month and the lowest temperature recorded at the particular location. From left to right at the top of the climatic diagram is the name of the location with the number of years of observations at this location below, the elevation of the site in meters, and the mean annual temperature in °C and precipitation in mm.

At these Syrian sites, essentially all precipitation occurs during winter and spring with an abrupt cutoff during the summer. Thus, plant growth during summer and autumn on these sites relies entirely on stored winter-spring moisture.

The climatic patterns which are most similar to that of Snowville are found in the high, dry, cold plateaus in Turkey, Iran, and Afghanistan. Fig. 2 shows climatic diagrams for representative plateau areas at Ankara, Ardabil, Meshed, and Kabul. These locations are colder than the lowland Mediterranean areas and have mean daily minimum temperatures below 0 °C for the coldest months.

The precipitation patterns of these highland sites are intermediate between those of the lowland Mediterranean areas and Snowville. Precipitation during the summer months is slightly greater in the plateau locations than the Mediterranean lowlands, but typically is less than that at Snowville.

Because of the cold winter temperature regime at Snowville and the high, dry, cold plateau areas, plant materials developed for these areas would likely be more cold resistant than cultivars developed for lowland Mediterranean areas. However, all three climatic areas receive the bulk of their precipitation during winter and early spring, and have a summer period with low rainfall and high temperatures. Thus, many of the principles involved in improving drought resistance in this discussion would likely be applicable to rangeland plant improvement in both the plateau and lowland Mediterranean areas. In particular, because drought occurs predictably during the same part of the growing season at all of these sites, there are good prospects for developing improved plant materials.

Defining the selection criteria

The effect of drought stress on range plants is particularly pronounced during germination, emergence, and early seedling growth (Wright 1971, 1975). McGinnies (1960), Knipe and Herbel (1960), and Wilson *et al.* (1970) reported that low soil moisture delayed germination, reduced total germination, and decreased the seedling growth of several range forage species. Planting failures could be minimized by the development of cultivars with superior establishment characteristics. However, seedling response is not necessarily indicative of mature plant response.

Because stored winter-spring moisture is critical in determining productivity on many semi-arid rangelands, growth during periods when moisture is available is essential for improved cultivars. This adaptation, drought avoidance, allows the plant to grow and mature during the portion of the season when water and temperature are most favorable for growth. Successful seedling establishment in semi-arid areas critically hinges on early root initiation and elongation. Rapid root extension allows the seedling to compete successfully with other species and with evaporative drying for the rapidly diminishing moisture in the upper soil layers (Harris, 1977).

After emergence, seedlings may not receive any more rain for 10 to 20 days. Consequently, than ability to survive desiccating conditions and resume growth after drought is another important attribute. Because of the extremely low precipitation and high evaporative demands during the summer, perennial forage plants generally become dormant. Thus, capacity to withstand summer drought is also important. Plant materials should also be able to resume growth in response to precipitation during late fall when evaporative demands are lower.

There are many possibilities for screening breeding populations for anatomical or morphological characteristics related to drought resistance. The association between individual morphological characters and drought was examined by Maximov (1929, 1931), Newton and Martin (1930), Shields (1950), Iljin (1957), Russell (1959), and Oppenheimer (1960). Their work documented the many interactions between morphological characters and the environment.

Because of these interactions, Wright (1971), Moss *et al.* (1974), and Boyer and McPherson (1975) stated that essentially none of these associations was useful as a reliable guide for indicating drought resistance. Ashton (1948) summarized the information on techniques for drought selection that were available 30 years ago. His statement is still applicable: 'In general, physical characters such as water requirements and transpiration rate, and anatomical characters have not been found to provide a simple and practical index of drought resistance in selection work.' Consequently, Wright (1964) recommended that until the fundamentals of drought resistance characteristics were known more precisely, plant improvement programs would have to rely on screening techniques that are based on plant response to drought stress, rather than on specific plant characteristics.

Depending on the specific objectives within a plant improvement program, additional plant traits besides drought resistance would likely be selected. These might include such traits as total and seasonal distribution of biomass production, forage quality, seed yield, compatability with associated species, and resistance to insects and disease. However, unless the genotypic and phenotypic variances and covariances are known and the relative economic weights of the traits are precisely defined, efficiency of selection probably would decrease with each additional selection characteristic which is added to the evaluation.

Assembling a broad genetic base

Genetic advance in a plant improvement program depends on assembling a diverse collection of germplasm from released cultivars, experimental breeding strains, collections from old plantings, and plant introductions. This diverse germplasm serves as a gene pool for variation for particular characteristics. Genetic variation for desirable plant characteristics or responses is fundamental to any plant improvement program and dictates potential progress.

Hurd (1971) emphasized the need for large populations because improvement in drought resistance probably involves many genes with small effects that are difficult to measure. Hanson (1972) also stressed the importance of genetic diversity when attempting to increase the frequency of genes for potentially valuable plant characteristics.

Development of screening techniques

After a broad-based germplasm pool has been assembled, plant improvement generally involves screening this large source population to isolate plants that have the desired combination of characters. Levitt (1964) and Cooper (1974) stressed the importance of reliable screening tests as an integral component of any plant improvement program. Hanson (1972) emphasized that progress in plant breeding has been impeded by the lack of appropriate screening procedures.

Ideally, plant screening techniques should: (i) assess plant performance at the critical developmental stage; (ii) be completed in a relatively short time; (iii) use relatively small quantities of plant material; and (iv) be capable of screening large populations. Although many techniques are available for examining plant water relationships, most are too laborious and time-consuming for use in a plant improvement program. Also, many techniques for evaluating plant water relations measure dynamic plant characteristics that change daily or throughout the season. Moss et al. (1974) suggested that selection criteria for improving plant performance under drought may require a compromise between impossibly complex measurements and convenient rapid screening techniques.

Roy and Murty (1970), and Mederski and Jeffers (1973), suggested that selection under conditions for optimum growth might yield lines that would also perform well under water stress conditions. In this case, improved response to drought would be an unidentified component of stability in performance over different environments. However, high yielding strains under conditions of adequate moisture are not always high yielding under drought stress (Burton,

1964; Hurd, 1968; Johnson *et al.,* 1968). Boyer and McPherson (1975) emphasized that to screen for characteristics that become apparent only during drought, selection must be done under desiccating conditions. This implies that the potentials for growth and drought resistance are under different genetic control. Thus, after drought resistance components have been identified, they must be incorporated into high yielding cultivars.

Breeding lines that use water efficiently in a dry environment may not do as well as other lines under more favorable conditions, apparently because of trade-offs between plant responses in different environments (Orians and Solbrig, 1977). Consequently, selection for wide adaptability, may in reality, be selection for mediocrity (Reitz, 1974). Probably the most promising route for plant improvement under drought stress involves selection under water-limiting conditions.

Field performance of plant lines is generally regarded as being the standard for evaluation of plant response under drought stress. However, field trials frequently require large expenditures of time and money. In addition, semi-arid areas are notorious for their fluctuations in environmental conditions from year to year and site to site. Because field screening can be time-consuming and uncertain (Wright, 1964), reliable laboratory or greenhouse screening techniques are also necessary. Laboratory procedures allow control of environmental conditions and isolation of the direct and indirect effects of drought.

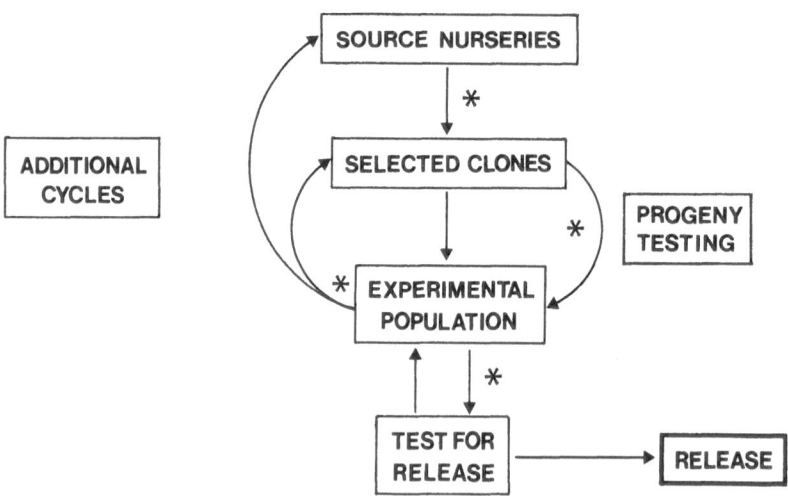

Fig. 3.   Diagram representing different phases of our plant improvement program. The asterisks denote where field and laboratory screening procedures can be effectively incorporated.

Application of screening procedures

Field and laboratory screening procedures can be effectively used during several phases of a plant breeding program (Fig. 3). In the grass breeding portion of our program, progenies are screened from genetically diverse parental lines in both laboratory and field trials. Data from each trial are then systematically included in an index for selection of parents in breeding improved cultivars.

Laboratory screening. In the laboratory, our breeding lines are subjected to controlled drought stress to permit screening for (i) seedling emergence under drought stress; (ii) seedling recovery after exposure to drought, and (iii) rate of seedling emergence and seedling vigor, using a planting depth of 7 cm.

(i) *Emergence under drought stress.* Germination and seedling emergence under drought stress are key factors in the establishment of range forage plants, and are assessed with a procedure which has been modified from Kaufmann (1969). The procedure, described by Johnson and Asay (1978), uses soil as the germination medium. A stainless steel frame, enclosed on the sides and bottom with a semi-permeable membrane, is filled with soil at field capacity and placed in a vessel containing a polyethylene glycol 6000 (PEG) solution. The frame is supported in the vessel so that the PEG solution reaches about 1 cm above the bottom edge of the frame. The vessel is then covered and sealed with a plastic bag for a 3-day equilibration. The amount of water moving through the membrane from the PEG solution to the soil is controlled by the strength of the PEG solution. After equilibration, 100 seeds are uniformly distributed on the soil in each frame and covered with about 0.5 cm of air-dry soil. The vessel is again sealed and after seven days at 25 °C, percentage emergence is determined.

We have screened 150 progeny lines each of Russian wild rye (*Elymus junceus* Fisch.) and crested wheatgrass (*Agropyron desertorum* [Fisch. ex Link] Schult., *A. cristatum* [L.] Gaertn., and *A. sibiricum* [Willd.] Beauv.) with this technique. Significant differences (P < 0.01) were found among progenies of both species under each of two stress levels and in the analyses of data from the two stress levels combined.

The genetic variance among the progeny lines was consistently more than 50 per cent of the total phenotypic variance. This variance, along with the wide range among the progenies, indicated that the material had excellent potential for improved germination and seedling emergence under drought stress.

However, the correlation between the two stress levels was not significant and the progeny by stress level interaction was significant (P < 0.01) for each species. These findings suggest that the relative performances of the progeny lines were not consistent over the two levels of soil moisture, and that the stress level must be carefully considered in interpreting the results. Nevertheless, some lines emerged consistently well at both stress levels.

(ii) *Seedling recovery after drought*. A technique similar to one proposed by Wright (1964) was used to screen the same crested wheatgrass progeny lines for their ability to recover after exposure to severe drought. Eighty-four seedlings of each entry were started in individualized plastic containers in a greenhouse. After three weeks, they were transferred to a growth chamber programmed to simulate the environmental regime encountered during the early summer on the semi-arid rangelands of the western U.S.A. In the latest trials, the entries were arranged in a lattice design to minimize the effects of temperature gradients within the chamber. After equilibration for one week, water was withheld for 17 days. This is a realistic rate of stress development, a consideration that is extremely important (Begg and Turner, 1976). The desiccated plants were then returned to the greenhouse and watered daily for three weeks. The recovery was rated by three independent observers. Ratings from the observers were closely correlated, and highly significant differences were found among the crested wheatgrass progeny lines. The magnitude of genetic variation and heritability values suggested that selection would genetically improve recovery of seedlings after drought.

(iii) *Seedling vigor*. The potential of Russian wild rye in U.S.A. has been severely limited by deficiencies in seedling vigor and other characteristics which are related to stand establishment. In our program to improve the seedling vigor of this species, seeds were planted 7 cm in the soil, following a similar procedure to Lawrence's (1963). The trials were conducted in a growth chamber programmed to simulate the same early summer, semi-arid environment used in earlier trials. Five seedling counts were made from 7 to 13 days after planting, and rate of emergence was computed for each line according to a method described by Maguire (1962). This value consisted of a summation of the five counts (expressed as per cent emergence), with each count divided by the number of days after planting. Highly significant differences were detected among progeny lines in rate of emergence, and the mean emergence rate of the progenies was significantly higher than that of commercially available check cultivars. We were also encouraged by the relative magnitude of the genetic variation among the progeny lines which was 64 per cent of the total phenotypic variance. Height and dry weight of the seedlings two weeks after planting followed similar trends, and provided additional evidence that selection could improve seedling vigor in Russian wild rye.

Field screening. Final evaluation of products from a breeding program should be based on seedling establishment under drought conditions in the field. Our breeding lines of crested wheatgrass and Russian wild rye were screened on a field site in northwestern Utah near Snowville, which receives an average annual precipitation of about 244 mm. In these plantings, a plot consisted of a 3 m row in which 500 seeds were planted 3.5 cm deep. The trials were planted during the late fall and, as was done in the laboratory, they were replicated three or four times. After germination during the following spring, per cent seedling emergence and seedling height were determined in June, dry matter yield in August, and per cent stand in the fall nearly a year after seeding.

Data from the field plantings were more consistent than expected, particularly for the Russian wild rye entries. Significant differences were found among the Russian wild rye progeny lines for all characters measured in the field. For example, per cent stand of the progenies ranged from 42 to 86 per cent and averaged 70 per cent, while the check cultivars averaged 58 per cent. The genetic variance among lines was 35 per cent of total phenotypic variance, and the coefficient of variation was 17 per cent. Data for rate of seedling emergence and seedling height followed similar trends, and indicated that selection under field conditions would also be fruitful.

Laboratory – field correlations. In general, the correlations between data from laboratory and field trials were low, ranging from -0.11 to $+0.34$. Data from seedling vigor trials in which seed was planted 7 cm deep, were correlated most closely with data from field trials. All but one of these correlations were statistically significant ($P < 0.05$) and ranged from 0.17 to 0.34. Data from the seedling vigor trials were also the most closely associated with seed weight. Correlation coefficients ranged from 0.24 for seedling dry weight to 0.57 for rate of germination.

These low correlations between the data from the laboratory and the field were somewhat disappointing, but not unexpected. Although the correlations are low, plant response to drought stress as evaluated by our current screening procedures may represent relatively few of the myriad of other interacting factors that affect plants in semi-arid rangeland. As a result, the small positive correlations may be the best we can expect.

Alternatively, the low correlations might indicate that the present screening procedures should be improved to approximate more closely the limiting environmental conditions on semi-arid rangelands. However, because our trials involved only one environment, additional trials involving several sites and different years would be more definitive.

Another approach that should help to assess our current screening procedures, is the comparison of performance of progenies from selected and unselected lines. We are evaluating these populations for shifts in laboratory and field performance, and should be able to assess progress, if any, in the use of our present screening techniques. The outcome should help to direct screening for improvement of drought resistance.

Other promising screening procedures

A field technique that appears promising for screening involves a line-source sprinkler installation (Hanks *et al.*, 1976). Sprinklers are spaced closely along an irrigation line so that water distribution is essentially constant along any line parallel to the sprinkler line. A continuous water variable is obtained by applying water so that the amount of water applied decreases linearly from a high rate at the point of origin to zero at some point a given distance from the sprinkler line. Thus, the system produces a water-application pattern that is uniform along the length of the irrigation line and continuously, but uniformly, variable at right angles to the sprinkler line.

This field installation seems ideal for evaluating the response of range seedlings to different water levels, an especially critical characteristic for successful seedling establishment on semi-arid rangeland. One disadvantage is that the technique cannot exclude precipitation. However, in semi-arid areas, experiments could be run during the summer months when precipitation is usually slight.

In many semi-arid situations, however, it is necessary to exclude precipitation from the plots, especially when mature perennial plants are evaluated. In this case, rainout shelters (Horton, 1962; Hiler, 1969; Teare *et al.*, 1973; Arkin *et al.*, 1975) could be used in combination with the sprinkler systems for strict control over the development of water stress conditions. These shelters can be placed, either manually or mechanically, over the field plots to exclude rainfall. Electronically controlled rainout shelters automatically protect the plots from precipitation, and also minimize disturbance to other environmental factors such as irradiation, temperature, and wind. Shelters typically cover small (12.2 × 12.2 m) areas and much larger ones might be necessary for plant breeding applications. However, the small shelters might be adequate if selections can be made at the seedling stage.

A notable exception to the limited use of plant characters in selecting for drought resistance was reported by Hurd (1969, 1971, 1974, 1976). Hurd and co-workers concluded that a rapidly penetrating root system with an extensive network of primary and secondary roots is essential for high wheat yields in a semi-arid environment.

In some semi-arid environments where plant growth must rely only on water stored in the soil, conservative water behavior has been suggested to be a desirable trait. Passioura (1972) demonstrated that increasing the resistance of the root system of wheat to water flow by reducing the number of seminal roots conserved water during the vegetative stage. This enhanced water availability during heading and resulted in seed yields that were higher than those from plants with normal root systems. However, in forage production where leaf rather than grain production is critical, water conservation for later growth stages is probably of less concern. Therefore, a rapidly penetrating and extensive root system probably should be of primary concern in programs to develop cultivars for semi-arid rangelands.

POSSIBILITIES FOR ENHANCED BIOLOGICAL NITROGEN FIXATION
ON SEMI-ARID RANGELANDS

Plant productivity on many semi-arid rangelands is often limited by lack of nitrogen (N). Although forage production on many types of rangelands can be increased by the addition of nitrogenous fertilizers, marginal economic returns have prohibited fertilization of all but a fraction of the world's rangelands. This is particularly true of semi-arid rangelands where plant response is restricted by low moisture availability throughout most of the growing season. Increasing costs for N will probably further limit its application to rangelands. Such limitations have focused increasing attention on biological fixation of N by use of legume-*Rhizobium* associations.

The influence of drought and temperature stress on nitrogen fixation

Although suggestions have been made to seed or interseed native or introduced legume species on semi-arid rangelands, plants growing in semi-arid climatic zones are affected seriously by both high temperature and drought stress. For most temperate legumes, N fixation is at a maximum between 20° and $35\,^\circ C$ (Gibson, 1971). Environmental rhizosphere limitations of N fixation by soybean, *Glycine max* (L.) Merr., included low and high rhizosphere temperatures, slow response to heating, and low reversibility (Hardy and Criswell, 1976). Between 14° and $34\,^\circ C$, the equilibrium response of soybean $N_2[C_2H_2]$ fixation to temperature was quadratic, with optima at 22° to $24\,^\circ C$; heating caused greater decreases than cooling.

Optimum temperatures for the development of leguminous plants, for nodulation and for N assimilation may not coincide, and those temperatures that promote formation of nodules on primary roots may differ from those that promote nodulation on secondary roots (Kornilov and Verteletskaya, 1952). For example, the optimum temperature for nodulation of the pea, *Pisum sativum* L., was not the same as that for N fixation (Lie, 1974). However, the response of N fixation to temperature can be markedly affected by the legume's area of origin. For example, legumes that evolved under tropical climates, withstood higher temperatures than those adapted to subtemperate climates (EK-Jander and Fahraeus, 1971).

Drought stress also seriously reduces N fixation in a number of legume-*Rhizobium* associations (Sprent, 1971a, b, 1972, 1976; Engin and Sprent, 1973; Minchin and Pate, 1975; Pankhurst and Sprent, 1975; Foulds, 1978). Soil moisture stress has particularly deleterious effects on shallow-rooted legumes (Ward *et al.,* 1966). Extremes of soil moisture also adversely affect root nodule bacteria (Engin and Sprent, 1973; Sprent, 1971a; Vincent, 1965). Plants of birdsfoot trefoil, *Lotus corniculatus* L., grown from seeds placed in soil of low average moisture content had small, white, and nonfunctional nodules with up to 55 per cent of the plants still not nodulated 77 days after seeding (McKee, 1961).

Two or more weeks of desiccation of inoculated seeds of alfalfa (*Medicago sativa* L.) and birdsfoot trefoil in a dry seed bed reduced inoculation and produced N deficiency in the legumes (Alexander and Chamblee, 1965). Mishustin and Shil'nikova (1971) stated that optimum soil moisture for inoculation was 60 to 70 per cent. They observed death of formed nodules with lack of moisture, although plant species differ in their critical moisture thresholds (Kornilov and Verteletskaya, 1952).

Pate (1976) summarized the effects of drought stress on symbiotic N fixation as:

• Loss of fixation activity on desiccation occurs in nodules with either determinate growth or indeterminate growth.

• Effects of water stress are reversible, provided that water loss from a nodule does not amount to more than 20 per cent of its maximum fresh weight.

• Irreversible structural damage occurs with severe drought damage as collapse of cells in the nodule cortex, and damage to plasmodesmatal connections in the nodule tissue takes place.

• In the field, fixation is found to be highest at or near field capacity, but is severely suppressed once flagging of the lower leaves has begun.

• In times of stress, a deficiency of water associated specifically with nodules of the upper zone of the root may be counteracted by lateral transfer to these nodules of water retrieved by roots in lower horizons of the soil profile.

• Osmotic damage to fixation may occur through the concentration of ions near the nodules.

Legumes for semi-arid areas

Recent research on the effects of temperature and drought stress on N fixation has involved *Rhizobium* associations with crop legumes that evolved and were growing in mesic environments. Legumes from environments with marked drought and high temperature stress, such as semi-arid rangelands, may have evolved unique adaptations. These adaptations may enable such legume-*Rhizobium* associations to fix more N under stress conditions than associations evolved in other environments. Alternatively, legumes with drought avoidance mechanisms may be more appropriate for semi-arid areas. Legume-*Rhizobium* associations could be selected that fix nitrogen rapidly when temperature and moisture are favorable. These associations could be either annuals which complete their life cycle before high stress or, being perennials, they could become dormant with the ability to remain viable during dormancy, such as those found by Pate (1958), may be particularly applicable in Mediterranean-type climates.

Whatever strategy or combination of strategies may prove the best, extrapolation of N fixation responses from legume-*Rhizobium* associations that evolved in low-stress environments may not accurately indicate the capabilities of associations evolved in drought- and temperature-stressed environments. Consequently, little is known about the N fixation capabilities of legume associations which grow under conditions of drought and temperature stress in semi-arid rangelands (Farnsworth *et al.*, 1978).

Annual legumes. Annual *Trifolium* and *Medicago* species have been successful in the Mediterranean-type climates of both Australia and California. For example, in California addition of phosphate along with seeded *Trifolium* species produced more than three-fold increases in forage and six-fold increases in protein as compared to on untreated land (Love, 1952; Williams *et al.*, 1956). Jones (1967) found that California grasslands seeded with annual *Trifolium* species produced forage yields equal to those from seeded grasslands fertilized with 45 to 90 kg N ha^{-1} in a moisture-deficient year. In a moisture-adequate year, annual *Trifolium*-grass mixtures produced more forage than did annual grass-dominated areas fertilized with 179 kg N ha^{-1}.

Because of the similarities of climate, similar results could probably be achieved in other Mediterranean-type areas with proper management. Work with annual legume species in Syrian rangeland is currently underway in Damascus at ACSAD (Arab Center for the Studies of Arid Zones and Dry Lands) (F.D. Iskander, pers. comm.).

Annual legumes are in the process of being used to integrate livestock production with farming systems in Mediterranean-type areas (Draz, 1977). Traditional ley farming in these areas involves a grain/fallow system. Fallow fields are usually not barren, but rather support weed 'crops,' which are sold for grazing by sheep. However, letting weeds grow during the fallow year results in abundant weed seeds. Severe weed infestations during the following year, compete with grain production and consequently result in lower yields.

A similar situation faced Australian farmers during the early 1900's. During the 1930's, the possibility of including nitrogen-fixing annual legumes (*Medicago* and *Trifolium* species) in the rotation system instead of fallow was examined. The annual legumes successfully competed against the weedy species and provided high quality grazing as well as an accumulation of nitrogen to support the subsequent wheat crop. Today, the cereal/annual legume rotation is an integral part of Australian agriculture, and is resulting in significant increases in the yields and quality of cereals.

CIMMYT (Centro Internacional de Mejoramiento de Maiz y Trigo) has a current program which involves the examination of annual legumes in North African grain rotation systems (Breth, 1975). The CIMMYT program suggests that annual legumes can be successful in the Mediterranean area. Scientists estimated that a cereal/legume rotation would probably be best-suited in areas with 350 to 500 mm of annual precipitation.

Annual *Medicago* species are native in Mediterranean areas and would be a logical species for use in a cereal/legume rotation in the more alkaline soils. Because of the abundance of annual medics and their likely genetic variation, a comprehensive plant breeding program is probably not necessary (Breth, 1975). Instead a short-term program involving collection, testing, and selection, would probably yield superior medic strains. With proper management, integration of the livestock and crop production systems could result in reduced grazing of overgrazed rangelands.

Perennial legumes. The relatively infrequent use of perennial legumes on semi-arid rangelands may be attributed largely to their low persistence under heavy grazing, and their bloat hazard to ruminants (Heinrichs, 1975). However, these negative attributes have probably resulted from the unavailability of adapted cultivars and the subsequent use of unadapted cultivars, which have been developed in mesic areas. Improper grazing management has also undoubtedly contributed to these negative attitudes. Whatever the case, improvement of perennial legume cultivars for semi-arid rangeland has received only limited attention.

Perennial legumes provide high quality forage and have increased forage yields in many mesic pasture situations. Even on less mesic sites such as the Northern Great Plains area of the U.S.A., increased short-term forage yields have been reported after the introduction of alfalfa on range sites (Lorenz and Rogler, 1962; Gomm, 1964; Miles, 1969). Evidence supporting long-term persistence of legumes on semi-arid range sites is extremely limited. However, Rumbaugh and Pederson (1979) have indicated that legumes can persist in semi-arid environments. They reported that under grazing, alfalfa plant density remained high on a semi-arid range site that receives about 280 mm annual precipitation. After 23 years, alfalfa yielded 121 per cent as much oven-dry forage as crested wheatgrass in an adjacent planting. Apparently, even on semi-arid range sites, legumes can provide long-term increases in forage quality and quantity.

Johnson and Rumbaugh (1980) examined acetylene reduction rates of excised root segments with attached nodules in legumes from two mountain grassland sites, one native sagebrush-dominated site, and three cultivated semi-arid range sites, formerly under big sagebrush, all in western U.S.A. Even during the most drought-stressed period of the semi-arid growing season, certain perennial range legumes fixed nitrogen. Alfalfa plants were particularly notable because they were nodulated and reduced acetylene in dry soils when other legumes were not active.

In addition, under the most favorable conditions of this study, nodules from alfalfa reduced acetylene most actively. These results suggest that, given more favorable temperature and moisture, nitrogen fixation capability may be considerable at least for some adapted range legumes. Use of adapted perennial legumes for enhancing nitrogen fixation on Mediterranean rangelands should not be overlooked.

Nitrogen availability to associated forage plants

If legumes can fix significant amounts of nitrogen in semi-arid range environments, more nitrogen could be made available to the non-associative nitrogen fixers and non-leguminous members of the plant community. Ideally the legume-fixed nitrogen would increase productivity and protein content of other forage plants in the community. This fixed nitrogen could be made available to other plants through leaf and stem fall, leaching of the vegetation by precipitation, root decomposition, and/or exudation from the roots (Dommergues, 1978). Because decomposition can be slow and because precipitation available for leaching is limited in semi-arid environments, root exudation may represent the most direct route of providing nitrogen for use by associated plants. Root exudation or loss of plant-synthesized organic compounds from the root surface into the soil may occur through both secretion and leakage (Hale *et al.*, 1978). Secretion is an active transport process which requires metabolic energy, whereas leakage is the loss of compounds by simple diffusion.

Hale *et al.* (1978) stated that drought stress affects the amount of root exudation. After a drying and wetting cycle. Vancura (1964) recovered organic compounds exuded from the roots of wheat and barley (*Hordeum vulgare* L.) seedlings that amounted to about 10 per cent of the total dry weight of the aboveground portions of the plant. Drought stress might increase exudation from legumes growing on semi-arid rangelands, and this could add considerable amounts of nitrogenous compounds to the soil for subsequent use by associated forage species.

SUMMARY AND CONCLUSIONS

As world population expands, semi-arid rangelands should become increasingly important as sources of food, and as locations for off-season grazing in rotation with croplands. Because vast expanses of semi-arid rangelands are producing far below their potential due to past overgrazing, range improvement through improved plant materials and proper grazing management is required. Inasmuch as drought stress is the overriding factor affecting plant growth on semi-arid rangelands, improved range plants must have drought resistance mechanisms.

The approach that our research team is using for plant improvement on semi-arid rangelands includes: (i) drought characterization; (ii) definition of selection criteria; (iii) assemblage of a broad genetic base; (iv) development of screening techniques, and (v) application of the screening procedures. Additionally, because plant productivity on many semi-arid rangelands is limited by nitrogen deficiency, both annual and perennial legumes should be examined to assess their potential for adding nitrogen to the range ecosystem.

As Boyer and McPherson (1975) stated, 'At this time, with our limited and inadequately integrated knowledge of plant performance under desiccating conditions, any suggestion of how to aim a plant improvement program must be tenuous at best.' Although this statement still applies, interdisciplinary programs must be initiated to test our best strategies for plant improvement under drought stress as a step towards the development of improved cultivars for the revegetation of semi-arid rangelands.

REFERENCES

Alexander, C. W. and Chamblee, D. S. 1965 Effect of sunlight and drying on the inoculation of legumes with *Rhizobium* species. Agron. J. 57, 550–553.

Arkin, G. F., Ritchie, J. T., Thompson, M. and Chaison, R. 1975 A Large Automated Rainout Shelter Installation for Crop Water Deficit Studies. Texas Agric. Exp. Sta. Misc. Publ. No. 1199. 8 pp.

Ashton, T. 1948 Techniques of Breeding for Drought Resistance in Crops. Commonwealth Bur. Plant Breed. Genet., Tech. Commun. No. 14. Aberystwyth.

Begg, J. E. and Turner, N. C. 1976 Crop water deficits. Adv. in Agron. 28, 161–217.

Blum, A. 1979 Genetic improvement of drought resistance in crop plants: A case history for sorghum. p. 429–446. In: Stress Physiology in Crop Plants. Mussel, H. and Staples, R. C., (eds.), John Wiley and Sons, New York.

Boyer, J. S. and McPherson, H. G. 1975 Physiology of water deficits in cereal crops. Adv. in Agron. 27, 1–23.

Breth, S. A. 1975 The return of medic. CIMMYT Today 3, 1–16.

Brown, R. W. 1977 Water relations of range plants. p. 97–140. In: Rangeland Plant Physiology. Sosebee, R. E. (ed.), Soc. Range Manage. Range Sci. Ser. No. 4. Denver, Colorado.

Burton, G. W. 1964 The geneticist's role in improving water-use efficiency by crops. p. 95–103. In: Research on Water. Amer. Soc. Agron. Spec. Publ. No. 4. Madison, Wisconsin.

Cooper, J. P. 1974 The use of physiological criteria in grass breeding. p. 95–102. Welsh Plant Breeding Station Report. Aberystwyth.

Dommergues, Y. R. 1978 The plant-microorganism system. p. 1–37. In: Interactions Between Non-Pathogenic Soil Microorganisms and Plants. Y. R. Dommergues and S. V. Krupa (eds.), Elsevier Sci. Publ. Co., New York.

Draz, O. 1977 Role of range management in the campaign against desertification: The Syrian experience as an applicable example for the Arabian Peninsula. UNDP/UNCOD/MISC/13. 37 p.

EK-Jander, J. and Fahraeus, G. 1971 Adaptation of *Rhizobium* to subarctic environment in Scandinavia. Plant Soil Spec. Vol. pp. 129–137.

Engin, M. and Sprent, J. I. 1973 Effects of water stress on growth and nitrogen fixing activity of *Trifolium repens*. New Phytol. 72, 117–126.

Farnsworth, R. B., Romney, E. M. and Wallace, A. 1978 Nitrogen fixation by microfloral-higher associations in arid to semiarid environments. p. 17–19. In: Nitrogen in Desert Ecosystems. West, N. E., Skujins, J. J. (eds.). Dowden, Hutchinson, and Ross, Inc. Stroudsburg, Pennsylvania.

Fischer, R. A. and Turner, N. C. 1978 Plant productivity in the arid and semiarid zones. Ann. Rev. Plant Physiol. 29, 277–317.

Foulds, W. 1978 Response to soil moisture supply in three leguminous species. II. Rate of $N_2[C_2H_2]$-fixation. New Phytol. 80, 847–555.

Gibson, A. H. 1971 Factors in the physical and biological environment affecting nodulation and nitrogen fixation by legumes. Plant Soil Spec. Vol. pp. 139–152.

Gomm, F. B. 1964 A comparison of two sweetclover strains and Ladak alfalfa alone and in mixture with crested wheatgrass for range and dryland seeding. J. Range Manage. 17, 19–22.

Hale, M. G., Moore, L. D. and Griffin, G. J. 1978 Root exudates and exudation. p. 163–203. In: Interactions Between Non-Pathogenic Soil MIcroorganisms and Plants. Y. R. Dommergues and S. V. Krupa (eds), Elsevier Sci. Publ. Co., New York.

Hanks, R. J., Keller, J., Rasmussen, V. P. and Wilson, G. D. 1976 Line source sprinkler for continuous variable irrigation-crop production studies. Soil Sci. Soc. Am. J. 40, 426–429.

Hanson, A. A. 1972 Breeding of grasses. p. 36–52. In: The Biology and Utilization of Grasses. V. A. Younger and C. M. McKell (eds.), Academic Press, New York.

Hardy, R. W. F. and Criswell, J. G. 1976 Assessment of environmental limitations of symbiotic $N_2[C_2H_2]$ fixation: temperature and pO_2, Agron. Abstr. p. 72.

Harris, G. A. 1977 Root phenology as a factor of competition among grass seedlings. p. 93–98. In: The Belowground Ecosystem: A Synthesis of Plant-Associated Processes. J. K. Marshall (ed.), Range Sci. Series No. 26, Colorado State Univ., Fort Collins, Colorado.

Heinrichs, D. H. 1975 Potentials of legumes for rangelands. p. 50–61. In: Improved Range Plants. Campbell, R. S. and Herbel, C. H. (eds.). Society for Range Manage. Range Symp. Series No. 1, Denver, Colorado.

Hiler, E. A. 1969 Quantitative evaluation of crop-drainage requirements. Trans. Amer. Soc. Agric. Eng. 12, 499–505.

Horton, M. L. 1962 'Rainout' shelter for corn. Iowa Farm Sci. 17, 16.

Hurd, E. A. 1968 Growth of roots of seven varieties of spring wheat at high and low moisture levels. Agron. J. 60, 201–205.

Hurd, E. A. 1969 A method of breeding for yield of wheat in semi-arid climates. Euphytica 18, 217–226.

Hurd, E. A. 1971 Can we breed for drought resistance? p. 77–88. In: Drought Injury and Resistance in Crops. Larson, K. L. and Eastin, J. D. (eds.). Crop Sci. Soc. Amer. Spec. Publ. No. 2. Madison, Wisconsin.

Hurd, E. A. 1974 Phenotype and drought tolerance in wheat. Agric. Meteorol. 14, 39–55.

Hurd, E. A. 1976 Plant breeding for drought resistance. p. 317–353. In: Water Deficits and Plant Growth Vol. IV. Soil Water Measurement, Plant Responses and Breeding for Drought Resistance. Kozlowski, T. T. (ed.). Academic Press, New York.

Iljin, W. S. 1957 Drought resistance in plants and physiological processes. Ann. Rev. Plant Physiol. 8, 257–274.

Johnson, D. A. 1980 Improvement of perennial herbaceous plants for drought.stressed western rangelands. (pp. 419–433). In: Adaptation of Plants to Water and High Temperature Stress. Turner, N. C. and Kramer, P. J., (eds.). Wiley-Interscience, New York.

Johnson, D. A. and Asay, K. H. 1978 A technique for assessing seedling emergence under drought stress. Crop Sci. 18, 520–522.

Johnson, D. A. and Rumbaugh, M. D. 1981 Nodulation and nitrogen fixation by certain rangeland legume species under field conditions. J. Range Manage. 34 (In press).

Johnson, V. A., Shafer, S. L. and Schmidt, J. W. 1968 Regression analysis of general adaptation in hard red winter wheat (*Triticum aestivum* L.). Crop Sci. 8, 187–191.

Jones, M. B. 1967 Forage and protein production by subclover-grass and nitrogen-fertilized California grasslands. Calif. Agric. 21, 4–7.

Kaufmann, M. R. 1969 Effects of water potential on germination of lettuce, sunflower, and citrus seeds. Can. J. Bot. 47, 1761–1764.

Knipe, D. and Herbel, C. H. 1960 The effects of limited moisture on germination and initial growth of six grass species. J. Range Manage. 13, 297–302.

Kornilov, A. A. and Verteletskaya, V. 1952 Penetration of sainfoin into dry steppe regions and the role of nodule bacteria. Microbiologiya 20, 423–428.

Lawrence, T. 1963 A comparison of methods of evaluating Russian wild ryegrass for seedling vigor. Can. J. Plant Sci. 43, 307–312.

Levitt, J. 1964 Drought. p. 57–66. In: Forage Plant Physiology and Soilrange Relationships. Amer. Soc. Agron. Spec. Publ. No. 5. Madison, Wisconsin.

Levitt, J. 1972 Response of Plants to Environmental Stress. Academic Press, New York. 697 pp.

Lie, T. A. 1974 Environmental effects on nodulation and symbiotic nitrogen fixation. p. 555–582. In: The Biology of Nitrogen Fixation. Quispel, A. (ed.). American Elsevier Publ. Co., New York.

Lorenz, R. J. and Rogler, G. A. 1962 A comparison of methods of renovating old stands of crested wheatgrass. J. Range Manage. 15, 215–219.

Love, R. M. 1952 Range improvement experiments on the Aurther E. Brown ranch, California. J. Range Manage. 5, 120–123.

Maguire, J. D. 1962 Speed of germination – aid in selection and evaluation for seedling emergence and vigor. Crop Sci. 2, 176–177.

Maximov, N. A. 1929 Internal factors of frost and drought resistance in plants. Protoplasma 7, 259–291.

Maximov, N. A. 1931 The physiological significance of the xeromorphic structure of plants. J. Ecol. 19, 273–282.

May, L. H. and Milthorpe, F. L. 1962 Drought resistance of crop plants. Commonwealth Bureau of Pastures and Field Crops, Hurley, U.K. Field Crop Abstracts 15, 171–179.

Mc Ginnies, W. G. 1957 Vegetation. p. 121–133. In: Arid Zone Research. Vol. IX. Guide Book to Research Data for Arid Zone Development. Dickson, B. T. (ed.). UNESCO, Paris.

McGinnies, W. J. 1960 Effects of moisture stress and temperature on germination of six range grasses. Agron. J. 52, 159–162.

McKee, G. W. 1961 Some effects of liming, fertilization, and soil moisture on seedling growth and nodulation in birdsfoot trefoil. Agron. J. 53, 237–240.

McKell, C. M. 1975 Shrubs – A neglected resource of arid lands. Science 187, 803–809.

Mederski, H. J. and Jeffers, D. J. 1973 Yield response of soybean varieties grown at two soil moisture stress levels. Agron. J. 65, 410–412.

Miles, A. D. 1969 Alfalfa as a range legume. J. Range Manage. 22, 205–207.

Minchin, F. R. and Pate, J. S. 1975 Effects of water, aeration, and salt regime on nitrogen fixation in a nodulated legume – definition of an optimum root environment. J. Exp. Bot. 26, 60–69.

Mishustin, E. N. and Shil'nikova, V. K. 1971 Biological fixation of atmospheric nitrogen. Penn. State Univ. Press. Univ. Park, Pennsylvania.

Moore, R. M. 1960 The management of native vegetation in arid or semi-arid regions. p. 173–190. In: Arid Zone Research. Vol. XV. Plant-Water Relationships in Arid and Semi-Arid Conditions. Reviews of Research. UNESCO, Paris.

Moss, D. N., Woolley, J. T. and Stone, J. F. 1974 Plant modification for more efficient water use: The challenge. Agric. Meteorol. 14, 311–320.

Naveh, A. 1972 The role of shrubs and shrub ecosystems in present and future Mediterranean land use. p. 414–428. In: Wildland Shrubs – Their Biology and Utilization. McKell, C. M., Blaisdell, J. P. and Goodin, J. R. (eds.). General Technical Report INT-1. U.S. Forest Service, Washington, D.C.

Newton, R. and Martin, W. M. 1930 Physiochemical studies on the nature of drought resistance in crop plants. Can. J. Res. 3, 336–427.

Oppenheimer, H. R. 1960 Adaptation to drought: Xerophytism. Arid Zone Res. 15, 105–138.

Oram, P. A. 1956 Pasture and fodder crops in rotations in Mediterranean agriculture. Working Party on the Development of the Grazing and Fodder Resources of the Near East, First Meeting. Cairo.

Orians, G. H. and Solbrig, O. T. 1977 A cost-income model of leaves and roots with special reference to arid and semiarid areas. Amer. Nat. 111, 677–690.

O'Toole, J. C. and Chang, T. T. 1979 Drought resistance in cereals – rice: A case study. p. 373–406. In: Stress Physiology in Crop Plants. Mussell, H. and Staples, R. C. (eds.). John Wiley and Sons, New York.

Pankhurst, C. E. and Sprent, J. I. 1975 Effects of water stress on the respiratory and nitrogen-fixing activity of soybean root nodules. J. Exp. Bot. 26, 287–304.

Parker, J. 1968 Drought-resistance mechanisms. p. 195–234. In: Water Deficits and Plant Growth. Vol. I. Development, Control, and Measurement. Kozlowski, T. T. (ed.). Academic Press, New York.

Passioura, J. B. 1972 The effect of root geometry on the yield of wheat growing on stored soil moisture. Aust. J. Agr. Res. 23, 745–752.

Pate, J. S. 1958 Nodulation studies in legumes. II. The influence of various environmental factors on symbiotic expression in the vetch (Vicia sativa L.) and other legumes. Austr. J. Biol. Sci. 11, 496–515.

Pate, J. S. 1976 Physiology of the reaction of nodulated legumes to environment. p. 335–360. In: Symbiotic Nitrogen Fixation in Plants. Nutman, P. S. (ed.). Cambridge Univ. Press. New York.

302

Reitz, L. P. 1974 Breeding for more efficient water use – is it real or a mirage. Agric. Meteorol. 14, 3–11.

Rossiter, R. C. 1969 Ecology of the Mediterranean annual-type pastures. Adv. in Agron. 18, 1–56.

Roy, N. N. and Murty, B. R. 1970 A selection procedure in wheat for stress environment. Euphytica 19, 509–521.

Rumbaugh, M. D. and Pedersen, M. W. 1979 Survival of alfalfa in five semiarid range seedings. J. Range Manage. 32, 48–51.

Russell, M. B. 1959 Drought tolerance of plants. Adv. in Agron. 11, 70–73.

Shantz, H. L. 1927 Drought resistance and soil moisture. Ecology 8, 145–157.

Shields, L. M. 1950 Leaf xeromorphy as related to physiological and structural influences. Bot. Rev. 16, 399–447.

Sprent, J. I. 1971a Effects of water stress on nitrogen fixation in root nodules. Plant Soil Spec. Vol. pp. 225–228.

Sprent, J. I. 1971b The effects of water stress on nitrogen-fixing root nodules. I. Effects on the physiology of detached soybean nodules. New Phytol. 70, 9–17.

Sprent, J. I. 1972 The effects of water stress on nitrogen-fixing root nodules. IV. Effects on whole plants of Vicia faba and Glycine max. New Phytol. 71, 603–611.

Sprent, J. I. 1976 Water deficits and nitrogen-fixing root nodules. p. 291–315. In: Water Deficits and Plant Growth. Vol. IV. Soil Water Measurement, Plant Responses, and Breeding for Drought Resistance. Kozlowski, T. T. (ed.). Academic Press, New York.

Stoddart, L. A., Smith, A. D. and Box, T. W. 1975 Range Management. McGraw-Hill Book Co., New York. 532 pp.

Sullivan, C. Y. and Ross, W. M. 1979 Selecting for drought and heat resistance in grain sorghum. p. 263–282. In: Stress Physiology in Crop Plants. Mussell, H., and Staples, R. C. (eds.). John Wiley and Sons, New York.

Teare, I. D., Schimmelpfenning, H. and Waldren, R. P. 1973 Rainout shelter and drainage lysimeters to quantitatively measure drought stress. Agron. J. 65, 544–547.

Townley-Smith, T. F. and Hurd, E. A. 1979 Testing and selecting for drought resistance in wheat. p. 447–464. In: Stress Physiology in Crop Plants. Mussell, H. and Staples, R. C. (eds.). John Wiley and Sons, New York.

Turner, N. C. 1979 Drought resistance and adaptation to water deficits in crop plants. p. 343–372. In: Stress Physiology in Crop Plants. Mussell, H. and Staples, R. C. (eds.). John Wiley and Sons, New York.

Turner, N. C. and Begg, J. E. 1978 Responses of pasture plants to water deficits. p. 50–66. In: Plant Relations in Pastures. Wilson, J. R. (ed.). CSIRO, Melbourne.

Vancura, V. 1964 Root exudates of plants. I. Analysis of root exudates of barley and wheat in their initial phases of growth. Plant Soil 21, 231–248.

Vincent, N. M. 1965 Environmental factors in the fixation of nitrogen by the legume. p. 384–435. In: Soil Nitrogen. Bartholemew, W. V. and Clark, F. E. (eds.). Amer. Soc. Agron., Madison, Wisconsin.

Walter, H. and Lieth, H. 1960 Klimadiagramm-Weltatlas. G. Fisher Verlag, Jena.

Ward, C. Y., Jones, J. N., Lillard, J. H., Moody, J. E., Brown, R. H. and Blaser, R. E. 1966 Effects of irrigation and cutting management on yield and botanical composition of selected legume-grass mixtures. Agron. J. 58, 181–184.

Williams, W. A., Love, R. M. and Conrad, J. P. 1956 Range improvement in California by seeding annual clovers, fertilization, and grazing management. J. Range Manage. 9, 28–33.

Wilson, A. M., Nelson, J. R. and Goebel, C. J. 1970 Effects of environment on the metabolism and germination of crested wheatgrass seeds. J. Range Manage. 23, 283–288.

Wright, L. N. 1964 Drought tolerance – program-controlled environmental evaluation among range grass genera and species. Crop Sci. 4, 472–474.

Wright, L. N. 1971 Drought influence on germination and seedling emergence. p. 19–44. In: Drought Injury and Resistance in Crops. Larson, K. L. and Eastin, J. D. (eds.). Crop Sci. Soc. Amer. Spec. Publ. No. 2. Madison, Wisconsin.

Wright, L. N. 1975 Improving range grasses for germination and seedling establishment under stress environments. p. 3–22. In: Improved Range Plants. Campbell, R. S. and Herbel, C. H. (eds.). Soc. Range Manage. Range Symp. Series No. 1. Denver, Colorado.

Epilogue: themes and variations

J. L. MONTEITH

The objective of agricultural research is to provide Man with a more stable supply of food from an environment which is inherently variable. In many types of climate, fluctuations of rainfall and temperature are the main sources of variability in yield but it was presumably the particular experience of Mediterranean farmers which was summed up long ago in one simple statement: *annus fructum fert, non tellus.* Expanding as well as translating, weather is the factor primarily responsible for differences in crop yields from year to year and changes in the composition and behaviour of the soil are less important when they are assessed on the same time scale. Even in a relatively constant and favourable climate, however, good yields cannot be maintained season after season without an input of nitrogen and other elements to replace those removed by harvesting. So it was appropriate at this workshop to consider the processes which determine the availability of nitrogen in agricultural soils as well as the more rapid fluctuations in water supply which are closely linked to the availability of nitrogen.

The concepts of stability and variability link all the contributions to this workshop and they are common to all the sciences. Physicists talk about 'damping' the response of a system to an external stimulus and chemists about the 'buffering' action of a solution. The wise farmer knows how to buffer his crops and his livestock against the excesses of climate, intuitively exploiting physiological mechanisms and forms of behaviour which have evolved over many thousands of years. During the workshop, I have tried to collect and categorise some of the attributes which already help to stabilise crop yields and which point the way towards further improvements in agricultural production.

Storage. The concept of storage was discussed in various contexts: storage of water in the soil profile; storage of carbohydrate in the stems of cereals, mobilised when the current supply of assimilate is restricted by drought; storage of grain or other foodstuffs for people and storage of fodder for livestock. In general, the more material that a system can store, the more stable it becomes, always provided there are no significant losses of material during storage.

Plasticity. We have used this word frequently without defining it. It describes the ability of a plant (or of any other biological system) to minimise the adverse effects of an unusually unfavourable environment and to exploit an unusually favourable environment. For example, it was suggested that breeders should be looking for wheat varieties with 'phenological plasticity' – the ability of a plant to alter its developmental timetable to match the weather.

One ingenious manifestation of plasticity is 'functional balance'. Plants which are short of water or nutrients tend to put a higher fraction of current assimilates into new roots to explore a larger volume of soil. In dull light on the other hand, leaves expand at the expense of the root system to trap more photons. The interaction between tops and roots creates a balance between the system collecting light and CO_2, and the system collecting water and nutrients, thus maximising the rate of dry matter production. Breeders should presumably be looking for varieties in which the balacing of root and shoot activity operates over the widest possible range of environmental conditions.

Feedback. Frequent reference has been made to the closure of stomata which helps plants to conserve water. Cowan and Farquhar have suggested that this type of feedback may optimise the use of water by plants and it is certainly consistent with the close correlation between water use and dry matter production which several speakers exploited. It is worth remembering, however, that the physiological evidence for a type of feedback which may hold the internal CO_2 concentration of leaves at an almost constant level is consistent with a system in which the 'water use efficiency' depends on the saturation deficit of the atmosphere. The drier the air, the more water must be used during the production of a given amount of dry matter.

There is also an important element of feedback and compromise in the behaviour of farmers whose decisions about the management of crops *next* year may be influenced by the weather *this* year as well as by a lifetime of climatic experience.

Reversibility. In the world of thermodynamics, no process can be perfectly reversible. In practice, we treat terrestrial processes as effectively reversible when they are driven by solar energy because we don't have to worry about the site at which entropy is increasing within the sun! Processes such as the production of fertilizers and pesticides or the operation of farm machinery need fossil fuel and are clearly irreversible. As world resources of energy and of materials become ever scarcer, the stability of agricultural systems is bound to be threatened by shortages of such things as fertilizers and pesticides.

One long-term solution to this problem lies in our ability to exploit organisms which grow and operate independently of human resources – bacteria which fix nitrogen, and insects, harmless to crops, which are the predators of agricultural pests. Agricultural scientists were using this type of 'biotechnology' long before the word was coined by their colleagues in the laboratory.

I hope ICARDA's programme will contain at least some work on all these aspects of buffering because they can help to bring greater agricultural stability to the region, an essential ingredient of the social and political stability which we all long for.

I now pass to a second topic frequently referred to by speakers and in discussion – the value of 'models' which can be used to define, to analyse, and to predict the response of a crop to its environment. Some confusion may have arisen because the word 'modelling' has been applied to at least three, if not four discrete activities and it may be helpful to distinguish them before we disperse.

In the first case, we talked about a 'model' for a research project, meaning a set of plans for experimental work to test new hypotheses or to confirm that general conclusions derived from the literature are valid for a particular crop in a specific environment. To avoid possible misunderstanding, such plans are better referred to as a research *scheme*. When a scheme has been implemented, the raw output consists of sets of figures describing the state of the crop at different stages of growth and corresponding features of the soil and of weather. This information can be handled in two ways. Any biological variable such as dry weight, representing an index of response, can be correlated with any number of appropriate environmental variables. The form of such a correlation is often referred to as a statistical or empirical 'model'. Again, the word 'model' seems inappropriate to me when no mechanisms are invoked and I see no good reason for abandoning the traditional word *analysis* to describe this type of exercise.

Field measurements can also be analysed in terms of a functional relationship or set of relationships, derived from a new hypothesis or known to be valid from other studies. For example, the analysis of dry matter production in terms of a 'water use efficiency' or some similar index, is based on the expectation that the uptake of carbon dioxide by a crop should be approximately proportional to its loss of water as vapour because both processes of gas exchange occur through the same system of valves. In this and similar cases, it is appropriate to apply the term *mechanistic model* to the equations which describe how the processes in a system depend on a set of mechanisms. The equations contain measurements as variables and parmeters which describe the state of the environment or of the crop at different stages of growth. The succes of the model may be judged by the extent to which the parameters stay constant or change in some rational and predictable way during the life of a crop.

Models of this type usually describe the behaviour of a crop at a fixed stage of growth and are therefore classified as 'static'. The study described by Kassam belongs to this category. The parameters derived from static models then become the building bricks of much more complex *simulation models* of a growing crop in which two or more aspects of response to the environment are combined in a computer program, using a language appropriate for a dynamic system (see chapter by van Keulen). Most models of this type are *deterministic* in the sense that environmental variables are assigned a fixed set of values. The next major step forward must be the development of *probablistic* or *stochastic* models of the type described by Smith and Harris.

The value of models to the crop ecologist, to the agronomist and to the breeder has been argued in many quarters. Unfortunately, the best efforts of modellers have often been frustrated by the Laws of Murphy, already referred to at this workshop. For example, the well-known tendency of simulation models to expand to unmanageable proportions is a direct consequence of Edington's Theory according to which 'The number of different hypotheses erected to explain a given biological phenomenon is inversely proportional to the available knowledge'. The so-called Harwood Law is relevant to many abortive attempts to 'test' models: 'Under the most rigorous controlled conditions of pressure, temperature, volume, humidity and other variables, the organism will do as it damn well pleases.'

A little more seriously, I believe that simulation modelling has not proved to be as useful as its exponents hoped a decade ago. Progress has not been impeded by lack of ingenuity in model building but rather by the large number of grey areas in our physiological knowledge which modellers have drawn attention to. Perhaps we should declare a moratorium on the more sophisticated forms of modelling until physiological work catches up; otherwise the word 'mimic' which some modellers innocently use will be singularly appropriate (Mimic: to ridicule by imitation - *Oxford Dictionary*).

We need to know much more about leaf growth and death in relation to the supply of water and nitrogen and to the state of the atmosphere; about the way in which roots respond to soil physical conditions and nutrient supply; about the mechanisms which determine how assimilate is partitioned between tops and roots; about the effect of weather on the quality of the harvested product; and about the impact of weather and soil conditions on pests and diseases. In all these areas of comparative ignorance, progress is limited not by the technical limitations of modelling but by the absence of a solid framework of principles around which robust models can be built.

At the beginning of this workshop, Dr Darling spoke of the urgent need for

ICARDA to attack some of the practical problems of agricultural production in the region for which it is responsible. With these problems in mind, we have reviewed a major area of agronomy and the proceedings of this workshop bring together many of the principles from which a strong and well-coordinated research programme can be developed.

In 1850, the Director of another newly formed institute – Rothamsted Experimental Station – published a paper* which formed an early basis for much of the work reviewed here. He presented figures for the ratio of water use to yield for several arable crops, drawing attention, perhaps for the first time, to the nitrogen – fixing ability of legumes. I have referred to this paper partly because the subject is so germane but also because it contains a splendid final paragraph in which Sir John Lawes makes a plea we should circulate to all the agencies which support national and international research in agriculture:

"We are convinced, indeed, that however important and useful miscellaneous agricultural analyses may be, the interest and progress of agriculture would be more surely and permanently served, if its great patron Societies were to permit to their scientific officers a wider range of discretion, and more liberal means for the selection and carrying out of definite questions of research. Results of this kind promise, it is true, but little prospect of immediate and direct practical application, but by their aid the uncertain dictates, whether of common experience, theory, or speculation, may, ere long, be replaced by the unerring guidance of principles; and then alone can it reasonably be anticipated that miscellaneous and departmental analyses may find their true interpretation and acquire a due and practical value."

* J. B. Lawes (1850): Experimental investigation into the amount of water given off by plants during their growth especially in relation to the fixation and source of their various constituents. J. Hortic. Soc. Lond. V, 3–28.

Recommendations

Recommendations which arose from the workshop were:

Land resource and climatic measurements

Experiments should be conducted on representative agro-ecological sites. The agro-ecological conditions should be characterised in terms of length of growing period, temperature and radiation regimes, physical and chemical properties of the soil, and potential biomass production. The climatic data set should include radiation, wet and dry bulb temperatures, wind speed, soil temperature, pan evaporation, and rainfall intensity measurements.

Soil water relationships

Extractable moisture should be measured in a range of soil types under crops of the region and under different management strategies. Emphasis should be on rotations, fertility levels and management practices such as fallowing and tillage.

The influence of soil moisture and atmospheric demand on crop/forage growth and yield should be studied. Studies on plant water status should be conducted to couple the influence of soil and atmosphere on yield.

Nitrogen dynamics

Assessment of the nitrogen fluxes and transformations should be made in carefully planned long-term crop rotation experiments. The animal component should be included.

This assessment under different agro-climatic environments, will need to quantify nitrogen inputs, particularly by nitrogen fixation, the nitrogen store and availability, plant uptake, and losses of nitrogen.

Additional activities should include base studies, using ^{15}N or other means, on interactions between the environment and nitrogen mineralisation, to generate an understanding of the underlying principles which govern the availability of nitrogen in soils.

Attempts should be made to increase nitrogen fixation through effectual Rhizobium inoculation techniques and plant selection.

311

Experimental procedure

Experimental sites should be properly cleaned by uniform cropping, and detailed baseline data, e.g. total nitrogen, organic matter and other soil physical and chemical characteristics which should be collected when the experiments begin.

Sites should be large enough to allow superimposition of a wide spectrum of experiments on farming systems and soil management within realistic existing and alternative rotation systems.

Improving the adaptability of crop and pasture plants to advance the exploitation of available water and nitrogen

The potential for improving the adaptation of the cereals and legumes should be developed in association with research on management systems to understand the dynamics of water and nitrogen, and improving the availability of these elements in the cropping and rangeland systems of·the region.

Recognition and characterisation of traits that may be associated with increased resistance to water stress and, with legumes, the fixation of greater quantities of nitrogen, should go hand in hand with the development of screening procedures to help plant breeders to incorporate these responses with the genetic material for release to national programs.

THE EDITORS

Index

320